SIMPLE GROUPS
OF
LIE TYPE

ROGER W. CARTER

A Wiley–Interscience Publication

Wiley Classics Library Edition Published 1989

WILEY

JOHN WILEY & SONS
London · New York · Sydney · Toronto

Copyright © 1972, 1989 by John Wiley & Sons Ltd

Library of Congress Catalog Card Number: 72-39228

British Library Cataloguing in Publication Data:

Carter, Roger W. (Roger William), *1934–*
  Simple groups of lie type.
  1. Simple lie groups
  I. Title
  512'.55

ISBN 0-471-50683-4 (pbk.)

Printed in the United States of America

10 9 8 7 6 5 4 3 2

Preface

Since Chevalley showed in 1955 how to construct analogues of the complex simple Lie groups over arbitrary fields, these 'simple groups of Lie type' and their twisted analogues have been the subject of detailed investigation from a number of different points of view. This book is intended to serve as an introduction to the theory of Chevalley groups. It is not an exhaustive account of these groups, but concentrates on the basic results in the structure theory of the Chevalley groups and the twisted groups.

The Chevalley groups are studied in this book as groups of automorphisms of Lie algebras. This approach implies a concentration on the adjoint Chevalley groups—indeed the universal Chevalley groups and other isogenous groups have only been touched on in the development. In developing the theory we have found it necessary to assume a certain familiarity with the theory of simple Lie algebras over the complex field. Fortunately several good accounts of this theory are now available, and the information which we need about the simple Lie algebras has been collected together in a survey chapter.

Considerable emphasis has been placed in the development upon the Weyl groups associated with the groups of Lie type, as it is becoming increasingly clear that many properties of the Chevalley groups which are independent of the field of definition can best be dealt with in terms of the Weyl group. Thus several chapters have been devoted entirely to the Weyl group. Most of the development is independent of the field of definition, although we concentrate attention from time to time on finite fields in order to determine the orders of the finite groups of Lie type. We also show that the classical linear and symplectic groups and some of the orthogonal and unitary groups are examples of Chevalley groups or their twisted analogues.

One of the most fruitful ways of thinking about the Chevalley groups is as split forms of simple algebraic groups, for many of the most striking properties of the Chevalley groups are special cases of general results on algebraic groups. The theory of algebraic groups thus sheds much light on the structure of the Chevalley groups. This theory, however, requires a considerable background knowledge of algebraic geometry, which is beyond the scope of this volume. A reader who masters the material in this book is nevertheless encouraged to follow up by acquiring a knowledge of the algebraic group approach. This approach is advantageous in that it includes in addition to the split forms (Chevalley groups) and quasi-split forms (Steinberg groups) also the non-split forms, and so includes for example those classical orthogonal and unitary groups which are not Chevalley or Steinberg groups. Our approach is, however, sufficient to include all the known finite groups of Lie type.

A considerable part of the theory of Chevalley and Steinberg groups can be developed in the more general context, introduced by Tits, of groups with a (B, N)-pair. We have followed this approach since it removes the necessity for repetition of several of the arguments. Groups with a (B, N)-pair operate naturally on geometrical systems called buildings, also introduced by Tits, and we have included a chapter in which these buildings are described and their connection with the groups established.

It was decided not to include any account of the representation theory of Chevalley groups, although elegant theories have been developed by Chevalley for groups over an algebraically closed field, and by Curtis for p-modular representations of groups over finite fields of characteristic p.

Finally, almost all the finite simple groups at present known may be regarded as groups of Lie type, and we have included a chapter which completes the contemporary picture regarding finite simple groups by giving a brief description of the sporadic simple groups not of Lie type.

Part of this book was written while the author held visiting appointments at the University of Chicago during the Autumn of 1968 and Simon Fraser University, British Columbia, during the Summer of 1969. Thanks are due to both these universities for their support. The book was based partly on lecture courses on Chevalley groups given at the University of Warwick and the University of Chicago, and on Weyl groups at Simon Fraser University and Warwick University. A number of helpful comments were made by members of the audience at these courses.

Thanks are due to C. W. Curtis for suggesting an improvement to my original approach to chapter 6.

Finally I wish to thank Mrs Susan Tall for typing the manuscript.

University of Warwick,   ROGER CARTER
May 1971

Contents

Preface to the Wiley Classics Edition

Only a few change have been made to the original version of this book. I am grateful to Brian Hartley for pointing out some errors in the proof of the simplicity of the twisted groups. The reader is invited to make the necessary corrections using the additional lemma printed on page 313. I have also taken the opportunity to complete the list of sporadic simple groups in the last chapter. A number of smaller amendments have also been made, but the book as a whole remains very much as before.

University of Warwick　　　　　　　　　　　　　Roger Carter
April 1988

CHAPTER 1

The Classical Simple Groups

1.1 Introduction

Since the first edition of this book was published in 1972 great progress has been made in the theory of finite simple groups. Above all, the classification of the finite simple groups was finally completed in 1981. This was an outstanding achievement made possible by collaboration between many mathematicians in different parts of the world.

Until about 1955 the only known finite simple groups were the cyclic groups of prime order, the alternating groups, the families of classical simple groups over a finite field discovered by Jordan [1] and investigated by Dickson [1] and Dieudonné [1,2], some finite analogues of the simple Lie groups of type G_2 and E_6 discovered by Dickson, and the five 'sporadic' simple groups of Mathieu [1,2].

Since that time the families of classical simple groups have been described in a unified way by means of the Lie theory, following the fundamental work of Chevalley [4].

Further families of simple groups of Lie type, which may be obtained by modifications of Chevalley's procedure, were later discovered by Steinberg, Tits, Hertzig, Suzuki and Ree. In fact all the infinite families of finite simple groups, with the exception of the cyclic and alternating groups, may be regarded as groups of Lie type over finite fields.

In addition, the five Mathieu groups have been supplemented by the discovery between 1965 and 1981 of 21 further 'sporadic' simple groups.

At the same time, classification theorems for simple groups were being proved of increasing strength. Following the Feit-Thompson theorem, proving that every non-cyclic simple group has even order, results were established by Brauer, Gorenstein, Suzuki, Walter, Aschbacher and many others classifying simple groups in many cases where the Sylow 2-subgroup or the centralizer of an involution was assumed to be known. These theorems culminated in the classification theorem for finite simple groups. According to this theorem every finite simple group is either a cyclic group of prime order, an alternating group of degree at least 5, a simple group of Lie type over a finite field, or one of the 26 sporadic simple groups.

Now the simple groups of Lie type have many structural features in common and our purpose in this volume is to describe some of the properties of these Lie families in a unified way, following Chevalley, Steinberg, Tits and others. In order to do this some knowledge of root systems and Euclidean reflection groups is required, and also some knowledge of the simple Lie algebras over the complex field. The results about reflection groups have been developed from first principles in chapter 2, and the results we shall require about Lie algebras have been collected together in chapter 3. With this introductory material, we begin in chapter 4 an exposition of the properties of Chevalley groups, regarded as groups of automorphisms of Lie algebras. The properties of the Lie families of simple groups are to a large extent independent of the field over which the groups are defined, and this field will usually be an arbitrary one. However we do specialize from time to time to the complex field or a finite field.

In the present introductory chapter we shall describe the classical groups of non-singular linear transformations of a vector space.

1.2 The Linear Groups

Let $\mathfrak{V}$ be a vector space of dimension n over a field K. The group of all non-singular linear transformations of $\mathfrak{V}$ into itself is called the general linear group $GL_n(K)$. The transformations of determinant 1 form a normal subgroup $SL_n(K)$, the special linear group. The factor group $GL_n(K)/SL_n(K)$ is isomorphic to the multiplicative group of non-zero elements of K. The centre Z of $GL_n(K)$ consists of all transformations of form $T(x) = \lambda x$ for $\lambda \in K$ with $\lambda \neq 0$. The factor group $GL_n(K)/Z$ is the projective general linear group $PGL_n(K)$. It operates on the projective space of dimension $n - 1$ associated with $\mathfrak{V}$. The centre of $SL_n(K)$ is the subgroup $Z \cap SL_n(K)$, and the factor group $SL_n(K)/Z \cap SL_n(K)$ is the projective special linear group $PSL_n(K)$.

The projective special linear groups are generally simple. In fact $PSL_n(K)$ is a simple group for all $n \geqslant 2$, except for the groups $PSL_2(2)$ and $PSL_2(3)$ (cf. Huppert [1], p. 182). The finite simple groups in this family are obtained by taking for K the Galois field $GF(q)$, where q is some power of a prime. The projective special linear group over this field is denoted by $PSL_n(q)$ and its order is given by

$$|PSL_n(q)| = \frac{1}{(n, q-1)} q^{n(n-1)/2}(q^2 - 1)(q^3 - 1) \ldots (q^n - 1).$$

1.3 The Symplectic Groups

We now suppose that the vector space $\mathscr{V}$ is endowed with a non-singular bilinear scalar product which associates with each pair x, y of elements of $\mathscr{V}$ an element (x, y) of K. We assume that this scalar product is skew-symmetric, so that

$$(y, x) = -(x, y)$$

for all x, $y \in \mathscr{V}$. A space endowed with a scalar product of this type is called a symplectic space.

Consider the non-singular linear transformations of $\mathscr{V}$ into itself which are isometries, i.e. which satisfy the condition

$$(Tx, Ty) = (x, y)$$

for all x, $y \in \mathscr{V}$. The isometries form a subgroup of $GL_n(K)$ called the symplectic group $Sp_n(K)$. Now this group is, to within isomorphism, independent of the choice of the scalar product. In fact any non-singular skew-symmetric scalar product can be represented, with respect to a suitable basis, by the matrix

$$A = \begin{pmatrix} 0 & 1 & & & & & & \\ -1 & 0 & & & & & & \\ & & & & & & 0 & \\ & & 0 & 1 & & & & \\ & & -1 & 0 & & & & \\ & & & & \ddots & & & \\ & & & & & \ddots & & \\ 0 & & & & & & & \\ & & & & & & 0 & 1 \\ & & & & & & -1 & 0 \end{pmatrix}$$

In particular, the dimension of any non-singular symplectic space is even. With respect to the above basis the symplectic transformations are represented by matrices T satisfying the condition

$$T'AT = A.$$

Now a symplectic transformation necessarily has determinant 1. (cf. Dieudonné [1]). The centre Z of $Sp_n(K)$ consists of the transformations

$Tx = \lambda x$, where $\lambda = \pm 1$. The factor group $Sp_n(K)/Z$ is called the projective symplectic group $PSp_n(K)$.

The projective symplectic groups are generally simple. In fact they are all simple except for $PSp_2(2)$, $PSp_2(3)$, $PSp_4(2)$. The symplectic transformations of a 2-dimensional space are just those of determinant 1, therefore we have

$$Sp_2(K) = SL_2(K),$$

$$PSp_2(K) = PSL_2(K).$$

The finite symplectic groups are denoted by $Sp_n(q)$, and the orders of the finite symplectic groups are given by the formula

$$| PSp_{2l}(q) | = \frac{1}{(2, q-1)} q^{l^2}(q^2 - 1)(q^4 - 1) \ldots (q^{2l} - 1).$$

1.4 The Orthogonal Groups

We assume now that $\mathfrak{V}$ is a vector space of dimension n over a field K of characteristic not equal to 2. (The orthogonal groups over fields of characteristic 2 must be defined in a different way, which we describe in section 1.6.) We assume that there is defined on $\mathfrak{V}$ a non-singular bilinear scalar product which is symmetric, so that

$$(y, x) = (x, y)$$

for all $x, y \in \mathfrak{V}$. This scalar product determines a quadratic form f given by

$$f(x) = (x, x).$$

Conversely, the quadratic form determines the scalar product by the formula

$$(x, y) = \tfrac{1}{2}(f(x+y) - f(x) - f(y)).$$

(We have used here the fact that the characteristic of K is not 2.) The non-singular linear transformations of $\mathfrak{V}$ which are isometries form a group $O_n(K, f)$, the orthogonal group associated with the quadratic form f. In contrast to the symplectic group, the structure of the group $O_n(K, f)$ does depend upon which quadratic form is taken.

The number of inequivalent quadratic forms which exist on $\mathfrak{V}$ is dependent on the field K. The most important invariant of a quadratic form is its index, which is defined in the following way. A subspace $\mathcal{U}$ of $\mathcal{V}$ is called isotropic if

$$(x, y) = 0$$

for all x, $y \in \mathcal{U}$. It was proved by Witt that the maximal isotropic subspaces of $\mathcal{V}$ all have the same dimension (cf. Dieudonné [1]), and this dimension is called the index of f. We shall denote the index of f by v. v cannot be greater than $\frac{1}{2}n$, and we say that a form f has maximal index if $v = \frac{1}{2}n$ when n is even and $v = \frac{1}{2}(n-1)$ when n is odd.

The determinant of an orthogonal transformation is ± 1 and the orthogonal transformations of determinant 1 form a subgroup $SO_n(K, f)$, the special orthogonal group of f. The centre Z of $O_n(K, f)$ consists of the transformations $Tx = \lambda x$ where $\lambda = \pm 1$, provided $n > 2$, and

$$Z \cap SO_n(K, f)$$

is the centre of $SO_n(K, f)$. Factoring by the centre we obtain the corresponding projective groups

$$PO_n(K, f) = O_n(K, f)/Z,$$

$$PSO_n(K, f) = SO_n(K, f)/Z \cap SO_n(K, f).$$

One might expect, by analogy with the symplectic groups, that the groups $PSO_n(K, f)$ are generally simple. This is not so, however, and it is necessary to descend to a smaller subgroup. We denote by $\Omega_n(K, f)$ the commutator subgroup of $O_n(K, f)$. Since $O_n(K, f)/SO_n(K, f)$ is abelian, $\Omega_n(K, f)$ is a subgroup of $SO_n(K, f)$. We define the corresponding projective group

$$P\Omega_n(K, f) = \Omega_n(K, f)/Z \cap \Omega_n(K, f).$$

Then the groups $P\Omega_n(K, f)$ are generally simple. In fact $P\Omega_n(K, f)$ is simple provided $n \geqslant 5$ and $v \geqslant 1$. It is not generally true that $P\Omega_n(K, f)$ is simple if $n = 4$, or if $n \geqslant 5$ and $v = 0$.

We now describe the families of finite simple orthogonal groups. Let K be the finite field $GF(q)$ and $\mathcal{V}$ be an n-dimensional vector space over K. We distinguish between the cases when n is odd and even. Suppose first that n is odd, and let $n = 2l + 1$. Then there are just two inequivalent non-singular symmetric scalar products on $\mathcal{V}$, which can be represented by the matrices A and ϵA, where ϵ is a non-square in K and

$$A = \begin{pmatrix} 1 & 0 & \cdots & & 0 \\ 0 & 0 & & & I_l \\ \cdot & & & & \\ \cdot & & & & \\ \cdot & & & & \\ 0 & I_l & & & 0 \end{pmatrix}.$$

These two scalar products give rise to the same orthogonal group $O_{2l+1}(q)$, and the order of the associated simple group is given by

$$| P\Omega_{2l+1}(q) | = \frac{1}{(2, q-1)} \, q^{l^2}(q^2-1)(q^4-1) \ldots (q^{2l}-1).$$

Note that this is the same as the order of the group $PSp_{2l}(q)$, although the two groups are in general not isomorphic.

Now suppose n is even and let $n = 2l$. There are again just two inequivalent non-singular symmetric scalar products on $\mathfrak{V}$, but this time they give rise to distinct orthogonal groups. These scalar products can be represented by the matrices

$$\begin{pmatrix} 0 & I_l \\ I_l & 0 \end{pmatrix}, \qquad \begin{pmatrix} 0 & & I_{l-1} & 0 & 0 \\ & & & \cdot & \cdot \\ & & & \cdot & \cdot \\ & & & \cdot & \cdot \\ I_{l-1} & & 0 & 0 & 0 \\ 0 & \cdots & 0 & 1 & 0 \\ 0 & \cdots & 0 & 0 & -\epsilon \end{pmatrix},$$

where ϵ is a non-square in K. The first of these gives rise to a quadratic form of maximal index l, and the second has a quadratic form of index $l-1$. We shall denote the corresponding orthogonal groups by $O_{2l}^+(q)$ and $O_{2l}^-(q)$ respectively. The orders of the associated simple groups are given by

$$| P\Omega_{2l}^+(q) | = \frac{1}{(4, q^l - 1)} \, q^{l(l-1)}(q^2 - 1)(q^4 - 1) \ldots (q^{2l-2} - 1)(q^l - 1),$$

$$| P\Omega_{2l}^-(q) | = \frac{1}{(4, q^l + 1)} \, q^{l(l-1)}(q^2 - 1)(q^4 - 1) \ldots (q^{2l-2} - 1)(q^l + 1).$$

1.5 The Unitary Groups

We now consider a vector space $\mathfrak{V}$ of dimension n over a field K and suppose $\lambda \to \bar{\lambda}$ is an automorphism of K of order 2. Suppose $\mathfrak{V}$ is endowed with a non-singular Hermitian scalar product (x, y). Thus (x, y) is linear in x, conjugate linear in y, and

$$(y, x) = \overline{(x, y)}.$$

This scalar product determines a Hermitian form f given by

$$f(x) = (x, x).$$

The values $f(x)$ of the form lie in the fixed field K_0 of the involutary automorphism of K. The non-singular linear transformations of $\mathfrak{V}$ which are isometries with respect to this scalar product form a group $U_n(K, f)$, the unitary group associated with the Hermitian form f. The structure of $U_n(K, f)$ again depends upon the choice of the Hermitian form f. The index ν of f is again defined as the dimension of the maximal isotropic subspaces of $\mathfrak{V}$.

The unitary transformations of determinant 1 form a subgroup $SU_n(K, f)$, the special unitary group of f. The centre Z of $U_n(K, f)$ consists of the transformations $Tx = \lambda x$ where $\lambda\bar{\lambda} = 1$, provided $n > 1$. The centre of $SU_n(K, f)$ is $Z \cap SU_n(K, f)$. Factoring by the centre we obtain the corresponding projective groups

$$PU_n(K, f) = U_n(K, f)/Z,$$

$$PSU_n(K, f) = SU_n(K, f)/Z \cap SU_n(K, f).$$

The groups $PSU_n(K, f)$ are usually simple. In fact $PSU_n(K, f)$ is always simple provided $n \geq 2$ and $\nu \geq 1$, with the exception of three finite unitary groups mentioned below. Again there are examples which show that $PSU_n(K, f)$ need not be simple if $\nu = 0$.

We now describe the finite unitary groups. A finite field K admitting an automorphism of order 2 must be a field $GF(q^2)$ for some prime-power

q, and the involutary automorphism is given by

$$\bar{\lambda} = \lambda^q.$$

By choosing a suitable basis for the vector space $\mathfrak{V}$ over $GF(q^2)$, any non-singular Hermitian scalar product can be represented by the matrix I_n. Thus there is essentially only one such scalar product, and this gives rise to a Hermitian form f of maximal index. The corresponding unitary group is denoted by $U_n(q^2)$, and the order of the associated projective special unitary group is given by

$$| PSU_n(q^2) | = \frac{1}{(n, q+1)} q^{n(n-1)/2}(q^2-1)(q^3+1)(q^4-1) \ldots (q^n-(-1)^n).$$

The three exceptional groups which are not simple are

$$PSU_2(2^2), \ PSU_2(3^2), \ PSU_3(2^2).$$

1.6 The Orthogonal Groups in Characteristic 2

The orthogonal groups over a field of characteristic 2 have to be defined in rather a different way from the orthogonal groups already considered. If K is a field of characteristic not 2 and (x, y) is a symmetric scalar product on a vector space $\mathfrak{V}$ over K, the corresponding quadratic form f is defined by $f(x) = (x, x)$, and therefore satisfies the condition

$$f(\lambda x + \mu y) = \lambda^2 f(x) + \mu^2 f(y) + 2\lambda\mu(x, y)$$

for all $\lambda, \ \mu \in K$.

Now suppose K is a field of characteristic 2. A quadratic form on $\mathfrak{V}$ is a function f with values in K satisfying the condition

$$f(\lambda x + \mu y) = \lambda^2 f(x) + \mu^2 f(y) + \lambda\mu(x, y)$$

for all $\lambda, \ \mu \in K$, where (x, y) is some bilinear scalar product on $\mathfrak{V}$. In particular, putting $\mu = 0$ we have

$$f(\lambda x) = \lambda^2 f(x)$$

and putting $\lambda = \mu = 1$ we have

$$(x, x) = 0$$

and

$$(y, x) = (x, y).$$

Thus (x, y) may be regarded as a symplectic scalar product on $\mathfrak{V}$. It is not assumed to be necessarily non-singular but by a suitable choice of basis for $\mathfrak{V}$ it can be represented by a matrix of form

$$
\begin{pmatrix}
0 & 1 & & & & & & & & \\
1 & 0 & & & & & & & & \\
& & 0 & 1 & & & & & 0 & \\
& & 1 & 0 & & & & & & \\
& & & & \ddots & & & & & \\
& & & & & 0 & 1 & & & \\
& & & & & 1 & 0 & & & \\
& & 0 & & & & & 0 & & \\
& & & & & & & & 0 & \\
& & & & & & & & & 0
\end{pmatrix}
$$

Let n be the dimension of $\mathfrak{V}$ and $2l$ be the rank of the above matrix. Let $\mathfrak{V}_0$ be the set of $x \in \mathfrak{V}$ such that $(x, y)=0$ for all $y \in \mathfrak{V}$. Then $\mathfrak{V}_0$ is a subspace of $\mathfrak{V}$ of dimension $d=n-2l$. On this subspace $\mathfrak{V}_0$ the quadratic form f satisfies the condition

$$f(\lambda x + \mu y) = \lambda^2 f(x) + \mu^2 f(y)$$

and f is said to be non-degenerate if no non-zero vector $x \in \mathfrak{V}_0$ satisfies $f(x)=0$. The dimension d of $\mathfrak{V}_0$ is called the defect of f.

The non-singular linear transformations T of $\mathfrak{V}$ which satisfy the condition

$$f(Tx)=f(x)$$

form the orthogonal group $O_n(K, f)$ associated with f. Since

$$(x, y)=f(x+y)+f(x)+f(y),$$

it is clear that

$$(Tx, Ty) = (x, y).$$

Thus each element of $O_n(K, f)$ is an isometry of the scalar product (x, y).

A vector $x \in \mathfrak{V}$ is called singular if $f(x) = 0$, and a subspace of $\mathfrak{V}$ is called totally singular if each vector in it is singular. Any two maximal totally singular subspaces of $\mathfrak{V}$ have the same dimension ν, and ν is called the index of f. As before ν is at most $\frac{1}{2}n$.

We shall again denote by $\Omega_n(K, f)$ the commutator subgroup of $O_n(K, f)$.

We shall consider only non-degenerate quadratic forms and distinguish between the cases when $d = 0$ and $d > 0$. Suppose first that f is a non-degenerate quadratic form of defect 0. Then (x, y) is a non-singular symplectic scalar product and so $O_n(K, f)$ is a subgroup of $Sp_n(K)$. (Note that $n = 2l$ is even.) The commutator subgroups $\Omega_n(K, f)$ are usually simple. In fact $\Omega_n(K, f)$ is simple provided $n \geqslant 6$ and $\nu \geqslant 1$. $\Omega_n(K, f)$ need not be simple if $n = 4$ or if $n \geqslant 6$ and $\nu = 0$. We now describe the finite groups of this type. A basis $e_1, e_2, \ldots, e_l, e_{-1}, e_{-2}, \ldots, e_{-l}$ can be chosen for $\mathfrak{V}$ such that f is one of the two following forms:

$$f(x) = x_1 x_{-1} + x_2 x_{-2} + \ldots + x_l x_{-l},$$

$$f(x) = x_1 x_{-1} + x_2 x_{-2} + \ldots + x_{l-1} x_{-(l-1)} + \alpha x_l^2 + x_l x_{-l} + \alpha x_{-l}^2,$$

where

$$x = \sum_l x_i e_i$$

and $\alpha t^2 + t + \alpha$ is an irreducible polynomial over $K = GF(q)$, where q is a power of 2. The indices of these forms f are l and $l-1$ respectively, and the respective orthogonal groups are denoted by $O_{2l}^+(q)$, $O_{2l}^-(q)$. The orders of these groups are given by the same formulae as the orders of the groups $O_{2l}^+(q)$, $O_{2l}^-(q)$ when q is odd.

Now suppose that f is a non-degenerate quadratic form of defect greater than 0. The defect $d = n - 2l$ is the dimension of the subspace $\mathfrak{V}_0$ of $\mathfrak{V}$, and on $\mathfrak{V}_0$ the form f satisfies the relation

$$f(\lambda x + \mu y) = \lambda^2 f(x) + \mu^2 f(y).$$

The set of values $f(x)$ for $x \in \mathfrak{V}_0$ therefore forms a subset of K which is a subspace over the subfield K^2 of K. Since $\mathfrak{V}_0$ has dimension d over K and

f is non-degenerate, $f(\mathcal{V}_0)$ has dimension d over K^2. In particular

$$d \leqslant |K : K^2|.$$

It can be shown (cf. Dieudonné [1]) that $O_n(K, f)$ is isomorphic to the subgroup of $Sp_{2l}(K)$ of transformations T satisfying the condition

$$f(Tx) + f(x) \in f(\mathcal{V}_0),$$

where x lies in some suitably chosen non-singular symplectic subspace of $\mathcal{V}$ of dimension $2l$.

If K is a perfect field, i.e. $K^2 = K$, it is clear from the above that $d = 1$ and

$$O_{2l+1}(K, f) \cong Sp_{2l}(K).$$

In particular there is one finite family of orthogonal groups $O_{2l+1}(q)$ of this kind, and $O_{2l+1}(q)$ is isomorphic to $Sp_{2l}(q)$.

If K is not perfect it can be shown that the commutator subgroup $\Omega_n(K, f)$ of $O_n(K, f)$ is simple provided $l \geqslant 1$ and $\nu \geqslant 1$ (with the possible exception of the case $l = 2$, $\nu = 2$).

The facts given above about the classical linear, symplectic, orthogonal and unitary groups will suffice for our present purposes. For further information about these classical groups the reader is referred to the books of Dieudonné [1, 2]. We shall show in later chapters that many of the simple classical groups, including all the finite families, can be interpreted as simple groups of Lie type.

CHAPTER 2

Weyl Groups

In the present chapter we shall describe some properties of certain finite groups of orthogonal transformations which are generated by reflections. Such reflection groups, called Weyl groups, play a crucial rôle in the theory of groups of Lie type—indeed there is a sense in which the Weyl group may be regarded as the 'skeleton' of the corresponding group of Lie type. In order to define the Weyl groups we introduce the concept of a root system. We do this axiomatically, but the way in which such root systems arise naturally in the theory of Lie algebras is explained in chapter 3.

2.1 Systems of Roots

Let $\mathfrak{V}$ be a Euclidean space of finite dimension l. For each non-zero vector r of $\mathfrak{V}$ we denote by w_r the reflection in the hyperplane orthogonal to r. This is the linear map defined by $w_r(r) = -r$ and $w_r(x) = x$ for all x with $(r, x) = 0$. If x is any vector in $\mathfrak{V}$ we have

$$w_r(x) = x - \frac{2(r, x)}{(r, r)} r.$$

Definition 2.1.1. A subset Φ of $\mathfrak{V}$ is called a system of roots in $\mathfrak{V}$ if the following axioms are satisfied:

(i) Φ is a finite set of non-zero vectors.
(ii) Φ spans $\mathfrak{V}$.
(iii) If $r, s \in \Phi$ then $w_r(s) \in \Phi$.
(iv) If $r, s \in \Phi$ then $2(r, s)/(r, r)$ is a rational integer.
(v)‡ If $r, \lambda r \in \Phi$, where $\lambda \in \mathbb{R}$, then $\lambda = \pm 1$.

We observe that if $r \in \Phi$ then $-r \in \Phi$ also. This follows from (iii) since $w_r(r) = -r$.

‡ Axiom (v) is sometimes omitted in the definition of a root system, but the present definition is the most convenient for our purposes.

Let Φ be a root system. We denote by $W(\Phi)$ the group generated by the reflections w_r for all $r \in \Phi$. $W(\Phi)$ is called the Weyl group of Φ. It is clearly a group of orthogonal transformations of $\mathfrak{P}$. Each element of W transforms Φ into itself, by axiom (iii). W operates faithfully on Φ, by axiom (ii). Since Φ is a finite set, W is therefore a finite group.

Now although Φ spans $\mathfrak{P}$, Φ is not linearly independent. Thus Φ contains a proper subset which is a basis for $\mathfrak{P}$. We shall show that Φ contains a subset Π satisfying the following conditions:

(i) Π is linearly independent.

(ii) Every root in Φ is a linear combination of roots in Π with coefficients which are either all non-negative or all non-positive.

A subset Π of Φ satisfying (i), (ii) is called a fundamental system of roots. If

$$\Pi = \{r_1, r_2, \ldots, r_l\}$$

and $r \in \Phi$ then we have

$$r = \sum_{i=1}^{l} \lambda_i r_i,$$

where $\lambda_i \in \mathbb{R}$ and either $\lambda_i \geqslant 0$ for all i or $\lambda_i \leqslant 0$ for all i.

In order to prove the existence of a fundamental system of roots we endow the space $\mathfrak{P}$ with a total ordering. Let $\mathfrak{P}^+$ be a subset of $\mathfrak{P}$ satisfying the conditions:

(i) If $v \in \mathfrak{P}^+$ and $\lambda > 0$ then $\lambda v \in \mathfrak{P}^+$.

(ii) If $v_1, v_2 \in \mathfrak{P}^+$ then $v_1 + v_2 \in \mathfrak{P}^+$.

(iii) For each $v \in \mathfrak{P}$ exactly one of the conditions $v \in \mathfrak{P}^+$, $-v \in \mathfrak{P}^+$, $v = 0$, holds.

Such a subset $\mathfrak{P}^+$ can be obtained, for example, by choosing a basis $v_1, v_2, \ldots, v_l$ of $\mathfrak{P}$ and taking for $\mathfrak{P}^+$ the set of all vectors

$$\sum_{l=1}^{l} \lambda_i v_i$$

in which the first non-zero coefficient λ_i is positive.

We now introduce an order relation $\succ$ by defining $v_1 \succ v_2$ whenever $v_1 - v_2 \in \mathfrak{P}^+$. This is a total ordering on $\mathfrak{P}$ compatible with addition and with scalar multiplication by positive elements of $\mathbb{R}$.

A subset of Φ is called a positive system of roots if it has the form $\Phi \cap \mathfrak{P}^+$ for some total ordering of $\mathfrak{P}$. If Π is a fundamental system of roots there is just one positive system of roots containing it. For we can certainly choose an ordering for which $\Pi \subset \mathfrak{P}^+$. But then any root

$r \in \Phi$ of form

$$\sum_{r_i \in \Pi} \lambda_i r_i$$

with $\lambda_i \geqslant 0$ for all i will satisfy $r \in \mathcal{Y}^+$, and each root for which $\lambda_i \leqslant 0$ for all i satisfies $-r \in \mathcal{Y}^+$. Thus $\Phi \cap \mathcal{Y}^+$ is determined by Π.

We now prove the existence of a fundamental system of roots in Φ.

PROPOSITION 2.1.2. *Every positive system of roots in Φ contains a fundamental system.*

PROOF. Let Φ^+ be a positive system in Φ. Then there is some total ordering of $\mathcal{Y}$ for which $\Phi^+ = \Phi \cap \mathcal{Y}^+$. Let Π be a subset of Φ^+ satisfying the conditions:

(i) Every root in Φ^+ is a linear combination of roots in Π with non-negative coefficients.

(ii) No subset of Π satisfies (i).

Such a subset Π certainly exists, since Φ^+ itself satisfies (i).

We show that a subset Π satisfying (i), (ii) is a fundamental system. To do this it is sufficient to prove that Π is linearly independent. However we prove first that $(r, s) \leqslant 0$ for any two distinct roots r, s of Π. Suppose, if possible, that $(r, s) > 0$. Then $w_r(s) = s - \lambda r$, where $\lambda > 0$. If $w_r(s) \in \Phi^+$ we have

$$w_r(s) = \sum_{r_i \in \Pi} \alpha_i r_i,$$

where each $\alpha_i \geqslant 0$. Thus

$$s = \lambda r + \sum_{r_i \in \Pi} \alpha_i r_i, \qquad \lambda > 0, \ \alpha_i \geqslant 0.$$

The coefficient of s on the right-hand side must be less than 1, otherwise

$$\lambda r + \sum_{r_i \in \Pi} \alpha_i r_i - s$$

would be in $\mathcal{Y}^+$. Thus s can be expressed as a linear combination with non-negative coefficients of the remaining roots in Π, a contradiction. On the other hand, if $-w_r(s) \in \Phi^+$ we have

$$-w_r(s) = \sum_{r_i \in \Pi} \alpha_i r_i$$

with $\alpha_i \geqslant 0$, whence

$$\lambda r = s + \sum_{r_i \in \Pi} \alpha_i r_i, \qquad \lambda > 0, \ \alpha_i \geqslant 0.$$

The coefficient of r on the right-hand side must be less than λ, thus r can be expressed as a linear combination with non-negative coefficients of the remaining roots in Π, which again gives a contradiction. Thus $(r, s) \leqslant 0$.

It is now easy to prove that Π is linearly independent. Any linear relation involving the elements of Π can be written in the form

$$\sum \alpha_i r_i = \sum \beta_i s_i, \qquad \alpha_i \geqslant 0, \ \beta_i \geqslant 0,$$

where r_i, s_i are distinct elements of Π. Let $v = \sum \alpha_i r_i = \sum \beta_i s_i$. Then

$$(v, v) = \sum_{i, j} \alpha_i \beta_j (r_i, s_j) \leqslant 0.$$

Thus $v = 0$. Since r_i, s_i are in $\mathfrak{P}^+$ this implies that $\alpha_i = 0$, $\beta_i = 0$ for all i. Hence Π is a fundamental system of roots. ∎

PROPOSITION 2.1.3. *Every positive system of roots in* Φ *contains just one fundamental system. Thus there is a one–one correspondence between positive systems and fundamental systems in* Φ.

PROOF. Suppose $\{r_1, r_2, \ldots, r_l\}$ and $\{s_1, s_2, \ldots, s_l\}$ are two fundamental systems in Φ^+. Then

$$r_i = \sum_{j=1}^{l} \alpha_{ij} s_j, \qquad s_i = \sum_{j=1}^{l} \beta_{ij} r_j,$$

where $\alpha_{ij} \geqslant 0$, $\beta_{ij} \geqslant 0$. (α_{ij}) and (β_{ij}) are inverse matrices. For each i there exists j with $\alpha_{ij} \neq 0$. Since

$$\sum_{m=1}^{l} \alpha_{im} \beta_{mk} = 0$$

for all $k \neq i$, we have $\beta_{jk} = 0$ for all $k \neq i$. Hence $\beta_{ji} \neq 0$. This implies similarly that $\alpha_{ik} = 0$ for $k \neq j$. Thus (α_{ij}) is a monomial matrix. By numbering $s_1, s_2, \ldots, s_l$ suitably we may assume (α_{ij}) is diagonal. Since $r_i, s_i \in \Phi^+$ we have $\alpha_{ii} > 0$ for each i. By 2.1.1 (v), $\alpha_{ii} = 1$ for each i. Thus $s_i = r_i$. ∎

COROLLARY 2.1.4. *If* $\Pi = \{r_1, r_2, \ldots, r_l\}$ *is a fundamental system in* Φ, *then* $(r_i, r_j) \leqslant 0$ *for all* $i \neq j$.

PROOF. This was true of the fundamental system constructed in 2.1.2, and every fundamental system arises in this way for some ordering.

We now choose a fundamental system Π and corresponding positive system Φ^+, and keep them fixed in the following discussion. (The relation between the different fundamental systems will be described later.) The roots in Φ^+ will be called positive roots and the remainder negative roots. The set of negative roots will be denoted by Φ^-.

LEMMA 2.1.5.   *Let* $r \in \Pi$. *Then* w_r *transforms* r *into* $-r$ *but every other positive root into a positive root.*

PROOF.   Let $s \in \Phi^+$ with $s \neq r$. Then

$$s = \sum_{r_i \in \Pi} \alpha_i r_i$$

with $\alpha_i \geqslant 0$. By 2.1.1 (v) there is some $\alpha_i > 0$ corresponding to $r_i \neq r$. The coefficient of r_i in $w_r(s)$ is therefore also positive. Thus $w_r(s) \in \Phi^+$.   ∎

PROPOSITION 2.1.6.   *Each root in* Φ *is a linear combination of roots in* Π *with rational integer coefficients.*

PROOF.   It is sufficient to prove this for roots in Φ^+. Let $r \in \Phi^+$. If $r \in \Pi$ the result is trivially true, so we assume $r \notin \Pi$. Then

$$r = \sum_{r_i \in \Pi} \lambda_i r_i,$$

where each $\lambda_i \geqslant 0$ and at least two λ_i are positive. Now there is some $r_i \in \Pi$ for which $(r_i, r) > 0$. For if all $(r_i, r) \leqslant 0$ we would have

$$(r, r) = \sum \lambda_i (r_i, r) \leqslant 0,$$

a contradiction. Choose $r_i \in \Pi$ with $(r_i, r) > 0$. Then

$$w_{r_i}(r) = r - \frac{2(r_i, r)}{(r_i, r_i)} r_i.$$

$w_{r_i}(r)$ differs from r in only one coefficient when expressed as a linear combination of roots in Π. Thus at least one coefficient of $w_{r_i}(r)$ is positive, and so $w_{r_i}(r) \in \Phi^+$.

We define $h(r)$, the height of r, by $h(r) = \sum \lambda_i$. Then $h(w_{r_i}(r)) < h(r)$ Thus for each positive root which is not fundamental there is another positive root of smaller height. Therefore the positive roots of minimal height are the fundamental roots, which have height 1.

We prove our result by induction on $h(r)$. It is true for roots of height 1. Given $r \in \Phi^+$ with $r \notin \Pi$, choose r_i as above. Then $w_{r_i}(r)$ is an integral combination of roots in Π, by induction. Thus r is also, since

$$r = w_{r_i}(r) + \frac{2(r_i, r)}{(r_i, r_i)} r_i$$

and $2(r_i, r)/(r_i, r_i)$ is a rational integer, by 2.1.1 (iv). ■

COROLLARY 2.1.7. *With respect to the basis Π of $\mathfrak{B}$, each element of W is represented by a matrix with rational integer coefficients.*

PROOF. If $r_i \in \Pi$, $w(r_i) \in \Phi$ so is an integral combination of the basis elements.

PROPOSITION 2.1.8. (i) *Every root in Φ is the image of some root in Π under some element of W.*
(ii) *W is generated by the fundamental reflections w_r for $r \in \Pi$.*

PROOF. Let W_0 be the subgroup of W generated by the reflections w_r for $r \in \Pi$. We show that every root in Φ has the form $w(s)$ for some $w \in W_0$, $s \in \Pi$. Let $r \in \Phi^+$. If $h(r) = 1$, r can certainly be expressed in the required form, so we use induction on $h(r)$. If $h(r) > 1$ there is a root $r_i \in \Pi$ such that $(r_i, r) > 0$. Then $w_{r_i}(r) \in \Phi^+$ and $h(w_{r_i}(r)) < h(r)$, as in the proof of 2.1.6. By induction $w_{r_i}(r) = w'(s)$ for some $w' \in W_0$, $s \in \Pi$. Hence $r = w_{r_i}w'(s)$ and $w_{r_i}w' \in W_0$. The negative roots can also be expressed in the required form since

$$-r = w_{r_i}w'(-s) = w_{r_i}w'w_s(s)$$

and $w_{r_i}w'w_s \in W_0$.
We now show that $W_0 = W$. Since W is generated by the reflections w_r for $r \in \Phi$, it is sufficient to show that $w_r \in W_0$. Now $r = w(s)$ for some $s \in \Pi$ and $w \in W_0$. It follows that $w_r = w w_s w^{-1}$. For

$$w w_s w^{-1}(x) = w \left(w^{-1}(x) - \frac{2(s, w^{-1}(x))}{(s, s)} s \right)$$

$$= x - \frac{2(r, x)}{(r, r)} r$$

$$= w_r(x).$$

Hence $w_r \in W_0$ and $W_0 = W$. ■

2.2 The Length Function

Having established that every element of the Weyl group W is a product of fundamental reflections w_r, $r \in \Pi$, we denote by $l(w)$ the minimal length of any expression of w of this form. Thus $l(1) = 0$ and $l(w_r) = 1$ for $r \in \Pi$. We shall show that $l(w)$ can also be described in quite a different way. We define for each $w \in W$ an integer $n(w)$ given by

$$n(w) = |\Phi^+ \cap w^{-1}(\Phi^-)| \ .$$

Thus $n(w)$ is the number of positive roots transformed by w into negative roots. We first derive some elementary properties of the function $n(w)$.

LEMMA 2.2.1.   *Let* $r \in \Pi$ *and* $w \in W$. *Then*
(i) $n(w_r w) = n(w) + 1$ *if* $w^{-1}(r) \in \Phi^+$,
(ii) $n(w_r w) = n(w) - 1$ *if* $w^{-1}(r) \in \Phi^-$,
(iii) $n(w w_r) = n(w) + 1$ *if* $w(r) \in \Phi^+$,
(iv) $n(w w_r) = n(w) - 1$ *if* $w(r) \in \Phi^-$.

PROOF.   By 2.1.5 w_r changes the sign of only two roots r and $-r$. Thus $n(w_r w) = n(w) \pm 1$ and $n(w w_r) = n(w) \pm 1$. Now $n(w_r w) = n(w) + 1$ if and only if $r \in w(\Phi^+)$, thus proving (i) and (ii), and $n(w w_r) = n(w) + 1$ if and only if $w(r) \in \Phi^+$, proving (iii) and (iv).   ∎

We now show that $l(w) = n(w)$.

THEOREM 2.2.2.   *The minimal length of an expression of* w *as a product of fundamental reflections is equal to the number of positive roots transformed by* w *into negative roots.*

PROOF.   Let w by an element of W with $l(w) = k$. Then w has an expression

$$w = w_{r_1} w_{r_2} \ldots w_{r_k}, \qquad r_i \in \Pi.$$

By 2.2.1 we have

$$n(w) \leqslant n(w_{r_1} w) + 1 \leqslant n(w_{r_2} w_{r_1} w) + 2 \leqslant \ldots \leqslant k.$$

Thus $n(w) \leqslant l(w)$.

Suppose, if possible, that $n(w) < k$. Then by 2.2.1 (iii) there is an integer $j \leqslant k - 1$ such that

$$w_{r_1} w_{r_2} \ldots w_{r_j}(r_{j+1}) \in \Phi^-.$$

It follows that there is some integer $i \leqslant j$ such that

$$w_{r_{i+1}} \ldots w_{r_j}(r_{j+1}) \in \Phi^+,$$

$$w_{r_i} w_{r_{i+1}} \ldots w_{r_j}(r_{j+1}) \in \Phi^-.$$

Since w_{r_i} changes the sign of only r_i, $-r_i$, we have

$$w_{r_{i+1}} \ldots w_{r_j}(r_{j+1}) = r_i.$$

It follows that

$$w_{r_i} = w_{r_{i+1}} \ldots w_{r_j} w_{r_{j+1}} w_{r_j} \ldots w_{r_{i+1}}$$

and so

$$w_{r_{i+1}} \ldots w_{r_{j+1}} = w_{r_i} \ldots w_{r_j}.$$

We can now make use of this relation to shorten the original expression for w.

$$w = w_{r_1} \ldots w_{r_k}$$

$$= w_{r_1} \ldots w_{r_i} w_{r_i} \ldots w_{r_j} w_{r_{j+2}} \ldots w_{r_k}$$

$$= w_{r_1} \ldots w_{r_{i-1}} w_{r_{i+1}} \ldots w_{r_j} w_{r_{j+2}} \ldots w_{r_k}.$$

Thus we have an expression for w as a produce of $k-2$ fundamental reflections. Thus $l(w) < k$, a contradiction. Hence $n(w) = l(w)$. ∎

COROLLARY 2.2.3. *If* $w \in W$ *satisfies* $w(\Pi) = \Pi$ *then* $w = 1$.

PROOF. If $w(\Pi) = \Pi$ then $w(\Phi^+) = \Phi^+$ and so $n(w) = 0$. Thus $l(w) = 0$ and so $w = 1$. ∎

We now discuss the way in which different fundamental systems in Φ are related to one another.

THEOREM 2.2.4. *If* Π *is a fundamental system in* Φ, *so is* $w(\Pi)$ *for any* $w \in W$. *Given any two fundamental systems* Π_1, Π_2 *in* Φ *there is exactly one element* $w \in W$ *such that* $w(\Pi_1) = \Pi_2$.

PROOF. Let Φ^+ be the positive system containing Π. Then $\Phi^+ = \Phi \cap \mathcal{P}^+$ for some total ordering on $\mathcal{P}$. $w(\mathcal{P}^+)$ also defines a total ordering on $\mathcal{P}$ and the positive roots with respect to this ordering are

$$w(\Phi^+) = \Phi \cap w(\mathcal{P}^+).$$

$w(\Pi)$ is clearly the fundamental system contained in $w(\Phi^+)$.

We now show that for any two fundamental systems Π_1, Π_2 there is some element $w \in W$ for which $w(\Pi_1) = \Pi_2$. Let Φ_1^+, Φ_2^+ be the positive systems containing Π_1, Π_2. We use induction on $n = |\Phi_1^+ \cap \Phi_2^-|$. If $n = 0$ we have $\Phi_1^+ = \Phi_2^+$ and so $\Pi_1 = \Pi_2$. Thus we may assume $n > 0$. Then $\Pi_1 \cap \Phi_2^-$ is not empty. For if every root in Π_1 were in Φ_2^+ the same would apply to every root in Φ_1^+. Let $r \in \Pi_1 \cap \Phi_2^-$. Then $w_r(\Phi_1^+)$ is the set of roots obtained from Φ_1^+ by replacing r by $-r$. Hence

$$|w_r(\Phi_1^+) \cap \Phi_2^-| = n - 1.$$

Now $w_r(\Pi_1)$ is the fundamental system contained in $w_r(\Phi_1^+)$ so, by induction, there exists $w' \in W$ with $w'w_r(\Pi_1) = \Pi_2$. Thus $w(\Pi_1) = \Pi_2$ where $w = w'w_r$.

Finally we prove the uniqueness of w. Suppose $w_1(\Pi_1) = \Pi_2$ and $w_2(\Pi_1) = \Pi_2$. Then $w_2^{-1}w_1(\Pi_1) = \Pi_1$ and so by 2.2.3 $w_2^{-1}w_1 = 1$. Thus there is a unique element of W transforming Π_1 into Π_2. ∎

COROLLARY 2.2.5. *The number of fundamental systems in* Φ *is equal to the order of* W.

PROOF. Let Π be a fixed fundamental system. Then $w(\Pi)$ gives each fundamental system exactly once as w runs through W. ∎

PROPOSITION 2.2.6. *Let* Φ^+ *be a positive system in* Φ *and* Φ^- *the corresponding negative system. Then there is a unique element* $w_0 \in W$ *such that* $w_0(\Phi^+) = \Phi^-$. *Moreover* w_0 *is an element of order* 2.

PROOF. Φ^+ is the positive system in Φ corresponding to some total ordering of $\mathcal{V}$, and Φ^- is the positive system corresponding to the reverse ordering. By 2.1.3 and 2.2.4 there exists a unique element $w_0 \in W$ with $w_0(\Phi^+) = \Phi^-$. Since $w_0^2(\Phi^+) = \Phi^+$ we have $w_0^2 = 1$, so that w_0 is an element of order 2. ∎

2.3 A Geometrical Interpretation of the Fundamental Systems

For each root $r \in \Phi$ we denote by H_r the hyperplane orthogonal to r. Thus $x \in H_r$ if and only if $(r, x) = 0$. Now the hyperplanes are closed subsets of $\mathcal{V}$, thus

$$\bigcup_{r \in \Phi} H_r$$

is closed and its complement

$$\mathcal{V} - \bigcup_{r \in \Phi} H_r$$

is open in $\mathfrak{P}$. It is disconnected, two points being in the same connected component if and only if they lie on the same side of each reflecting hyperplane. The connected components of

$$\mathfrak{P} - \bigcup_{r \in \Phi} H_r$$

are called chambers.

Let C be any chamber in $\mathfrak{P}$ and $\delta(C)$ be the boundary of C. Then the hyperplanes H_r such that $H_r \cap \delta(C)$ is not contained in any proper subspace of H_r are called the bounding hyperplanes, or walls, of C.

Let $\Pi = \{r_1, r_2, \ldots, r_l\}$ be a fundamental system in Φ. Then the set C of vectors $x \in \mathfrak{P}$ with $(v_i, x) > 0$ for $i = 1, 2, \ldots, l$ is a chamber. For if H_r is any reflecting hyperplane, we may assume $r \in \Phi^+$, and so $(r, x) > 0$ for each $x \in C$. Thus all vectors in C lie on the same side of each reflecting hyperplane. We show that the bounding hyperplanes of C are

$$H_{r_1}, H_{r_2}, \ldots, H_{r_l}.$$

Now $H_{r_i} \cap \delta(C)$ consists of the vectors x for $(r_i, x) = 0$ and $(r_j, x) \geqslant 0$ for all $j \neq i$. Since $r_1, r_2, \ldots, r_l$ are linearly independent, $H_{r_i} \cap \delta(C)$ is not contained in any proper subspace of H_{r_i}. On the other hand, let r be a positive root which is not fundamental. Then

$$r = \sum_{i=1}^{l} \lambda_i r_i,$$

where each $\lambda_i \geqslant 0$ and at least two λ_i are positive. If $x \in H_r \cap \delta(C)$ then

$$\sum_{i=1}^{l} \lambda_i (r_i, x) = 0,$$

whence $(r_i, x) = 0$ whenever $\lambda_i > 0$. Thus $H_r \cap \delta(C)$ is contained in a proper subspace of H_r. Therefore the bounding hyperplanes of C are $H_{r_1}, H_{r_2}, \ldots, H_{r_l}$ and the roots $r_1, r_2, \ldots, r_l$ may be characterized as those roots in Φ which are orthogonal to the bounding hyperplanes of C and point into C (viz., r_i lies on the same side of the hyperplane H_{r_i} as C does).

The following proposition describes the relation between the set of chambers and the set of fundamental systems.

PROPOSITION 2.3.1. *The roots orthogonal to the bounding hyperplanes of a chamber and pointing into the chamber form a fundamental system. Moreover, every fundamental system arises in this way from some chamber.*

B

PROOF. Let w be an element of W. Since $w(H_r)=H_{w(r)}$, w permutes the reflecting hyperplanes. It follows that if C is a chamber so is $w(C)$. Thus the Weyl group operates on the set of chambers. We show that this operation is transitive.

Let Π be a fundamental system in Φ and C be the chamber defined by $(r_i, x)>0$ for all $r_i \in \Pi$. Let C' be any chamber and v be a vector in C'. Let Φ^+ be the positive system containing Π and let $\Phi^+=\Phi \cap \mathcal{V}^+$, where Φ^+ is the set of positive elements with respect to a total ordering $\prec$. Consider the set of transforms $w(v)$ of v by elements of W. Let v' be the greatest of all these transforms with respect to the ordering $\prec$. Then we have

$$w_{r_i}(v')=v'-\frac{2(r_i, v')}{(r_i, r_i)}\,r_i, \qquad r_i \in \Pi,$$

and since $w_{r_i}(v') \preccurlyeq v'$ we have $(r_i, v') \geqslant 0$. Since this holds for all $r_i \in \Pi$, v' must be in the closure $\bar{C}$ of C.

Let $v'=w(v)$. Then v' is in the chamber $w(C')$. However, the only chamber intersecting $\bar{C}$ is C. Thus $w(C')$, which intersects $\bar{C}$, must be equal to C. It follows that W operates transitively on the set of chambers.

Now since $C'=w^{-1}(C)$, the roots orthogonal to the bounding hyperplanes of C' and pointing into C' form the set $w^{-1}(\Pi)$, which is a fundamental system. Conversely, any fundamental system has this form for some $w \in W$ by 2.2.4. ∎

COROLLARY 2.3.2. *Given any two chambers C_1, C_2 there is a unique element $w \in W$ such that $w(C_1)=C_2$.*

PROOF. Let Π_1, Π_2 be the fundamental systems corresponding to C_1, C_2, as in 2.3.1. Then $w(\Pi_1)=\Pi_2$, and w is unique by 2.2.4.

COROLLARY 2.3.3. *The number of chambers is equal to the order of W.*

PROPOSITION 2.3.4. *Let C be a chamber in $\mathcal{V}$. For each $v \in \mathcal{V}$ there is a unique vector $v' \in \bar{C}$ such that v can be transformed into v' by some element of W.*

PROOF. Every vector in $\mathcal{V}$ lies in the closure of some chamber. Since W operates transitively on the set of chambers, the given vector v can be transformed into $\bar{C}$ by some element of W. Suppose $w_1(v)=v_1$, $w_2(v)=v_2$, where v_1, $v_2\in\bar{C}$. Then $w_2 w_1^{-1}(v_1)=v_2$.

We shall show that if $w(v_1)=v_2$, where v_1, $v_2\in\bar{C}$, then $v_1=v_2$. We prove this by induction on $l(w)$, the result being clear if $l(w)=0$. Suppose $l(w)>0$. Then there is some $r\in\Pi$ such that $w(r)\in\Phi^-$. Hence

$$0\leqslant(v_1,r)=(v_2,w(r))\leqslant 0.$$

It follows that $(v_1,r)=0$, whence $w_r(v_1)=v_1$. But now $ww_r(v_1)=v_2$ and $l(ww_r)=l(w)-1$ by 2.2.1. Thus $v_1=v_2$ by induction.

COROLLARY 2.3.5. *If C is a chamber in $\mathcal{V}$, its closure $\bar{C}$ contains just one element of each orbit of $\mathcal{V}$ under W.*

When we are considering a fixed fundamental system, as is the case in most of the discussion to follow, the corresponding chamber will be called the fundamental chamber. Corollary 2.3.5 shows that its closure is a 'fundamental region' for $\mathcal{V}$ under the action of W.

2.4 Definitions by Generators and Relations

In the present section we describe two quite distinct ways in which a Weyl group can be defined by generators and relations. Let Π be a fundamental system in Φ. Then the fundamental reflections w_r, $r\in\Pi$, generate W by 2.1.8. Let $r,s\in\Pi$ and m_{rs} be the order of $w_r w_s$. In particular $m_{rr}=1$ for each $r\in\Pi$. Then $(w_r w_s)^{m_{rs}}=1$ is a set of relations in W, and we show it is in fact a system of defining relations.

THEOREM 2.4.1. *W is defined as an abstract group by generators w_r, $r\in\Pi$, subject to relations $(w_r w_s)^{m_{rs}}=1$.*

PROOF. The following beautiful proof of this theorem is due to R. Steinberg. Suppose the result is false and let

$$w_{r_1}w_{r_2}\ldots w_{r_k}=1, \qquad r_i\in\Pi,$$

be a relation of minimal length k which is not a consequence of the given relations. Now each reflection has determinant -1, thus $\det w_r=-1$ and $\det(w_{r_1}w_{r_2}\ldots w_{r_k})=(-1)^k$. It follows that k is even, and we write

$k = 2m$. Thus

$$w_{r_1}w_{r_2} \ldots w_{r_m}w_{r_{m+1}} = w_{r_{2m}}w_{r_{2m-1}} \ldots w_{r_{m+2}}$$

and so $l(w_{r_1}w_{r_2} \ldots w_{r_{m+1}}) < m+1$. By 2.2.1 there is an integer $j \leqslant m$ such that $w_{r_1}w_{r_2} \ldots w_{r_j}(r_{j+1}) \in \Phi^-$. Therefore there is some integer $i \leqslant j$ for which

$$w_{r_{i+1}} \ldots w_{r_j}(r_{j+1}) \in \Phi^+,$$

$$w_{r_i}w_{r_{i+1}} \ldots w_{r_j}(r_{j+1}) \in \Phi^-.$$

Since $r_i \in \Pi$ this implies

$$w_{r_{i+1}} \ldots w_{r_j}(r_{j+1}) = r_i.$$

Thus

$$w_{r_i} = w_{r_{i+1}} \ldots w_{r_j}w_{r_{j+1}}w_{r_j} \ldots w_{r_{i+1}}$$

and therefore

$$w_{r_{i+1}} \ldots w_{r_{j+1}} = w_{r_i} \ldots w_{r_j}.$$

This is a relation of length $2(j-i+1)$. If this length is less than $2m$, the relation is a consequence of the given relations. By using this relation and the relations $w_i^2 = 1$ we have

$$w_{r_1} \ldots w_{r_k} = w_{r_1} \ldots w_{r_{i-1}}w_{r_{i+1}} \ldots w_{r_j}w_{r_{j+2}} \ldots w_{r_k}.$$

The relation $w_{r_1} \ldots w_{r_{i-1}}w_{r_{i+1}} \ldots w_{r_j}w_{r_{j+2}} \ldots w_{r_k} = 1$ has length less than k, so can be deduced from the given relations, and hence

$$w_{r_1} \ldots w_{r_k} = 1$$

can be deduced from the given relations also. This is a contradiction, and so we must have $2(j-i+1) = 2m$, whence $j = m$, $i = 1$. It follows that $w_{r_2} \ldots w_{r_m}(r_{m+1}) = r_1$.

However the relation $w_{r_1} \ldots w_{r_k} = 1$ is equivalent to the relation $w_{r_2} \ldots w_{r_k}w_{r_1} = 1$. Therefore $w_{r_3} \ldots w_{r_{m+1}}(r_{m+2}) = r_2$. Hence

$$w_{r_2} = w_{r_3} \ldots w_{r_{m+1}}w_{r_{m+2}}w_{r_{m+1}} \ldots w_{r_3}$$

and so

$$w_{r_3} \ldots w_{r_{m+2}} = w_{r_2} \ldots w_{r_{m+1}}.$$

Now this relation cannot be deduced from the given relations. For, by using it, $w_{r_1} \ldots w_{r_k}$ can be proved equal to the shorter expression $w_{r_1}w_{r_3} \ldots w_{r_{m+1}}w_{r_{m+3}} \ldots w_{r_k}$ and $w_{r_1}w_{r_3} \ldots w_{r_{m+1}}w_{r_{m+3}} \ldots w_{r_k} = 1$ can be deduced from the given relations. Thus we have another relation of length $2m$ which cannot be deduced from the given relations, and may be written in the form

$$w_{r_3}w_{r_2}w_{r_3}w_{r_4} \ldots w_{r_{m+1}}w_{r_{m+2}}w_{r_{m+1}} \ldots w_{r_4} = 1.$$

As before we obtain $w_{r_2} \ldots w_{r_m}(r_{m+1}) = r_3$. However we already have $w_{r_2} \ldots w_{r_m}(r_{m+1}) = r_1$, therefore $r_1 = r_3$.

We may now write the relation $w_{r_1} \ldots w_{r_k} = 1$ in equivalent forms by cyclic permutation of $r_1, \ldots, r_k$, and deduce that

$$r_1 = r_3 = \ldots = r_{2m-1},$$

$$r_2 = r_4 = \ldots = r_{2m}.$$

Putting $r_1 = r$ and $r_2 = s$, the relation $w_{r_1} \ldots w_{r_k} = 1$ becomes $(w_r w_s)^m = 1$. Thus m must be a multiple of the order m_{rs} of $w_r w_s$. But then $(w_r w_s)^m = 1$ is a consequence of $(w_r w_s)^{m_{rs}} = 1$, which is one of the given relations. Thus we have a contradiction, and the theorem is proved. ∎

Definition 2.4.2. A group defined by generators and relations $G = \langle a_i; (a_i a_j)^{m_{ij}} = 1 \rangle$, where $m_{ii} = 1$, is called a Coxeter group.

Thus every Weyl group is a Coxeter group. Most but not all of the finite Coxeter groups are Weyl groups (cf. Bourbaki [1]).

The second definition of W by generators and relations involves the set of all reflections w_r, $r \in \Phi$. The proof is also due to Steinberg.

THEOREM 2.4.3. *W is defined as an abstract group by generators w_r, $r \in \Phi$, subject to relations $w_r^2 = 1$ and $w_r w_s w_r = w_{w_r(s)}$.*

PROOF. We show first that the abstract group defined by the given generators and relations is generated by the elements w_r for $r \in \Pi$. This is certainly true of the Weyl group W. In fact if r is any root in Φ we have

$$r = w_{r_1} w_{r_2} \ldots w_{r_k}(s)$$

for suitable roots $r_1, \ldots, r_k, s \in \Pi$, by 2.1.8.

Thus

$$w_r = w_{r_1} \ldots w_{r_k} w_s w_{r_k} \ldots w_{r_1}.$$

We show that this relation, which expresses an arbitrary reflection as a product of fundamental reflections, is a consequence of the given relations. This is clear, since the given relations imply

$$w_{r_k} w_s w_{r_k} = w_{s_1}, \quad \text{where} \quad s_1 = w_{r_k}(s),$$

$$w_{r_{k-1}} w_{s_1} w_{r_{k-1}} = w_{s_2}, \quad \text{where} \quad s_2 = w_{r_{k-1}} w_{r_k}(s),$$

and hence

$$w_{r_1} \ldots w_{r_k} w_s w_{r_k} \ldots w_{r_1} = w_r.$$

Thus the group defined by the given generators and relations is generated by the w_r for $r \in \Pi$.

Now any relation involving the generators w_r, $r \in \Phi$, can be expressed, using the given relations, in the form

$$w_{r_1} w_{r_2} \ldots w_{r_k} = 1, \qquad r_i \in \Pi.$$

We show by induction on k that such a relation is a consequence of the given relations. This is clear if $k = 0$, so assume $k > 0$. Since

$$l(w_{r_1} w_{r_2} \ldots w_{r_k}) < k,$$

there exists an integer $j \leqslant k - 1$ such that

$$l(w_{r_1} \ldots w_{r_j} w_{r_{j+1}}) = l(w_{r_1} \ldots w_{r_j}) - 1.$$

Then by 2.2.1 we have $w_{r_1} \ldots w_{r_j}(r_{j+1}) \in \Phi^-$. Hence there exists an integer $i \leqslant j$ such that

$$w_{r_{i+1}} \ldots w_{r_j}(r_{j+1}) \in \Phi^+,$$

$$w_{r_i} w_{r_{i+1}} \ldots w_{r_j}(r_{j+1}) \in \Phi^-.$$

Since $r_i \in \Pi$ this implies that $w_{r_{i+1}} \ldots w_{r_j}(r_{j+1}) = r_i$, and therefore

$$w_{r_i} = w_{r_{i+1}} \ldots w_{r_j} w_{r_{j+1}} w_{r_j} \ldots w_{r_{i+1}}.$$

Now this relation is clearly a consequence of the given relations. Thus so is the relation

$$w_{r_i} \ldots w_{r_j} = w_{r_{i+1}} \ldots w_{r_{j+1}}.$$

However the relation $w_{r_1} \ldots w_{r_k} = 1$ can be proved equivalent, using this relation, to the shorter relation

$$w_{r_1} \ldots w_{r_{i-1}} w_{r_{i+1}} \ldots w_{r_j} w_{r_{j+2}} \ldots w_{r_k} = 1.$$

This can be deduced from the given relations by induction, so

$$w_{r_1} \ldots w_{r_k} = 1$$

can be deduced from the given relations also, and the theorem is proved. ∎

2.5 Parabolic Subgroups of a Weyl Group

We shall now discuss certain subgroups of a Weyl group which are themselves Weyl groups of certain subsystems of the root system Φ. Let Π by a fundamental system in Φ and Φ^+ be the corresponding positive system. Let J be a subset of Π. We define $\mathfrak{V}_J$ to be the subspace of $\mathfrak{V}$ spanned by J; Φ_J to be $\Phi \cap \mathfrak{V}_J$; and W_J to be the subgroup of W generated by the fundamental reflections w_r with $r \in J$.

PROPOSITION 2.5.1. Φ_J is a system of roots in $\mathfrak{V}_J$. J is a fundamental system in Φ_J. The Weyl group of Φ_J is W_J.

PROOF. It is clear that Φ_J spans $\mathfrak{V}_J$. If $r, s \in \Phi_J$, then

$$w_r(s) = s - \frac{2(r, s)}{(r, r)} r$$

is in Φ_J also. Thus Φ_J is a system of roots in $\mathfrak{V}_J$.

Now J is a linearly independent set and every root in Φ_J is a linear combination of elements of J. Since J is a subset of Π the coefficients in such a linear combination must be all non-negative or all non-positive. Thus J is a fundamental system in Φ_J. The Weyl group of Φ_J is generated by its fundamental reflections w_r, $r \in J$, so is W_J. ∎

The subgroups W_J and their conjugates in W are called parabolic subgroups of W.

PROPOSITION 2.5.2. The subgroups W_J for distinct subsets J of Π are all distinct.

PROOF. Suppose J, K are distinct subsets of Π and that $W_J = W_K$. Assume, without loss of generality, that there is a root r in K but not in J. Then

$$w_r(x) - x = -\frac{2(r, x)}{(r, r)} r.$$

Now $w_r \in W_J$ and so $w_r(x) - x \in \mathfrak{V}_J$ for each $x \in \mathfrak{V}$. Choosing x so that $(r, x) \neq 0$, we have $r \in \mathfrak{V}_J$. This is a contradiction since the fundamental roots are linearly independent. ∎

LEMMA 2.5.3. *Let v be a vector in $\bar{C}$, the closure of the fundamental chamber, and let w be an element of W such that $w(v)=v$. Then $w \in W_J$, where J is the set of roots in Π orthogonal to v.*

PROOF. We use induction on $l(w)$. If $l(w)=0$ the result is clear. If $l(w)>0$ there is a root $r \in \Pi$ such that $w(r) \in \Phi^-$. Then

$$0 \leqslant (r, v)=(w(r), v) \leqslant 0$$

since $v \in \bar{C}$. Hence $(r, v)=0$ and $w_r(v)=v$. But now $ww_r(v)=v$ and

$$l(ww_r)=l(w)-1$$

by 2.2.1. Thus $ww_r \in W_J$ by induction, whence $w \in W_J$.

COROLLARY 2.5.4. *Let v be any vector in $\mathfrak{V}$ and w be an element of W such that $w(v)=v$. Then w is a product of reflections corresponding to roots orthogonal to v.*

PROOF. Choose a fundamental system so that v is in the closure of the fundamental chamber, and apply 2.5.3. ∎

THEOREM 2.5.5. *Let $w \in W$ and $\mathfrak{U}$ be the subspace of $\mathfrak{V}$ of all vectors fixed by w. Then w is a product of reflections corresponding to roots in the orthogonal complement $\mathfrak{U}^\perp$ of $\mathfrak{U}$.*

PROOF. Let $v_1, \ldots, v_k$ be a basis for $\mathfrak{U}$. We must show that w is a product of reflections corresponding to roots orthogonal to $v_1, \ldots, v_k$. We shall prove more generally that if w fixes any finite set of vectors $v_1, \ldots, v_k$ then w is a product of reflections corresponding to roots orthogonal to $v_1, \ldots, v_k$. We proceed by induction on k, the result being true for $k=1$ by 2.5.4.

Choose a fundamental system Π so that v_k is in the closure of the fundamental chamber, and let J be the set of fundamental roots orthogonal to v_k. Consider the decomposition

$$\mathfrak{V}=\mathfrak{V}_J \oplus \mathfrak{V}_{\bar{J}}$$

and let $v_i=v_i'+v_i''$, where $v_i' \in \mathfrak{V}_J$, $v'' \in \mathfrak{V}_{\bar{J}}$. Now $w \in W_J$ by 2.5.3, and therefore $w(v_i'')=v_i''$. Since $w(v_i)=v_i$ we have also $w(v_i')=v_i'$. Now W_J is a Weyl group operating on $\mathfrak{V}_J$ and the element w of W_J fixes

$$v_1', v_2', \ldots, v_{k-1}'.$$

Thus w is a product of reflections corresponding to roots in $\mathfrak{P}_J$ orthogonal to $v_1', \ldots, v_{k-1}'$, by induction. These roots are all orthogonal to

$$v_1, v_2, \ldots, v_k,$$

so the theorem is proved. ∎

THEOREM 2.5.6. *Let J, K be subsets of Π. Then*:
(i) *the subgroup of W generated by W_J, W_K is $W_{J \cup K}$*
(ii) $W_J \cap W_K = W_{J \cap K}$.

PROOF. The statement (i) is clear from the definitions of W_J, W_K. In statement (ii) it is clear that $W_{J \cap K} \subseteq W_J \cap W_K$. Let $w \in W_J \cap W_K$. We must show that $w \in W_{J \cap K}$. Now $w(v) = v$ for each $v \in \mathfrak{P}_J^\perp$, also for each $v \in \mathfrak{P}_K^\perp$. Thus w fixes every vector in $\mathfrak{P}_J^\perp + \mathfrak{P}_K^\perp$.

Now

$$\mathfrak{P}_J^\perp + \mathfrak{P}_K^\perp \subseteq \mathfrak{P}_{J \cap K}^\perp.$$

Also

$$\dim (\mathfrak{P}_J^\perp + \mathfrak{P}_K^\perp) = \dim \mathfrak{P}_J^\perp + \dim \mathfrak{P}_K^\perp - \dim (\mathfrak{P}_J^\perp \cap \mathfrak{P}_K^\perp)$$
$$= \dim \mathfrak{P}_J^\perp + \dim \mathfrak{P}_K^\perp - \dim (\mathfrak{P}_J + \mathfrak{P}_K)^\perp$$
$$= l - \dim \mathfrak{P}_J + l - \dim \mathfrak{P}_K - l + \dim (\mathfrak{P}_J + \mathfrak{P}_K)$$
$$= l - \dim (\mathfrak{P}_J \cap \mathfrak{P}_K) = l - \dim \mathfrak{P}_{J \cap K}$$
$$= \dim \mathfrak{P}_{J \cap K}^\perp.$$

Hence $\mathfrak{P}_J^\perp + \mathfrak{P}_K^\perp = \mathfrak{P}_{J \cap K}^\perp$.
Now w fixes every vector in $\mathfrak{P}_{J \cap K}^\perp$, so is a product of reflections corresponding to roots in $\mathfrak{P}_{J \cap K}$ by 2.5.5. Thus $w \in W_{J \cap K}$.

COROLLARY 2.5.7. *The subgroups W_J form a lattice of 2^l subgroups in W.*

PROOF. This follows from 2.5.2 and 2.5.6. ∎

We now show that there is a natural way of choosing a system of representatives of the left cosets wW_J of W_J in W. Let D_J be the set of elements $w \in W$ such that $w(r) \in \Phi^+$ for all $r \in J$. D_J is a subset, although not generally a subgroup, of W.

THEOREM 2.5.8. *Let J be a subset of Π. Then each element of W has a unique expression of the form $w = d_J w_J$, where $d_J \in D_J$ and $w_J \in W_J$. Furthermore, we have $l(w) = l(d_J) + l(w_J)$.*

PROOF. We show first that each element of W can be expressed in the form $w = d_J w_J$, where $d_J \in D_J$, $w_J \in W_J$ and $l(w) = l(d_J) + l(w_J)$. If $l(w) = 0$, $1 = 1.1$ is the required factorization, thus we assume $l(w) > 0$ and use induction on $l(w)$. If $w \in D_J$ then $w = w.1$ is the required factorization. If $w \notin D_J$ there is a root $r \in J$ such that $w(r) \in \Phi^-$. Then

$$l(w w_r) = l(w) - 1$$

by 2.2.1. By induction we have $w w_r = d_J w_J$ with $l(d_J) + l(w_J) = l(w w_r)$. Thus $w = d_J w_J w_r$, where $d_J \in D_J$, $w_J w_r \in W_J$ and $l(d_J) + (l(w_J) + 1) = l(w)$. Now $l(w_J w_r) \leqslant l(w_J) + 1$; however if $l(w_J w_r) < l(w_J) + 1$ there would be an expression for w as a product of fundamental reflections of length less than $l(d_J) + l(w_J) + 1$, a contradiction. Hence $l(w_J w_r) = l(w_J) + 1$ and $l(d_J) + l(w_J w_r) = l(w)$.

We now prove the uniqueness of the factorization $w = d_J w_J$. Suppose we have

$$d_J w_J = d'_J w'_J,$$

where d_J, $d'_J \in D_J$ and w_J, $w'_J \in W_J$. Then $d'_J = d_J w_J (w'_J)^{-1}$. Suppose $w_J (w'_J)^{-1} \neq 1$. By 2.5.1 W_J is the Weyl group of the root system Φ_J with fundamental system J. Thus there exists $r \in J$ such that $w_J (w'_J)^{-1}(r) \in \Phi_J^-$. Thus $d_J w_J (w'_J)^{-1}(r) \in \Phi^-$. However $d'_J(r) \in \Phi^+$, and so we have a contradiction. This shows that $w'_J = w_J$ and $d'_J = d_J$. ∎

COROLLARY 2.5.9. *In each coset $w W_J$ there is a unique element of D_J. The length of this element is smaller than the length of any other element in $w W_J$.*

The elements of D_J are called distinguished coset representatives of W_J in W.

2.6 The Coxeter Complex

The parabolic subgroups of a Weyl group W can be described geometrically in terms of the Coxeter complex of W. This is a family of subsets of $\mathcal{V}$ defined as follows. We introduce an equivalence relation on $\mathcal{V}$ writing $x \sim y$ if, for each hyperplane H_r, $r \in \Phi$, the points x, y are either both in

H_r or both not in H_r but on the same side of H_r. The points in $\mathcal{V}$ fall into equivalence classes with respect to this relation, and the set of these equivalence classes is the Coxeter complex.

If, for each root $r \in \Phi^+$, we define

$$H_r^+ = \{v; \ (r, v) > 0\},$$
$$H_r^- = \{v; \ (r, v) < 0\},$$
$$H_r = H_r^0 = \{v; \ (r, v) = 0\},$$

then the Coxeter complex is the collection of all subsets of $\mathcal{V}$ of the form

$$\underset{r \in \Phi^+}{\cap} H_r^{\epsilon_r}, \qquad \epsilon_r = +, \ -, \ \text{or} \ 0.$$

Observe that every element of the Coxeter complex lies in the closure of some chamber. Choosing a fundamental system Π, the elements of the Coxeter complex contained in the closure of the fundamental chamber are those of the form

$$C_J = \left\{ v; \ \begin{matrix} (v, r) = 0 \ \text{for} \ r \in J \\ (v, r) > 0 \ \text{for} \ r \in \Pi - J \end{matrix} \right\},$$

where J is any subset of Π.

Let w be an element of W. Then w transforms each reflecting hyperplane into another, so transforms each set $H_r^{\epsilon_r}$ into a set $H_s^{\epsilon_s}$. Thus if $\mathfrak{K}$ is an element of the Coxeter complex so is $w(\mathfrak{K})$. We therefore have an operation of the Weyl group on the Coxeter complex.

PROPOSITION 2.6.1. *The stabilizer of C_J in W is W_J.*

PROOF. It is clear that an element $w \in W_J$ stabilizes C_J. In fact, since C_J is contained in the hyperplane H_r for all $r \in J$, w fixes each vector in C_J.

Suppose w is an element in W such that $w(C_J) = C_J$. We may write w in the form $w = d_J w_J$, $d_J \in D_J$, $w_J \in W_J$, as in 2.5.8. Then

$$C_J = w(C_J) = d_J w_J(C_J) = d_J(C_J).$$

Suppose $d_J \neq 1$. Then there is a root $r \in \Pi$ such that $d_J(r) \in \Phi^-$. r cannot be in J, by definition of D_J. Thus $r \in \Pi - J$. Let v be a vector in C_J. Then $(v, r) > 0$ and so $(d_J(v), -d_J(r)) < 0$. However $d_J(v) \in C_J$, so lies in the closure $\bar{C}$ of the fundamental chamber, and $-d_J(r) \in \Phi^+$. Thus $(d_J(v), -d_J(r)) \geq 0$ and we have a contradiction. Hence $d_J = 1$ and $w \in W_J$. ∎

COROLLARY 2.6.2. *Every element of W which transforms C_J into itself transforms every vector in C_J into itself.*

PROPOSITION 2.6.3. *Each element of the Coxeter complex can be transformed into exactly one element C_J by an element of the Weyl group.*

PROOF. Let $\mathfrak{K}$ be an element of the Coxeter complex. $\mathfrak{K}$ is contained in the closure of some chamber C' and there is an element $w \in W$ such that $w(C') = C$, the fundamental chamber. Hence $w(\mathfrak{K}) \subseteq \bar{C}$, and so $w(\mathfrak{K}) = C_J$ for some subset J of Π.

In order to prove the uniqueness of J it is sufficient to show that $w(C_J) = C_{\bar{J}}$ implies $J = \bar{J}$. We write $w = d_J w_J$ as in 2.5.8. Then $d_J(C_J) = C_{\bar{J}}$. Suppose $d_J \neq 1$. Then there is a root $r \in \Pi - J$ such that $d_J(r) \in \Phi^-$. Let v be a vector in C_J. Then $(v, r) > 0$ and $(d_J(v), -d_J(r)) < 0$. But $d_J(v) \in C_{\bar{J}}$ so is in the closure of the fundamental chamber. Also $-d_J(r) \in \Phi^+$, there-therefore $(d_J(v), -d_J(r)) \geqslant 0$ and we have a contradiction. Thus $d_J = 1$ and $C_J = C_{\bar{J}}$. ∎

We may now obtain the required geometrical description of the parabolic subgroups of the Weyl group.

PROPOSITION 2.6.4. *The parabolic subgroups of W are the stabilizers in W of the elements of the Coxeter complex.*

PROOF. Let $\mathfrak{K}$ be an element of the Coxeter complex. Then $\mathfrak{K} = w(C_J)$ for some $w \in W$ and some subset J of Π. Now $w'(\mathfrak{K}) = \mathfrak{K}$ if and only if $w^{-1}w'w(C_J) = C_J$, which holds if and only if $w' \in wW_Jw^{-1}$ by 2.6.1. Thus the stabilizer of $\mathfrak{K}$ is wW_Jw^{-1}, a parabolic subgroup of W. Conversely, any parabolic subgroup of W has form wW_Jw^{-1}, so is the stabilizer of some element of the Coxeter complex. ∎

CHAPTER 3

Simple Lie Algebras

The simple groups with which we are concerned in this volume are defined as groups of automorphisms of Lie algebras. Before introducing them we need certain introductory material on Lie algebras, in particular a knowledge of the structure of simple Lie algebras over the complex field. Each such Lie algebra determines a root system and a Weyl group, to which the results of the preceding chapter apply. In the present chapter we summarize the properties of Lie algebras which we shall need. Proofs of all the properties which we describe can be found, for example, in Jacobson's book [1].

3.1 Lie Algebras and Subalgebras

A Lie algebra is a vector space $\mathfrak{L}$ over a field K on which a product operation $[xy]$ is defined satisfying the following axioms:

(i) $[xy]$ is bilinear for x, $y \in \mathfrak{L}$.

(ii) $[xx] = 0$ for $x \in \mathfrak{L}$.

(iii) $[[xy]z] + [[yz]x] + [[zx]y] = 0$ for x, y, $z \in \mathfrak{L}$.

Axiom (iii) is called the Jacobi identity. We note that $[[xy]z]$ is not necessarily equal to $[x[yz]]$, thus Lie multiplication is not in general associative. As a simple consequence of axioms (i), (ii) we have

$$0 = [x+y, x+y] = [xx] + [xy] + [yx] + [yy] = [xy] + [yx]$$

Thus $[yx] = -[xy]$ and Lie multiplication is anticommutative. In the present work we shall be concerned only with finite-dimensional Lie algebras.

Let $\mathfrak{L}$ be a Lie algebra and $\mathfrak{M}$, $\mathfrak{N}$ be subspaces of $\mathfrak{L}$. We define $[\mathfrak{M}\mathfrak{N}]$ to be the subspace of $\mathfrak{L}$ spanned by all elements of form $[xy]$ for $x \in \mathfrak{M}$, $y \in \mathfrak{N}$. Since $[yx] = -[xy]$ it is clear that $[\mathfrak{N}\mathfrak{M}] = [\mathfrak{M}\mathfrak{N}]$. Thus multiplication of subspaces is commutative.

A subalgebra of $\mathfrak{L}$ is a subspace $\mathfrak{M}$ such that $[\mathfrak{M}\mathfrak{M}] \subseteq \mathfrak{M}$, and an ideal of $\mathfrak{L}$ is a subspace $\mathfrak{M}$ such that $[\mathfrak{M}\mathfrak{L}] \subseteq \mathfrak{M}$. Since $[\mathfrak{M}\mathfrak{L}] = [\mathfrak{L}\mathfrak{M}]$ there is no distinction in the theory of Lie algebras between left ideals and right ideals. Every ideal is two-sided.

For each element x of a Lie algebra $\mathfrak{L}$ we define a map ad x of $\mathfrak{L}$ into itself by

$$\text{ad } x.y = [xy], \qquad y \in \mathfrak{L}.$$

ad x is a linear map, and also satisfies the condition

$$\text{ad } x.[yz] = [x[yz]] = [[xy]z] + [y[xz]] = [\text{ad } x.y, z] + [y, \text{ad } x.z].$$

A linear map δ of $\mathfrak{L}$ into itself satisfying

$$\delta[yz] = [\delta y, z] + [y, \delta z], \qquad y, z \in \mathfrak{L},$$

is called a derivation of $\mathfrak{L}$. Thus ad x is a derivation of $\mathfrak{L}$ for each x. We note further that

$$\text{ad } x.\text{ad } y - \text{ad } y.\text{ad } x = \text{ad } [xy].$$

For, given $z \in \mathfrak{L}$, we have

$$(\text{ad } x.\text{ad } y - \text{ad } y.\text{ad } x).z = [x[yz]] - [y[xz]] = [[xy]z] = \text{ad } [xy].z.$$

An important rôle in the theory of Lie algebras is played by a scalar product called the Killing form. For each x, $y \in \mathfrak{L}$ we define the scalar product (x, y) by

$$(x, y) = \text{tr } (\text{ad } x.\text{ad } y).$$

As the trace of the linear map ad $x.\text{ad } y$, (x, y) is an element of the field K. The scalar product defined in this way is certainly bilinear, and is also symmetric, since tr $(\theta\phi) = \text{tr } (\phi\theta)$ for any two linear maps θ, ϕ of $\mathfrak{L}$ into itself.

After these introductory definitions we turn to a consideration of the simple Lie algebras over the complex field $\mathbb{C}$. A Lie algebra is said to be simple if it has no ideals other than itself and the zero subspace. The 1-dimensional Lie algebra over any field is certainly simple and is called a trivial algebra. We are concerned with simple non-trivial Lie algebras, and we begin with an example which illustrates clearly the main features of the general theory. We observe first that any associative algebra can be made into a Lie algebra by defining the Lie multiplication by

$$[xy] = xy - yx.$$

For $[xy]$ is clearly bilinear, $[xx] = 0$ and

$$[[xy]z] + [[yz]x] + [[zx]y] = (xy - yx)z - z(xy - yx) + (yz - zy)x$$
$$- x(yz - zy) + (zx - xz)y - y(zx - xz)$$
$$= 0.$$

Consider the algebra of all $(l+1) \times (l+1)$ matrices over $\mathbb{C}$. This algebra has dimension $(l+1)^2$ and may be made into a Lie algebra as described above. The matrices of trace 0 form a subalgebra of this Lie algebra of dimension $(l+1)^2 - 1 = l(l+2)$. For we have

$$\mathrm{tr}\ (x+y) = \mathrm{tr}\ x + \mathrm{tr}\ y = 0,$$

$$\mathrm{tr}\ (\lambda x) = \lambda\ \mathrm{tr}\ x = 0, \qquad \lambda \in \mathbb{C},$$

$$\mathrm{tr}\ [xy] = \mathrm{tr}\ (xy - yx) = 0.$$

The Lie algebra of all $(l+1) \times (l+1)$ matrices of trace 0 is in fact simple.

3.2 The Cartan Decomposition

The classification of the simple Lie algebras over $\mathbb{C}$ was obtained by W. Killing [1] and E. Cartan [1]. This classification is achieved by decomposing such an algebra with respect to a certain type of subalgebra, now called a Cartan subalgebra. A subalgebra $\mathfrak{H}$ of the Lie algebra $\mathfrak{L}$ is called a Cartan subalgebra if it satisfies the following two conditions:

(i) $[[[\mathfrak{H}\mathfrak{H}]\mathfrak{H}]\ldots] = 0$ for some r.
$\overset{\longleftarrow r \longrightarrow}{}$

Subalgebras satisfying this condition are called nilpotent.

(ii) If $[xh] \in \mathfrak{H}$ for all $h \in \mathfrak{H}$ then $x \in \mathfrak{H}$.

This condition means that $\mathfrak{H}$ is not contained as an ideal in any larger subalgebra of $\mathfrak{L}$.

It can be shown that any Lie algebra over $\mathbb{C}$ has Cartan subalgebras and any two Cartan subalgebras are isomorphic. In fact, given any two Cartan subalgebras of $\mathfrak{L}$, there is an automorphism of $\mathfrak{L}$ which transforms one into the other. The dimension of the Cartan subalgebras of $\mathfrak{L}$ is called the rank of $\mathfrak{L}$, and will usually be denoted by l.

If the algebra $\mathfrak{L}$ is simple over $\mathbb{C}$ the Cartan subalgebras actually satisfy $[\mathfrak{H}\mathfrak{H}] = 0$, although this is not true in general for non-simple algebras. Thus for a simple Lie algebra, Lie multiplication inside a Cartan subalgebra is trivial.

Let $\mathfrak{L}$ be simple over $\mathbb{C}$ and $\mathfrak{H}$ be a Cartan subalgebra of $\mathfrak{L}$. Then $\mathfrak{L}$ can be decomposed into a direct sum of $\mathfrak{H}$ with a number of 1-dimensional subspaces all invariant under multiplication by $\mathfrak{H}$. Thus

$$\mathfrak{L} = \mathfrak{H} \oplus \mathfrak{L}_{r_1} \oplus \mathfrak{L}_{r_2} \oplus \ldots \oplus \mathfrak{L}_{r_k},$$

where dim $\mathfrak{L}_{r_i} = 1$ and $[\mathfrak{H}\mathfrak{L}_{r_i}] = \mathfrak{L}_{r_i}$, for each i. This is called a Cartan decomposition of $\mathfrak{L}$.

For example, if $\mathfrak{L}$ is the algebra of all $(l+1) \times (l+1)$ matrices of trace 0, it is easy to see that the diagonal matrices of trace 0 form a Cartan subalgebra $\mathfrak{H}$. Then we have

$$\mathfrak{L} = \mathfrak{H} \oplus \sum_{i \neq j} \mathbb{C}e_{ij},$$

where e_{ij} is the elementary matrix with 1 in the (i, j) position and 0 elsewhere. This direct decomposition is a Cartan decomposition. For let $h = \text{diag}\,(\lambda_0, \lambda_1, \ldots, \lambda_l)$. Then

$$[he_{ij}] = he_{ij} - e_{ij}h = (\lambda_i - \lambda_j)e_{ij}.$$

Hence the 1-dimensional subspace $\mathbb{C}e_{ij}$ is invariant under $\mathfrak{H}$.

3.3 The Roots of a Simple Lie Algebra

Let $\mathfrak{L}$ be a simple Lie algebra over $\mathbb{C}$ and

$$\mathfrak{L} = \mathfrak{H} \oplus \mathfrak{L}_{r_1} \oplus \ldots \oplus \mathfrak{L}_{r_k}$$

be a Cartan decomposition of $\mathfrak{L}$. In each 1-dimensional subspace $\mathfrak{L}_r$ we choose a non-zero element e_r. Then, for each $h \in \mathfrak{H}$, $[he_r]$ is a scalar multiple of e_r, and we write

$$[he_r] = r(h)e_r.$$

The map $r : \mathfrak{H} \rightarrow \mathbb{C}$ defined in this way is certainly linear, so is an element of the dual space of $\mathfrak{H}$. The maps $r_1, r_2, \ldots, r_k$ from $\mathfrak{H}$ into $\mathbb{C}$ are called the roots of $\mathfrak{L}$ and the subspaces $\mathfrak{L}_{r_1}, \mathfrak{L}_{r_2}, \ldots, \mathfrak{L}_{r_k}$ are called the root-spaces of $\mathfrak{L}$ (relative to the given Cartan subalgebra $\mathfrak{H}$). This terminology originated from the fact that $r(h)$ is a root of the characteristic equation of the map ad h. The roots $r_1, r_2, \ldots, r_k$ are in fact all distinct and all non-zero. Thus the zero map is not a root.

Although the roots are defined as elements of the dual space of $\mathfrak{H}$ they can, by considering the Killing form, be regarded as elements of $\mathfrak{H}$ itself. It can be shown that the Killing form of a Lie algebra $\mathfrak{L}$ is non-singular if and only if $\mathfrak{L}$ is semi-simple, i.e. has no proper ideal in which the Lie multiplication is trivial. In particular every simple non-trivial algebra is semi-simple, and so the Killing form of $\mathfrak{L}$ remains non-singular when restricted to the Cartan subalgebra of $\mathfrak{H}$ (although the Killing form of $\mathfrak{H}$ itself is identically zero). Thus each element of the dual space of $\mathfrak{H}$

is expressible in the form $h \to (x, h)$ for a unique element $x \in \mathfrak{H}$. The element x associated with the map $h \to r(h)$ may be identified with the root r. Thus r can be regarded either as an element of $\mathfrak{H}$ or an element of its dual space; the relation between these two being given by

$$r(h) = (r, h), \qquad h \in \mathfrak{H}.$$

Considering the roots as elements of $\mathfrak{H}$, let Φ be the finite subset of $\mathfrak{H}$ obtained in this way. It can be shown that Φ spans $\mathfrak{H}$, and that if we choose any subset of Φ which is a basis for $\mathfrak{H}$ then each element of Φ is a linear combination of the roots in this subset with *rational* coefficients. Also (r, s) is rational for all $r, s \in \Phi$. We denote by $\mathfrak{H}_{\mathbb{R}}$ the set of all elements of $\mathfrak{H}$ which are linear combinations of elements of Φ with real coefficients. By the preceding remarks it is evident that $\mathfrak{H}_{\mathbb{R}}$ is a real vector space of the same dimension as the complex dimension of $\mathfrak{H}$. Also one can show that, if $x \in \mathfrak{H}_{\mathbb{R}}$, then $(x, x) \geqslant 0$ and $(x, x) = 0$ only if $x = 0$. Thus the Killing form is positive definite on $\mathfrak{H}_{\mathbb{R}}$ and so $\mathfrak{H}_{\mathbb{R}}$ may be regarded as a Euclidean space. In particular we can define the length of an element $x \in \mathfrak{H}_{\mathbb{R}}$ by

$$|x| = \sqrt{(x, x)}$$

and the angle θ between $x, y \in \mathfrak{H}_{\mathbb{R}}$ by

$$(x, y) = |x| \, |y| \cos \theta.$$

Now it is shown in the theory of simple Lie algebras that the subset Φ of the Euclidean space $\mathfrak{H}_{\mathbb{R}}$ forms a system of roots in the sense defined in 2.1.1. In particular, $2(r, s)/(r, r)$ is a rational integer for all $r, s \in \Phi$. We give an interpretation of this integer. Suppose r, s are linearly independent. Since Φ is finite there exist intergers $p, q \geqslant 0$ such that $ir + s \in \Phi$ for $-p \leqslant i \leqslant q$ but $-(p+1)r + s$ and $(q+1)r + s$ are not in Φ. The sequence of roots

$$-pr + s, \ldots, s, \ldots, qr + s$$

will be called the r-chain of roots through s. Now the reflection w_r in the hyperplane orthogonal to r can be shown to permute the elements of Φ. In fact it has the effect of inverting each r-chain of roots. In particular it transforms $-pr + s$ into $qr + s$ and so $-pr + s$, $qr + s$ are mirror images in the hyperplane orthogonal to r. Hence

$$((-pr + s) + (qr + s), r) = 0.$$

It follows that

$$\frac{2(r, s)}{(r, r)} = p - q.$$

For each pair of roots r, $s \in \Phi$ we define

$$A_{rs} = \frac{2(r, s)}{(r, r)}.$$

Thus A_{rs} is a rational integer which satisfies $A_{rs} = p - q$ and

$$w_r(s) = s - A_{rs}r.$$

By 2.1.2 the root system Φ contains a subsystem Π which is a system of fundamental roots. We shall denote such a subsystem by

$$\Pi = \{p_1, p_2, \ldots, p_l\}.$$

Every root in Φ is an integral combination of roots in Π with coefficients which are all non-negative or all non-positive and we denote by Φ^+, Φ^- the sets of positive and negative roots with respect to the fundamental system Π.

We illustrate the situation by means of an example. Let $\mathfrak{L}$ be the Lie algebra of all $(l+1) \times (l+1)$ matrices of trace 0. We have seen that the diagonal matrices in $\mathfrak{L}$ form a Cartan subalgebra $\mathfrak{H}$, and that

$$\mathfrak{L} = \mathfrak{H} \oplus \sum_{i \neq j} \mathbb{C}e_{ij}$$

is a Cartan decomposition. Let h be the diagonal matrix

$$\text{diag } (\lambda_0, \lambda_1, \ldots, \lambda_l).$$

Then

$$[he_{ij}] = (\lambda_i - \lambda_j)e_{ij}$$

and so the root corresponding to the subspace $\mathbb{C}e_{ij}$ is the map $h \to \lambda_i - \lambda_j$ of $\mathfrak{H}$ into $\mathbb{C}$. Let $p_1, p_2, \ldots, p_l$ be the roots defined by

$$p_1 : h \to \lambda_0 - \lambda_1,$$
$$p_2 : h \to \lambda_1 - \lambda_2,$$
$$p_l : h \to \lambda_{l-1} - \lambda_l.$$

Then $p_1, p_2, \ldots, p_l$ form a system of fundamental roots, and the other roots have form

$$\pm (p_{i+1} + \ldots + p_j), \qquad i < j.$$

The positive roots of this Lie algebra are therefore the sums of consecutive fundamental roots.

3.4 The Dynkin Diagram

Let p_i, p_j be distinct fundamental roots of a simple Lie algebra $\mathfrak{L}$ and θ_{ij} be the angle between them. Since $-p_i + p_j$ is not a root, p_j is the first member of the p_i-chain of roots through it. Using the relation

$$\frac{2(r, s)}{(r, r)} = p - q,$$

derived in section 3.3, and noting that $p = 0$, we see that $(p_i, p_j) \leqslant 0$. Thus the angle between two distinct fundamental roots is obtuse.

There are only a few possibilities for the value of this angle. For $2(p_i, p_j)/(p_i, p_i)$ and $2(p_j, p_i)/(p_j, p_j)$ are both integers, and so

$$\frac{4(p_i, p_j)^2}{(p_i, p_i)(p_j, p_j)} = 4 \cos^2 \theta_{ij}$$

is an integer also. Since $0 \leqslant \cos^2 \theta_{ij} \leqslant 1$, we have $4 \cos^2 \theta_{ij} = 0$, 1, 2, 3 or 4. Since θ_{ij} is obtuse, θ_{ij} is one of $\pi/2$, $2\pi/3$, $3\pi/4$, $5\pi/6$ or π. The fact that p_i, p_j are linearly independent excludes the possibility $\theta_{ij} = \pi$. Thus $\theta_{ij} = \pi/2$, $2\pi/3$, $3\pi/4$ or $5\pi/6$. We define an integer n_{ij} by $n_{ij} = 4 \cos^2 \theta_{ij}$. Thus $n_{ij} = 0$, 1, 2 or 3 if $i \neq j$. n_{ij} admits a factorization

$$n_{ij} = \frac{2(p_i, p_j)}{(p_i, p_i)} \cdot \frac{2(p_j, p_i)}{(p_j, p_j)}$$

into a product of two non-positive integers. We consider this factorization in the different cases which can arise.

(a) If $n_{ij} = 1$ the factorization must be $1 = -1 \cdot -1$. Thus $(p_i, p_i) = (p_j, p_j)$ and the roots p_i, p_j have the same length.

(b) If $n_{ij} = 2$ the factorization must be $2 = -1 \cdot -2$. Thus one of p_i, p_j is $\sqrt{2}$ times as long as the other.

(c) If $n_{ij} = 3$ the factorization must be $3 = -1 \cdot -3$. Thus one of p_i, p_j is $\sqrt{3}$ times as long as the other.

(d) If $n_{ij} = 0$ we obtain no information about the relative lengths of p_i, p_j.

We observe that the p_i-chain of roots through p_j has length 1, 2, 3 or 4. For

$$\frac{2(p_i, p_j)}{(p_i, p_i)} = p - q,$$

where p, q are the integers defined as before by the p_i-chain of roots through p_j. We have seen that

$$\frac{2(p_i, p_j)}{(p_i, p_i)} = 0, \ -1, \ -2 \text{ or } -3;$$

furthermore $p=0$ since p_j begins the p_i-chain through it. Hence $q \leqslant 3$, and so the p_i-chain through p_j contains at most four roots.

The same argument shows in fact that any r-chain has at most four roots. For let s be the first root in some r-chain. Then

$$\frac{2(r, s)}{(r, r)} = p - q$$

and, as before, this must take one of the values 0, -1, -2, -3. Since $p=0$ we have $q \leqslant 3$, and so the r-chain has at most four roots.

We now define the Dynkin diagram of the Lie algebra $\mathfrak{L}$. This is a graph with l nodes, one associated with each fundamental root p_i, such that the ith node is joined to the jth node by a bond of strength n_{ij}.

For example, in the Lie algebra of all $(l+1) \times (l+1)$ matrices of trace 0 it can be shown that the fundamental roots $p_1, p_2, \ldots, p_l$ all have the same length. Consecutive roots p_i, p_{i+1} are inclined at an angle $2\pi/3$ whereas fundamental roots which are not consecutive are orthogonal to one another. Thus the Dynkin diagram of this algebra is

$$
\begin{array}{ccccc}
1 & 2 & & l\text{-}1 & l \\
\circ\!\!&\!\!\!-\!\!\!&\!\!\circ\!-\!-\!-\!-\!\circ\!\!&\!\!-\!\!&\!\!\circ
\end{array}
$$

Now the possible Dynkin diagrams of simple Lie algebras can be enumerated, using the classical results of Killing and Cartan. It can be shown that the Dynkin diagram of a simple Lie algebra must be a connected graph. Furthermore the only connected graphs which can be Dynkin diagrams of simple Lie algebras are the ones in the following list

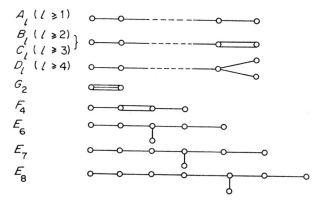

The diagrams are usually named as shown, the suffix denoting the rank (i.e. the number of nodes in the graph). The reason that the second type

of diagram is given two different names is that the Dynkin diagram does not always determine the simple Lie algebra to within isomorphism. Consider the problem of recovering the configuration formed by the fundamental roots from a knowledge of the Dynkin diagram. In the diagram

it is evident that the corresponding fundamental roots p_1, p_2, ..., p_{l-1} all have the same length, but p_l is either $\sqrt{2}$ times shorter or $\sqrt{2}$ times longer than the remainder. If p_l is shorter the system of fundamental roots is said to have type B_l, and if p_l is longer the system has type C_l. If $l=2$ there is no distinction between B_l, C_l as we can obtain either by numbering the nodes suitably. For the same reason there is only one fundamental root system of type G_2 and one of type F_4, since the diagrams are symmetric. In all remaining cases the Dynkin diagram contains only single bonds—thus all the fundamental roots have the same length and the configuration formed by the fundamental roots is uniquely determined by the diagram.

Now it is possible to recover the complete system of roots (as linear combinations of the fundamental roots) from a knowledge of the relative lengths of the fundamental roots and the angles between them. For if we know the configuration formed by the fundamental roots, we know the fundamental reflections w_r, $r \in \Pi$. Since the fundamental reflections generate W, the Weyl group is known. Finally, each root is the image of some fundamental root under an element of the Weyl group, thus the complete root system is determined.

3.5 The Existence and Isomorphism Theorems

We have seen that every simple Lie algebra over $\mathbb{C}$ determines a root system, and we now consider which root systems arise from simple Lie algebras, and to what extent a simple Lie algebra is determined by its root system.

A root system Φ is said to be indecomposable if it cannot be decomposed into two non-empty complementary subsets Φ_1, Φ_2 such that $(r, s) = 0$ for all $r \in \Phi_1$, $s \in \Phi_2$. Two root systems Φ_1, Φ_2 are said to be equivalent if there exists a bijection $\alpha : \Phi_1 \to \Phi_2$ such that

$$(\alpha(r), \alpha(s)) = \lambda(r, s), \qquad r, s \in \Phi_1,$$

where λ is some positive real number independent of r, s.

Now the root system determined by a simple Lie algebra is indecomposable. This follows from the fact that the Dynkin diagram of a simple Lie algebra is connected, using 2.1.8.

THEOREM 3.5.1 (Existence theorem). *Let* Φ *be an indecomposable root system. Then there exists a simple Lie algebra over* $\mathbb{C}$ *which has a root system equivalent to* Φ.

A proof of the existence theorem using the concepts we have outlined can be found in Tits [15].

We now consider the relation between two simple Lie algebras which have equivalent root systems. We first describe some properties concerning the multiplication of the root spaces in a Cartan decomposition. Let

$$\mathfrak{L} = \mathfrak{H} \oplus \sum_{r \in \Phi} \mathfrak{L}_r$$

be a Cartan decomposition of $\mathfrak{L}$. Then, for any pair of roots $r, s \in \Phi$ we have:

(i) $[\mathfrak{L}_r \mathfrak{L}_s] = \mathfrak{L}_{r+s}$,   if $r+s \in \Phi$.
(ii) $[\mathfrak{L}_r \mathfrak{L}_s] = 0$,   if $r+s \notin \Phi$, $r+s \neq 0$.
(iii) $[\mathfrak{L}_r \mathfrak{L}_{-r}] = \mathbb{C}r$.
(iv) $[\mathfrak{H} \mathfrak{L}_r] = \mathfrak{L}_r$.

In (iii) r is interpreted as an element of $\mathfrak{H}$. Instead of considering the root $r \in \mathfrak{H}$ it is often convenient to take a scalar multiple h_r of r, defined by

$$h_r = \frac{2r}{(r, r)}.$$

Since we have

$$[h_r e_s] = \frac{2(r, s)}{(r, r)} e_s = A_{rs} e_s,$$

it is evident that ad h_r transforms e_s into an integral multiple of itself. By property (iii) above we can find, for each $e_r \neq 0 \in \mathfrak{L}_r$, an element $e_{-r} \in \mathfrak{L}_{-r}$ such that $[e_r e_{-r}] = h_r$.

We can now state the isomorphism theorem for simple Lie algebras.

THEOREM 3.5.2 (Isomorphism theorem). *Let* $\mathfrak{L}$, $\mathfrak{L}'$ *be simple Lie algebras over* $\mathbb{C}$ *with Cartan subalgebras* $\mathfrak{H}$, $\mathfrak{H}'$ *of the same dimension* l. *Let* $p_1, p_2, \ldots, p_l$; *and* $p'_1, p'_2, \ldots, p'_l$ *be sets of fundamental roots for*

$\mathfrak{L}$, $\mathfrak{L}'$ and let

$$A_{ij}=\frac{2(p_i,p_j)}{(p_i,p_i)}, \qquad A'_{ij}=\frac{2(p'_i,p'_j)}{(p'_i,p'_i)}.$$

Let

$$h_{p_i}=\frac{2p_i}{(p_i,p_i)},$$

and let $e_{p_i}\in\mathfrak{L}_{p_i}$, $e_{-p_i}\in\mathfrak{L}_{-p_i}$ be chosen so that $[e_{p_i}e_{-p_i}]=h_{p_i}$. Define $h_{p'_i}$, $e_{p'_i}$ $e_{-p'_i}$ similarly in $\mathfrak{L}'$.

Suppose $A_{ij}=A'_{ij}$ for all i,j. Then there is a unique isomorphism $\theta:\mathfrak{L}\to\mathfrak{L}'$ such that $\theta(h_{p_i})=h_{p'_i}$, $\theta(e_{p_i})=e_{p'_i}$, $\theta(e_{-p_i})=e_{-p'_i}$.

In particular any two simple Lie algebras over $\mathbb{C}$ with equivalent root systems are isomorphic.

A proof of the isomorphism theorem can be found in Jacobson [1], p. 127.

3.6 Description of the Simple Lie Algebras

It follows from what has been said in sections 3.4 and 3.5 that the simple Lie algebras over $\mathbb{C}$ are the ones shown in the 'standard list' exhibited below. We have given for each algebra the dimension, the rank, the number N of positive roots, the order of the Weyl group, and the Dynkin diagram.

$\mathfrak{L}$	dim $\mathfrak{L}$	rank $\mathfrak{L}$	N	$\lvert W\rvert$	Dynkin diagram
$A_l(l\geqslant 1)$	$l(l+2)$	l	$\tfrac{1}{2}l(l+1)$	$(l+1)!$	
$B_l(l\geqslant 2)$	$l(2l+1)$	l	l^2	$2^l.l!$	
$C_l(l\geqslant 3)$	$l(2l+1)$	l	l^2	$2^l.l!$	
$D_l(l\geqslant 4)$	$l(2l-1)$	l	$l(l-1)$	$2^{l-1}.l!$	
G_2	14	2	6	12	
F_4	52	4	24	$2^7.3^2$	
E_6	78	6	36	$2^7.3^4.5$	
E_7	133	7	63	$2^{10}.3^4.5.7$	
E_8	248	8	120	$2^{14}.3^5.5^2.7$	

The matrix (A_{ij}) defined in 3.5.2 is called the Cartan matrix of $\mathfrak{L}$. The isomorphism theorem shows that the Cartan matrix determines the Lie algebra $\mathfrak{L}$. The Cartan matrices of the individual simple algebras are shown below.

$$
A_l: \begin{pmatrix}
2 & -1 & & & & & & \\
-1 & 2 & -1 & & & & & \\
& -1 & 2 & -1 & & & & \\
& & -1 & & & & & \\
& & & & & -1 & & \\
& & & & -1 & 2 & -1 & \\
& & & & & -1 & 2 & -1 \\
& & & & & & -1 & 2
\end{pmatrix},
$$

$$
B_l: \begin{pmatrix}
2 & -1 & & & & & & \\
-1 & 2 & -1 & & & & & \\
& -1 & 2 & -1 & & & & \\
& & -1 & & & & & \\
& & & & & -1 & & \\
& & & & -1 & 2 & -1 & \\
& & & & & -1 & 2 & -1 \\
& & & & & & -2 & 2
\end{pmatrix},
$$

$$
C_l: \begin{pmatrix}
2 & -1 & & & & & & \\
-1 & 2 & -1 & & & & & \\
& -1 & 2 & -1 & & & & \\
& & -1 & & & & & \\
& & & & & -1 & & \\
& & & & -1 & 2 & -1 & \\
& & & & & -1 & 2 & -2 \\
& & & & & & -1 & 2
\end{pmatrix},
$$

$$
D_l: \begin{pmatrix}
2 & -1 & & & & & & \\
-1 & 2 & -1 & & & & & \\
& -1 & 2 & -1 & & & & \\
& & -1 & & & & & \\
& & & & & -1 & & \\
& & & & -1 & 2 & -1 & -1 \\
& & & & & -1 & 2 & 0 \\
& & & & & -1 & 0 & 2
\end{pmatrix},
$$

G_2: $\begin{pmatrix} 2 & -1 \\ -3 & 2 \end{pmatrix},$

F_4: $\begin{pmatrix} 2 & -1 & 0 & 0 \\ -1 & 2 & -1 & 0 \\ 0 & -2 & 2 & -1 \\ 0 & 0 & -1 & 2 \end{pmatrix},$

E_6: $\begin{pmatrix} 2 & -1 & 0 & 0 & 0 & 0 \\ -1 & 2 & -1 & 0 & 0 & 0 \\ 0 & -1 & 2 & -1 & -1 & 0 \\ 0 & 0 & -1 & 2 & 0 & 0 \\ 0 & 0 & -1 & 0 & 2 & -1 \\ 0 & 0 & 0 & 0 & -1 & 2 \end{pmatrix},$

E_7: $\begin{pmatrix} 2 & -1 & 0 & 0 & 0 & 0 & 0 \\ -1 & 2 & -1 & 0 & 0 & 0 & 0 \\ 0 & -1 & 2 & -1 & 0 & 0 & 0 \\ 0 & 0 & -1 & 2 & -1 & -1 & 0 \\ 0 & 0 & 0 & -1 & 2 & 0 & 0 \\ 0 & 0 & 0 & -1 & 0 & 2 & -1 \\ 0 & 0 & 0 & 0 & 0 & -1 & 2 \end{pmatrix},$

E_8: $\begin{pmatrix} 2 & -1 & 0 & 0 & 0 & 0 & 0 & 0 \\ -1 & 2 & -1 & 0 & 0 & 0 & 0 & 0 \\ 0 & -1 & 2 & -1 & 0 & 0 & 0 & 0 \\ 0 & 0 & -1 & 2 & -1 & 0 & 0 & 0 \\ 0 & 0 & 0 & -1 & 2 & -1 & -1 & 0 \\ 0 & 0 & 0 & 0 & -1 & 2 & 0 & 0 \\ 0 & 0 & 0 & 0 & -1 & 0 & 2 & -1 \\ 0 & 0 & 0 & 0 & 0 & 0 & -1 & 2 \end{pmatrix}.$

We also give a description of the indecomposable root systems. We begin with the systems of rank 1 and 2, viz., systems of type A_1, A_2, B_2, G_2. Figure 1 shows the roots expressed as integral combinations of fundamental roots.

In order to describe the root systems of higher rank it is convenient to use an orthonormal basis of the vector space containing the roots.

(i) *Type A_l.* Let $e_0, e_1, \ldots, e_l$ be an orthonormal basis of a Euclidean space of dimension $l+1$, and let $\mathcal{V}$ be the subspace of vectors

$$\sum_{i=0}^{l} \lambda_i e_i \text{ with } \sum_{i=0}^{l} \lambda_i = 0.$$

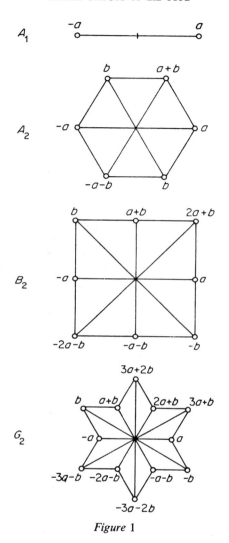

Figure 1

Then the following vectors in $\mathcal{V}$ form a fundamental system of type A_l.

$$\underset{e_0-e_1}{\circ}\quad\underset{e_1-e_2}{\circ}\ \text{-}\text{-}\text{-}\ \underset{e_{l-2}-e_{l-1}}{\circ}\quad\underset{e_{l-1}-e_l}{\circ}$$

The full system of roots with the above fundamental system is given by

$$\Phi = \{e_i - e_j;\ i \neq j,\ i, j = 0, 1, \ldots, l\}.$$

(ii) *Type B_l.* Let $e_1, e_2, \ldots, e_l$ be an orthonormal basis of a Euclidean space $\mathcal{V}$. The following vectors form a fundamental system of type B_l.

The full system of roots with the above fundamental system is given by

$$\Phi = \left\{ \begin{array}{l} \pm e_i \pm e_j; \ i \neq j, \ i, j = 1, 2, \ldots, l \\ \pm e_i; \ i = 1, 2, \ldots, l \end{array} \right\}.$$

(iii) *Type C_l.* Let $e_1, e_2, \ldots, e_l$ be an orthonormal basis for $\mathcal{V}$. Then the following vectors form a fundamental system of type C_l.

The full system of roots with this fundamental system is given by

$$\Phi = \left\{ \begin{array}{l} \pm e_i \pm e_j; \ i \neq j, \ i, j = 1, 2, \ldots, l \\ \pm 2e_i; \ i = 1, 2, \ldots, l \end{array} \right\}.$$

(iv) *Type D_l.* Let $e_1, e_2, \ldots, e_l$ be an orthonormal basis for $\mathcal{V}$. The following vectors form a fundamental system of type D_l.

The full system of roots with this fundamental system is given by

$$\Phi = \{ \pm e_i \pm e_j; \ i \neq j, \ i, j = 1, 2, \ldots, l \}.$$

(v) Type G_2. This has already been described.

(vi) Type F_4. Let e_1, e_2, e_3, e_4 be an orthonormal basis for $\mathcal{V}$. The following vectors form a fundamental system of type F_4.

The full system of roots is

$$\Phi = \left\{ \begin{array}{l} \pm e_i \pm e_j; \ i \neq j, \ i, j = 1, 2, 3, 4 \\ \pm e_i; \ i = 1, 2, 3, 4 \\ \frac{1}{2}(\pm e_1 \pm e_2 \pm e_3 \pm e_4) \end{array} \right\}.$$

(vii) It is convenient to describe next the root system of type E_8. The systems E_7, E_6 are then easily obtainable as subsystems.

Let e_1, e_2, e_3, e_4, e_5, e_6, e_7, e_8 be an orthonormal basis for $\mathfrak{H}$. The following vectors form a fundamental system of type E_8.

$$e_1-e_2 \quad e_2-e_3 \quad e_3-e_4 \quad e_4-e_5 \quad e_5-e_6 \quad e_6+e_7 \quad -\tfrac{1}{2}\sum_{i=1}^{8} e_i$$

$$e_6-e_7$$

The full root system is

$$\Phi = \left\{ \begin{array}{l} \pm e_i \pm e_j;\ i \neq j,\ i,j = 1,2,3,4,5,6,7,8 \\[2mm] \tfrac{1}{2}\sum_{i=1}^{8} \epsilon_i e_i;\ \epsilon_i = \pm 1,\ \prod_{i=1}^{8} \epsilon_i = 1 \end{array} \right\}.$$

(viii) Let e_i ($i = 1, 2, \ldots 8$) be as in (vii). Then we have a fundamental system of type E_7 given by

$$e_2-e_3 \quad e_3-e_4 \quad e_4-e_5 \quad e_5-e_6 \quad e_6+e_7 \quad -\tfrac{1}{2}\sum_{i=1}^{8} e_i$$

$$e_6-e_7$$

These vectors lie in the subspace of elements

$$\sum_{i=1}^{8} \lambda_i e_i \text{ satisfying } \lambda_1 = \lambda_8.$$

The full root system is

$$\Phi = \left(\begin{array}{l} \pm e_i \pm e_j;\ i \neq j,\ i,j = 2,3,4,5,6,7 \\[2mm] \pm(e_1 + e_8) \\[2mm] \tfrac{1}{2}\sum_{i=1}^{8} \epsilon_i e_i;\ \epsilon_i = \pm 1,\ \epsilon_1 = \epsilon_8 = 1,\ \prod_{i=1}^{8} \epsilon_i = 1 \\[2mm] -\tfrac{1}{2}\sum_{i=1}^{8} \epsilon_i e_i;\ \epsilon_i = \pm 1,\ \epsilon_1 = \epsilon_8 = 1,\ \prod_{i=1}^{8} \epsilon_i = 1 \end{array} \right).$$

(ix) Let $e_i(i=1, 2, \ldots 8)$ be as in (vii). Then we have a fundamental system of type E_6 given by

$$
\begin{array}{cccccc}
e_3-e_4 & e_4-e_5 & e_5-e_6 & e_6+e_7 & -\frac{1}{2}\sum_{i=1}^{8} e_i \\
\circ & \!\!\!\!-\!\!\!\! & \circ & \!\!\!\!-\!\!\!\! & \circ & \!\!\!\!-\!\!\!\! & \circ & \!\!\!\!-\!\!\!\! & \circ
\end{array}
$$

$$
\begin{array}{c}
| \\
\circ \\
e_6-e_7
\end{array}
$$

These vectors lie in the 6-dimensional subspace of elements

$$\sum_{i=1}^{8} \lambda_i e_i \text{ satisfying } \lambda_1=\lambda_2=\lambda_8.$$

The full root system is

$$
\Phi = \left(
\begin{array}{l}
\pm e_i \pm e_j;\ i \neq j,\ i,j=3, 4, 5, 6, 7 \\[2mm]
\frac{1}{2}\sum_{i=1}^{8} \epsilon_i e_i;\ \epsilon_i = \pm 1,\ \epsilon_1=\epsilon_2=\epsilon_8=1,\ \prod_{i=1}^{8} \epsilon_i=1 \\[2mm]
-\frac{1}{2}\sum_{i=1}^{8} \epsilon_i e_i;\ \epsilon_i = \pm 1,\ \epsilon_1=\epsilon_2=\epsilon_8=1,\ \prod_{i=1}^{8} \epsilon_i=1
\end{array}
\right).
$$

For further information about these indecomposable root systems, the reader is referred to Jacobson's book [1].

Let Φ be any indecomposable root system. For each root $r \in \Phi$ define

$$h_r = \frac{2r}{(r, r)}.$$

h_r is called the co-root corresponding to r (cf. 3.5.1). Let Φ^* be the set of co-roots h_r for all $r \in \Phi$.

PROPOSITION 3.6.1. Φ^* *is also a root system. Moreover if Φ contains roots of two different lengths, r is a short root of Φ if and only if h_r is a long root of Φ^*.*

PROOF. We show that Φ^* satisfies the axioms for a root system (2.1.1). Since h_r is a scalar multiple of r we have $w_r = w_{h_r}$. Thus

$$w_{h_r}(h_s) = w_r \frac{2s}{(s, s)} = \frac{2w_r(s)}{(w_r(s), w_r(s))} = h_{w_r(s)}.$$

Also we have

$$\frac{2(h_r, h_s)}{(h_r, h_r)} = \frac{2(s, r)}{(s, s)},$$

which is an integer. Thus Φ^* is a root system.

Now $(h_r, h_r) = 4/(r, r)$ and so $(r, r) < (s, s)$ if and only if $(h_r, h_r) > (h_s, h_s)$. Thus r is a short root in Φ if and only if h_r is a long root in Φ^*. ∎

Φ^* is called the dual root system of Φ. It is clear that $\Phi^{**} = \Phi$. It is readily verified that the dual of a root system of type B_l is a system of type C_l and that G_2 and F_4 are self-dual systems. Since $B_2 = C_2$ this is a self-dual system also. The duality is trivial for systems whose roots all have the same length.

We remark that if Π is a fundamental system in Φ then Π^* is a fundamental system in Φ^*.

We conclude the present chapter by giving two lemmas which will be useful in the development to follow.

LEMMA 3.6.2. *Any positive root $r \in \Phi^+$ can be expressed as a sum of fundamental roots*

$$r = p_{i_1} + p_{i_2} + \ldots + p_{i_k}$$

in such a way that $p_{i_1} + p_{i_2} + \ldots + p_{i_a}$ is a root for all $a \leqslant k$.

PROOF. Let

$$r = \sum_{i=1}^{l} n_i p_i$$

be the expression of r as an integral combination of fundamental roots. Then

$$(r, r) = \left(r, \sum_{i=1}^{l} n_i p_i \right) = \sum_{i=1}^{l} n_i (r, p_i).$$

Now $(r, r) > 0$ and $n_i \geqslant 0$ for each i. Thus there is some i for which $(r, p_i) > 0$. Suppose r is not a fundamental root. Then r, p_i are linearly independent, and r cannot be the first member of the p_i-chain of roots through it, thus $r - p_i$ is a root. By repeating this process we obtain the required expression for r as a sum of fundamental roots. ∎

LEMMA 3.6.3. *Let r, s be roots such that $r + s$ is a root. Then the integral combinations of r, s which are roots (i.e. the elements of Φ of form $ir + js$ with $i, j \in \mathbb{Z}$) form a root system of type A_2, B_2 or G_2.*

PROOF. The elements of Φ of form $ir + js$, with $i, j \in \mathbb{Z}$, satisfy the axioms 2.1.1 for a root system in the 2-dimensional space they generate. This 2-dimensional system is indecomposable, since it contains two independent non-orthogonal roots, so must have type A_2, B_2 or G_2. ∎

CHAPTER 4

The Chevalley Groups

We now begin the development of the theory of the Chevalley groups, making use of the properties of the simple Lie algebras over $\mathbb{C}$ described in the last chapter. The information given there about these simple Lie algebras may be regarded as classical; however we shall require certain additional facts about them which necessitate a closer look at the Cartan decomposition. We shall show that if $\mathfrak{L}$ is a simple Lie algebra over $\mathbb{C}$ it is possible to choose a basis for $\mathfrak{L}$, adapted to a Cartan decomposition, such that the constants of multiplication with respect to this basis are all rational integers.

4.1 Properties of the Structure Constants

Let

$$\mathfrak{L} = \mathfrak{H} \oplus \sum_{r \in \Phi} \mathfrak{L}_r$$

be a Cartan decomposition of $\mathfrak{L}$. Let

$$h_r = \frac{2r}{(r, r)}$$

be the co-root corresponding to the root $r \in \Phi$. For each root r let e_r be a non-zero element of $\mathfrak{L}_r$. If e_r is already chosen for $r \in \Phi^+$ there is a unique element $e_{-r} \in \mathfrak{L}_{-r}$ such that $[e_r e_{-r}] = h_r$, and we shall suppose e_{-r} chosen in this way. The set

$$\{h_r, r \in \Pi; \ e_r, r \in \Phi\}$$

is a basis for $\mathfrak{L}$. It consists of the fundamental co-roots h_r together with the set of all root vectors e_r. The elements of this basis multiply together as follows:

$$
\begin{aligned}
[h_r h_s] &= 0, & r, s &\in \Pi, \\
[h_r e_s] &= A_{rs} e_s, & r &\in \Pi, s \in \Phi \\
[e_r e_{-r}] &= h_r, & r &\in \Phi, \\
[e_r e_s] &= 0, & r, s &\in \Phi, r+s \notin \Phi.
\end{aligned}
$$

If r, s, $r+s \in \Phi$ then $[e_r e_s]$ is a scalar multiple of e_{r+s} since $[\mathfrak{L}_r \mathfrak{L}_s] = \mathfrak{L}_{r+s}$. We define $N_{r,\,s}$ by

$$[e_r e_s] = N_{r,\,s} e_{r+s}.$$

The elements $N_{r,\,s}$ for r, $s \in \Phi$ are called the structure constants of $\mathfrak{L}$. They clearly depend upon the choice of the root vectors e_r. Our first task is to consider the relations between them.

(i) Since $[e_s e_r] = -[e_r e_s]$, it is clear that $N_{s,\,r} = -N_{r,\,s}$ for all $r, s \in \Phi$.

(ii) Suppose r_1, r_2, r_3 are three roots such that $r_1 + r_2 + r_3 = 0$. By the Jacobi identity we have

$$[[e_{r_1} e_{r_2}] e_{r_3}] + [[e_{r_2} e_{r_3}] e_{r_1}] + [[e_{r_3} e_{r_1}] e_{r_2}] = 0.$$

Thus

$$N_{r_1,\,r_2}[e_{-r_3} e_{r_3}] + N_{r_2,\,r_3}[e_{-r_1} e_{r_1}] + N_{r_3,\,r_1}[e_{-r_2} e_{r_2}] = 0.$$

Hence

$$N_{r_1,\,r_2} h_{r_3} + N_{r_2,\,r_3} h_{r_1} + N_{r_3,\,r_1} h_{r_2} = 0.$$

It follows that

$$\frac{2 N_{r_1,\,r_2} r_3}{(r_3, r_3)} + \frac{2 N_{r_2,\,r_3} r_1}{(r_1, r_1)} + \frac{2 N_{r_3,\,r_1} r_2}{(r_2, r_2)} = 0.$$

Using the fact that $r_1 + r_2 + r_3 = 0$ we have

$$\left(\frac{N_{r_2,\,r_3}}{(r_1, r_1)} - \frac{N_{r_1,\,r_2}}{(r_3, r_3)} \right) r_1 + \left(\frac{N_{r_3,\,r_1}}{(r_2, r_2)} - \frac{N_{r_1,\,r_2}}{(r_3, r_3)} \right) r_2 = 0.$$

Now r_1, r_2 are linearly independent. For otherwise $r_2 = \pm r_1$ and $r_3 = -2r_1$ or 0, a contradiction. Therefore the coefficients of r_1, r_2 in the above equation must be zero, and we have

$$\frac{N_{r_1,\,r_2}}{(r_3, r_3)} = \frac{N_{r_2,\,r_3}}{(r_1, r_1)} = \frac{N_{r_3,\,r_1}}{(r_2, r_2)}.$$

(iii) Suppose r, $s \in \Phi$ are linearly independent. By the Jacobi identity we have

$$[[e_r e_{-r}] e_s] + [[e_{-r} e_s] e_r] + [[e_s e_r] e_{-r}] = 0.$$

Hence

$$[h_r e_s] + N_{-r,\,s}[e_{-r+s} e_r] + N_{s,\,r}[e_{r+s} e_{-r}] = 0.$$

(It is convenient here and throughout the exposition to assume that $N_{r,\,s}=0$ if $r,\,s\in\Phi$ but $r+s$ is not a root.) Thus

$$A_{rs}e_s+N_{-r,\,s}N_{-r+s,\,r}e_s+N_s,\,_rN_{r+s,\,-r}e_s=0.$$

Using the relations obtained in (ii) above, we have

$$A_{rs}+N_{r-s,\,-r}N_{-r+s,\,r}\frac{(-r+s,\,-r+s)}{(s,\,s)}+N_s,\,_rN_{-r,\,-s}\frac{(s,\,s)}{(r+s,\,r+s)}=0.$$

We may rewrite this as

$$N_{r,\,s}N_{-r,\,-s}\frac{(s,\,s)}{(r+s,\,r+s)}-N_{r,\,-r+s}N_{-r,\,r-s}\frac{(-r+s,\,-r+s)}{(s,\,s)}=A_{rs}.$$

We define $M_{r,\,s}$ for $r,\,s\in\Phi$ by

$$M_{r,\,s}=N_{r,\,s}N_{-r,\,-s}\frac{(s,\,s)}{(r+s,\,r+s)}.$$

Then the above equation becomes

$$M_{r,\,s}-M_{r,\,-r+s}=A_{rs}.$$

We now consider the r-chain of roots through s and apply this equation repeatedly. Let this r-chain be

$$-pr+s,\,\ldots,\,s,\,\ldots,\,qr+s.$$

Then we have

$$M_{r,\,s}-M_{r,\,-r+s}=A_{rs},$$

$$M_{r,\,-r+s}-M_{r,\,-2r+s}=A_{r,\,-r+s}=A_{rs}-2,$$

$$M_{r,\,-(p-1)r+s}-M_{r,\,-pr+s}=A_{r,\,-(p-1)r+s}=A_{rs}-2(p-1),$$

$$M_{r,\,-pr+s}=A_{r,\,-pr+s}=A_{rs}-2p.$$

(The term $M_{r,\,-(p+1)r+s}$ does not appear in the last equation since $-(p+1)r+s$ is not a root.) Adding these equations we obtain

$$M_{r,\,s}=(p+1)A_{rs}-p(p+1).$$

However, $A_{rs}=p-q$, as in section 3.3, thus

$$M_{r,\,s}=-(p+1)q.$$

Hence

$$N_{r,\,s}N_{-r,\,-s}=-(p+1)q\frac{(r+s,\,r+s)}{(s,\,s)}.$$

c

This expression can be simplified by the use of the following lemma.

LEMMA 4.1.1. *Suppose $r, s, r+s \in \Phi$. Then*

$$\frac{(r+s, r+s)}{(s, s)} = \frac{p+1}{q}.$$

PROOF. We consider separately the different possibilities for the r-chain through s, bearing in mind that the length of any r-chain is at most 4. We use the information given in section 3.4 about the relative lengths of roots inclined at a given angle. The various possibilities for the r-chain are shown below.

$\underset{\substack{s \qquad r+s}}{\circ\!\!-\!\!\circ}$	$p=0 \quad q=1 \quad (s,s) = (r+s, r+s)$
$\underset{\substack{s \quad r+s \; 2r+s}}{\circ\!\!-\!\!\circ\!\!-\!\!\circ}$	$p=0 \quad q=2 \quad (s,s) = 2(r+s, r+s)$
$\underset{\substack{-r+s \quad s \quad r+s}}{\circ\!\!-\!\!\circ\!\!-\!\!\circ}$	$p=1 \quad q=1 \quad (s,s) = \tfrac{1}{2}(r+s, r+s)$
$\underset{\substack{s \quad r+s}}{\circ\!\!-\!\!\circ\!\!-\!\!\circ\!\!-\!\!\circ}$	$p=0 \quad q=3 \quad (s,s) = 3(r+s, r+s)$
$\underset{\substack{\quad\;\; s \quad r+s}}{\circ\!\!-\!\!\circ\!\!-\!\!\circ\!\!-\!\!\circ}$	$p=1 \quad q=2 \quad (s,s) = (r+s, r+s)$
$\underset{\substack{\qquad\quad s \quad r+s}}{\circ\!\!-\!\!\circ\!\!-\!\!\circ\!\!-\!\!\circ}$	$p=2 \quad q=1 \quad (s,s) = \tfrac{1}{3}(r+s, r+s)$

The result of the lemma is clearly valid in all cases. ∎

Applying this result we obtain

$$N_{r, s}N_{-r, -s} = -(p+1)^2.$$

(iv) Finally we consider four roots r_1, r_2, r_3, r_4 such that

$$r_1 + r_2 + r_3 + r_4 = 0$$

and such that no pair are equal and opposite. By the Jacobi identity we have

$$[[e_{r_1}e_{r_2}]e_{r_3}] + [[e_{r_2}e_{r_3}]e_{r_1}] + [[e_{r_3}e_{r_1}]e_{r_2}] = 0.$$

Thus

$$N_{r_1, r_2}N_{r_1+r_2, r_3} + N_{r_2, r_3}N_{r_2+r_3, r_1} + N_{r_3, r_1}N_{r_3+r_1, r_2} = 0.$$

Now by the formulae obtained in (ii) above we have

$$\frac{N_{r_1+r_2, r_3}}{(r_4, r_4)} = \frac{N_{r_3, r_4}}{(r_1+r_2, r_1+r_2)}$$

and there are corresponding formulae obtained by permuting the roots. Using these formulae we obtain

$$\frac{N_{r_1, r_2}N_{r_3, r_4}}{(r_1+r_2, r_1+r_2)} + \frac{N_{r_2, r_3}N_{r_1, r_4}}{(r_2+r_3, r_2+r_3)} + \frac{N_{r_3, r_1}N_{r_2, r_4}}{(r_3+r_1, r_3+r_1)} = 0.$$

As usual we note that a term may be 0 if the corresponding vector is not a root. However, suppose $r_1+r_2 \in \Phi$. Then $r_3+r_4 \in \Phi$ since

$$r_3+r_4 = -(r_1+r_2).$$

Thus the first term in the above formula is non-zero. Hence at least one of the other terms is non-zero. Thus either r_2+r_3 and r_1+r_4 are roots, or r_3+r_1 and r_2+r_4 are roots, or both.

We summarize the results we have obtained in the following theorem.

THEOREM 4.1.2. *The structure constants of a simple Lie algebra $\mathfrak{L}$ over $\mathbb{C}$ satisfy the following relations:*

(i) $N_{s, r} = -N_{r, s}, \qquad r, s \in \Phi.$

(ii) $\dfrac{N_{r_1, r_2}}{(r_3, r_3)} = \dfrac{N_{r_2, r_3}}{(r_1, r_1)} = \dfrac{N_{r_3, r_1}}{(r_2, r_2)}$

if $r_1, r_2, r_3 \in \Phi$ satisfy $r_1+r_2+r_3=0$.

(iii) $N_{r, s}N_{-r, -s} = -(p+1)^2, \qquad r, s \in \Phi.$

(iv) $\dfrac{N_{r_1, r_2}N_{r_3, r_4}}{(r_1+r_2, r_1+r_2)} + \dfrac{N_{r_2, r_3}N_{r_1, r_4}}{(r_2+r_3, r_2+r_3)} + \dfrac{N_{r_3, r_1}N_{r_2, r_4}}{(r_3+r_1, r_3+r_1)} = 0$

if $r_1, r_2, r_3, r_4 \in \Phi$ satisfy $r_1+r_2+r_3+r_4=0$ and if no pair are opposite.

4.2 The Chevalley Basis

Formula (iii) of Theorem 4.1.2 suggests that, by a sufficiently careful choice of the root vectors e_r, it might be possible to arrange that

$$N_{r, s} = \pm(p+1)$$

for all relevant pairs r, s of roots. This is in fact so, and to prove it we use the isomorphism theorem for simple Lie algebras, stated in 3.5.2. In the notation of 3.5.2 we define $\mathfrak{L}' = \mathfrak{L}$ and $p'_i = -p_i$ for $i=1, 2, \ldots, l$. Then it is clear that $A_{ij} = A'_{ij}$. We also define $e_{p_i} = -e_{-p_i}$ and $e_{-p'_i} = -c_{p_i}$.

Then
$$[e_{p_i'}e_{-p_i'}] = [e_{-p_i}e_{p_i}] = -h_{p_i} = h_{-p_i} = h_{p_i'},$$

as required. Thus there is an isomorphism θ of $\mathfrak{L}$ to itself, i.e. an automorphism of $\mathfrak{L}$, such that

$$\theta(e_{p_i}) = -e_{-p_i},$$
$$\theta(e_{-p_i}) = -e_{p_i},$$
$$\theta(h_{p_i}) = -h_{p_i}.$$

Now θ^2 transforms e_{p_i}, e_{-p_i}, h_{p_i} into themselves, so must be the identity by the uniqueness part of 3.5.2. Thus θ has order 2.

Now by 3.6.2 each root $r \in \Phi^+$ can be expressed as a sum of fundamental roots

$$r = r_1 + r_2 + \ldots + r_k$$

in such a way that $r_1 + r_2 + \ldots + r_a$ is a root for all $a \leqslant k$. It follows that $[[e_{r_1}e_{r_2}] \ldots e_{r_k}] \in L_r$ and is a non-zero scalar multiple of e_r. The image of this element under θ is $[[-e_{-r_1}, -e_{-r_2}] \ldots -e_{-r_k}]$, which is a non-zero scalar multiple of e_{-r}. Thus e_r is transformed by θ into a scalar multiple of e_{-r}.

Let $\theta(e_r) = \lambda e_{-r}$. Then, since θ has order 2, $\theta(e_{-r}) = \lambda^{-1}e_r$. Hence

$$\theta(\mu e_r) = \mu \lambda e_{-r} = \mu^2 \lambda (\mu^{-1} e_{-r}).$$

Now it is possible to choose $\mu \in \mathbb{C}$ such that $\mu^2 = -\lambda^{-1}$. With such a choice of μ we have $\theta(\mu e_r) = -\mu^{-1}e_{-r}$ and $[\mu e_r, \mu^{-1}e_{-r}] = h_r$. We now alter our choice of the root vectors. By choosing μe_r as the root vector in $\mathfrak{L}_r$ and $\mu^{-1}e_{-r}$ as the root vector in $\mathfrak{L}_{-r}$, and then renaming these e_r, e_{-r} respectively, we see that it is possible to choose $e_r \in \mathfrak{L}_r$, $e_{-r} \in \mathfrak{L}_{-r}$ such that $[e_r e_{-r}] = h_r$ and $\theta(e_r) = -e_{-r}$.

Now

$$[e_r e_s] = N_{r,\,s} e_{r+s}$$

whenever r, s, $r+s \in \Phi$, and so applying θ we have

$$[-e_{-r}, -e_{-s}] = -N_{r,\,s} e_{-r-s}.$$

It follows that $N_{-r,\,-s} = -N_{r,\,s}$. However $N_{r,\,s} N_{-r,\,-s} = -(p+1)^2$, and so $N_{r,\,s} = \pm(p+1)$.

We can now state Chevalley's basis theorem for simple Lie algebras.

THEOREM 4.2.1. *Let $\mathfrak{L}$ be a simple Lie algebra over $\mathbb{C}$ and*

$$\mathfrak{L} = \mathfrak{H} \oplus \sum_{r \in \Phi} \mathfrak{L}_r$$

be a Cartan decomposition of $\mathfrak{L}$. *Let* $h_r \in \mathfrak{H}$ *be the co-root corresponding to the root r. Then, for each root* $r \in \Phi$, *an element* e_r *can be chosen in* $\mathfrak{L}_r$ *such that*

$$[e_r e_{-r}] = h_r,$$

$$[e_r e_s] = \pm(p+1)e_{r+s},$$

where p is the greatest integer for which $s - pr \in \Phi$.

The elements $\{h_r, r \in \Pi; e_r, r \in \Phi\}$ *form a basis for* $\mathfrak{L}$, *called a Chevalley basis. The basis elements multiply together as follows:*

$$[h_r h_s] = 0,$$

$$[h_r e_s] = A_{rs} e_s,$$

$$[e_r e_{-r}] = h_r,$$

$$[e_r e_s] = 0 \qquad\qquad \textit{if } r + s \notin \Phi,$$

$$[e_r e_s] = N_{r,\,s} e_{r+s} \qquad \textit{if } r + s \in \Phi,$$

where $N_{r,\,s} = \pm(p+1)$.

The multiplication constants of the algebra with respect to the Chevalley basis are all integers.

PROOF. Since A_{rs} and $N_{r,\,s}$ are integers, it only remains to prove that each co-root h_r is a linear combination of the fundamental co-roots with integer coefficients. This follows from the fact that the co-roots h_r for $r \in \Pi$ form a fundamental system of the dual root system (cf. 3.6.1). ∎

Now a simple Lie algebra $\mathfrak{L}$ has many different Chevalley bases, and we consider the amount of freedom available in the choice of such a basis. In the first place, every Chevalley basis is defined relative to some Cartan subalgebra. If a Cartan subalgebra $\mathfrak{H}$ of $\mathfrak{L}$ is prescribed, then the root spaces $\mathfrak{L}_r$ are determined. The elements of the basis which lie in $\mathfrak{H}$ are defined relative to some fundamental system Π in Φ. If Π is prescribed then the fundamental co-roots h_r, $r \in \Pi$, are determined. The fundamental root vectors, e_r, $r \in \Pi$, may be chosen as arbitrary non-zero elements of the root spaces $\mathfrak{L}_r$, $r \in \Pi$. The remaining positive root vectors e_r, $r \in \Phi^+$, are now determined *to within a sign* by $[e_r e_s] = \pm(p+1)e_{r+s}$, using 3.6.2. The relation $[e_r e_{-r}] = h_r$ then determines the basis vectors e_r for $r \in \Phi^-$.

Systems of structure constants

Now every Chevalley basis determines a system of structure constants $N_{r,\,s}$ given by $[e_r e_s] = N_{r,\,s} e_{r+s}$. Since $N_{r,\,s} = \pm (p+1)$, where p is defined as before, the absolute value of $N_{r,\,s}$ is determined. However, different Chevalley bases will give different values for the signs of the $N_{r,\,s}$. We consider the extent to which these signs can be chosen arbitrarily in determining a system of structure constants. We show that for certain ordered pairs (r, s) of roots the sign of $N_{r,\,s}$ may be chosen arbitrarily and that the remaining structure constants are then determined.

Suppose we are given a total ordering on the space containing the roots, as in section 2.1. An ordered pair (r, s) of roots will be called a *special pair* if $r+s \in \Phi$ and $0 \prec r \prec s$. An ordered pair (r, s) is called *extra special* if (r, s) is a special pair and if for all special pairs (r_1, s_1) with $r+s = r_1 + s_1$ we have $r \leqslant r_1$. Then every root in Φ^+ which is the sum of two roots in Φ^+ can be expressed uniquely as the sum of an extraspecial pair. Since by 3.6.2 every root in Φ^+ which is not in Π has this property, the extraspecial pairs are in $1-1$ correspondence with the roots in $\Phi^+ - \Pi$.

Now given a Chevalley basis $\{h_r, \ r \in \Pi; \ e_r, \ r \in \Phi\}$, we may change the sign of any subset of $\{e_r, \ r \in \Phi^+ - \Pi\}$, and changing the signs of the e_r for $r \in \Phi^-$ in a corresponding way to preserve the relation $[e_r e_{-r}] = h_r$, we obtain another Chevalley basis. Thus the signs of the structure constants $N_{r,\,s}$ for extraspecial pairs (r, s) can be chosen arbitrarily. On the other hand, we show that if the $N_{r,\,s}$ are given for the extraspecial pairs they are determined for all pairs.

PROPOSITION 4.2.2. *The signs of the structure constants $N_{r,\,s}$ may be chosen arbitrarily for extraspecial pairs (r, s), and then the structure constants for all pairs are uniquely determined.*

PROOF. Let $\{h_r, \ r \in \Pi; \ e_r, \ r \in \Phi\}$ and $\{h_r, \ r \in \Pi; \ e'_r, \ r \in \Phi\}$ be two Chevalley bases giving rise to systems of structure constants $(N_{r,\,s})$, $(N'_{r,\,s})$. Thus $[e_r e_s] = N_{r,\,s} e_{r+s}$ and $[e'_r e'_s] = N'_{r,\,s} e'_{r+s}$.

Let $e'_r = \lambda_r e_r$, where $\lambda_r \neq 0 \in \mathbb{C}$. Then $\lambda_r \lambda_s N_{r,\,s} = \lambda_{r+s} N'_{r,\,s}$. Now suppose that $N_{r,\,s} = N'_{r,\,s}$ for all extraspecial pairs (r, s). Then $\lambda_r \lambda_s = \lambda_{r+s}$ for such pairs. By 3.6.2 it follows that if $r \in \Phi^+$ and $r = n_1 p_1 + \ldots + n_l p_l$, then

$$\lambda_r = \lambda_{p_1}^{n_1} \lambda_{p_2}^{n_2} \ldots \lambda_{p_l}^{n_l}.$$

The same equation holds for negative roots since $\lambda_{-r} = \lambda_r^{-1}$. As a consequence of this we have $\lambda_r \lambda_s = \lambda_{r+s}$ for all pairs (r, s) of roots, and so

$N_{r,\,s}=N'_{r,\,s}$ for all pairs. Thus the structure constants are uniquely determined by their values on the extraspecial pairs.   ∎

It is of interest to point out that the values of the structure constants $N_{r,\,s}$ can be derived from the structure constants on the extraspecial pairs by the relations (i), (ii), (iii), (iv) of 4.1.2. To prove this consider the set of ordered pairs (r, s) of roots such that $r+s \in \Phi$. If (r, s) is such a pair, so are the following twelve pairs of roots:

$$(r, s), \qquad\qquad (s, r),$$
$$(s, -r-s), \qquad\qquad (-r-s, s),$$
$$(-r-s, r), \qquad\qquad (r, -r-s),$$
$$(-r, -s), \qquad\qquad (-s, -r),$$
$$(-s, r+s), \qquad\qquad (r+s, -s),$$
$$(r+s, -r), \qquad\qquad (-r, r+s).$$

Since $r+s+(-r-s)=0$, either two of r, s, $-r-s$ are positive or one is positive. It follows that of the above twelve ordered pairs of roots, exactly one is a special pair. Moreover relations (i), (ii), (iii) of 4.1.2 determine all the structure constants in terms of the $N_{r,\,s}$ for special pairs (r, s).

It remains to show that the $N_{r,\,s}$ for special pairs can be expressed in terms of the $N_{r,\,s}$ for extraspecial pairs. Suppose (r, s) is a pair of roots which is special but not extraspecial. Then there is a unique extraspecial pair (r_1, s_1) such that $r_1+s_1=r+s$. Since $r+s+(-r_1)+(-s_1)=0$, relation (iv) of 4.1.2 gives

$$\frac{N_{r,\,s}N_{-r_1,\,-s_1}}{(r+s, r+s)}+\frac{N_{s,\,-r_1}N_{r,\,-s_1}}{(s-r_1, s-r_1)}+\frac{N_{-r_1,\,r}N_{s,\,-s_1}}{(-r_1+r, -r_1+r)}=0.$$

Now the roots r, s, r_1, s_1 are ordered by

$$0 \prec r_1 \prec r \prec s \prec s_1.$$

Thus the special pairs associated with the pairs

$$(-r_1, -s_1), \; (s, -r_1), \; (r, -s_1), \; (-r_1, r), \; (s, -s_1)$$

respectively are

$$(r_1, s_1), \; (r_1, s-r_1), \; (s_1-r, r), \; (r-r_1, r_1), \; (s_1-s, s).$$

However for each of these special pairs $(\bar r, \bar s)$ we have $\bar r + \bar s \prec r + s$. Thus, by using relations (i), (ii), (iii), (iv) of 4.1.2, $N_{r,\,s}$ can be expressed in terms of $N_{r_1,\,s_1}$ and various terms $N_{\bar r,\,\bar s}$, where $(\bar r, \bar s)$ is a special pair with

$\bar{r}+\bar{s} \prec r+s$. By using induction on the sum $r+s$ it can be seen that $N_{r,\,s}$ is determined by the values of the structure constants on the extraspecial pairs.

4.3 The Exponential Map

We shall now show how to construct certain automorphisms of a Lie algebra, using the exponential map. We recall that an automorphism of a Lie algebra $\mathfrak{L}$ is a non-singular linear map θ of $\mathfrak{L}$ into itself such that

$$[\theta x,\ \theta y] = \theta [x,\ y].$$

The set of all automorphisms of $\mathfrak{L}$ clearly forms a group.

LEMMA 4.3.1. *Let $\mathfrak{L}$ be a Lie algebra over a field of characteristic 0 and δ be a derivation of $\mathfrak{L}$ which is nilpotent, i.e. satisfies $\delta^n = 0$ for some n. Then*

$$\exp \delta = 1 + \delta + \frac{\delta^2}{2!} + \ldots + \frac{\delta^{n-1}}{(n-1)!}$$

is an automorphism of $\mathfrak{L}$.

PROOF. $\exp \delta$ is a non-singular linear map, its inverse being $\exp(-\delta)$. Now we have

$$\delta[xy] = [\delta x, y] + [x, \delta y], \qquad x, y \in \mathfrak{L}.$$

Thus

$$\delta^r [xy] = \sum_{i=0}^{r} \binom{r}{i} [\delta^i x, \delta^{r-i} y].$$

$$\frac{\delta^r}{r!} [xy] = \sum_{i=0}^{r} \left[\frac{\delta^i}{i!} x, \frac{\delta^{r-i}}{(r-i)!} y \right]$$

$$= \sum_{\substack{i,j \\ i+j=r}} \left[\frac{\delta^i}{i!} x, \frac{\delta^j}{j!} y \right].$$

Therefore

$$\exp \delta[xy] = \sum_{r \geqslant 0} \sum_{\substack{i,j \\ i+j=r}} \left[\frac{\delta^i}{i!} x, \frac{\delta^j}{j!} y \right]$$

$$= \sum_{i \geqslant 0} \sum_{j \geqslant 0} \left[\frac{\delta^i}{i!} x, \frac{\delta^j}{j!} y \right]$$

$$= [\exp \delta.x, \exp \delta.y]. \qquad \blacksquare$$

Now let $\mathfrak{L}$ be a simple Lie algebra over $\mathbb{C}$ with Cartan decomposition

$$\mathfrak{L} = \mathfrak{H} \oplus \sum_{r \in \Phi} \mathfrak{L}_r$$

and Chevalley basis $\{h_r, r \in \Pi; e_r, r \in \Phi\}$. Then the map ad e_r is a derivation of $\mathfrak{L}$ (cf. section 3.1) and this derivation is in fact nilpotent. For we have

$$\text{ad } e_r.\mathfrak{H} = \mathfrak{L}_r, \qquad (\text{ad } e_r)^2.\mathfrak{H} = 0,$$

$$\text{ad } e_r.\mathfrak{L}_r = 0,$$

$$\text{ad } e_r.\mathfrak{L}_{-r} \subseteq \mathfrak{H}, \qquad (\text{ad } e_r)^3.\mathfrak{L}_{-r} = 0.$$

$(\text{ad } e_r)^{q+1}.\mathfrak{L}_s = 0$ if r, s are linearly independent, since $(q+1)r+s$ is not a root, Thus

$$(\text{ad } e_r)^n.\mathfrak{L} = 0$$

for all sufficiently large values of n.

Let $\zeta \in \mathbb{C}$. Then ad $(\zeta e_r) = \zeta$ ad e_r is also a nilpotent derivation of $\mathfrak{L}$. Thus exp $(\zeta \text{ ad } e_r)$ is an automorphism of $\mathfrak{L}$. We write

$$x_r(\zeta) = \exp (\zeta \text{ ad } e_r).$$

We now consider the effect of the automorphism $x_r(\zeta)$ on the elements of the Chevalley basis. We have

$$x_r(\zeta).e_r = e_r,$$

$$x_r(\zeta).e_{-r} = e_{-r} + \zeta h_r - \zeta^2 e_r,$$

$$x_r(\zeta).h_r = h_r - 2\zeta e_r.$$

Also, if r, s are linearly independent,

$$x_r(\zeta).h_s = h_s - A_{sr}\zeta e_r,$$

$$x_r(\zeta).e_s = e_s + N_{r,s}\zeta e_{r+s} + \frac{1}{2!} N_{r,s}N_{r,r+s}\zeta^2 e_{2r+s}$$

$$+ \ldots + \frac{1}{q!} N_{r,s}N_{r,r+s} \ldots N_{r,(q-1)r+s}\zeta^q e_{qr+s}.$$

We write

$$M_{r,s,i} = \frac{1}{i!} N_{r,s}N_{r,r+s} \ldots N_{r,(i-1)r+s}.$$

Then

$$x_r(\zeta).e_s = \sum_{i=0}^{q} M_{r,s,i}\zeta^i e_{ir+s}.$$

(We define $M_{r, s, 0} = 1$.) Using the fact that $N_{r, s} = \pm (p+1)$ we see that

$$M_{r, s, i} = \pm \frac{(p+1)(p+2)\ldots(p+i)}{i!} = \pm \binom{p+i}{i}.$$

In particular $M_{r, s, i}$ is an integer. Thus *the automorphism $x_r(\zeta)$ transforms each element of the Chevalley basis into a linear combination of basis elements, the coefficients being non-negative integral powers of ζ with rational integer coefficients.*

It is this property which enables us to define automorphisms of this type over an arbitrary field.

4.4 Algebras and Groups Over an Arbitrary Field

Let $\mathfrak{L}$ be a simple Lie algebra over $\mathbb{C}$ with Chevalley basis

$$\{h_r, r \in \Pi; \ e_r, r \in \Phi\}.$$

We denote by $\mathfrak{L}_\mathbb{Z}$ the subset of $\mathfrak{L}$ of all linear combinations of the basis elements with coefficients in the ring $\mathbb{Z}$ of rational integers. $\mathfrak{L}_\mathbb{Z}$ is an additive abelian group. By 4.2.1 the Lie product of two basis vectors lies in $\mathfrak{L}_\mathbb{Z}$, thus $\mathfrak{L}_\mathbb{Z}$ is closed under Lie multiplication. $\mathfrak{L}_\mathbb{Z}$ is therefore a Lie algebra over $\mathbb{Z}$.

Now let K be any field. We form the tensor product of the additive group of K with the additive group of $\mathfrak{L}_\mathbb{Z}$, and define

$$\mathfrak{L}_K = K \otimes \mathfrak{L}_\mathbb{Z}.$$

Then $\mathfrak{L}_K$ is an additive abelian group. Let 1_K be the unit element of K. Then every element of $\mathfrak{L}_K$ can be written in the form

$$\sum_{r \in \Pi} \lambda_r (1_K \otimes h_r) + \sum_{r \in \Phi} \mu_r (1_K \otimes e_r),$$

where $\lambda_r, \mu_r \in K$. We write

$$\bar{h}_r = 1_K \otimes h_r \qquad \bar{e}_r = 1_K \otimes e_r.$$

Then $\mathfrak{L}_K$ is a vector space over K with basis

$$\{\bar{h}_r, r \in \Pi; \ \bar{e}_r, r \in \Phi\}.$$

We now define a Lie multiplication on $\mathfrak{L}_K$.

PROPOSITION 4.4.1. *Let x, y be any two elements of the Chevalley* *basis of* $\mathfrak{L}$. *Then the multiplication on* $\mathfrak{L}_K$ *defined by*

$$[1_K \otimes x, 1_K \otimes y] = 1_K \otimes [xy]$$

and extended by linearity makes $\mathfrak{L}_K$ *into a Lie algebra over* K. *The multiplication constants of* $\mathfrak{L}_K$ *with respect to the basis* $\{\bar{h}_r, r \in \Pi; \bar{e}_r, r \in \Phi\}$ *are the multiplication constants of* $\mathfrak{L}$ *with respect to the basis*

$$\{h_r, r \in \Pi; e_r, r \in \Phi\}$$

interpreted as elements of the prime subfield of K.

PROOF. This is clear, since by 4.2.1 the multiplication constants of $\mathfrak{L}$ are in $\mathbb{Z}$. ∎

Having introduced the Lie algebra $\mathfrak{L}_K$, we shall define automorphisms of $\mathfrak{L}_K$ analogous to the automorphisms $x_r(\zeta)$ of $\mathfrak{L}$. Let $A_r(\zeta)$ be the matrix representing $x_r(\zeta)$ with respect to the Chevalley basis of $\mathfrak{L}$. We have seen that the coefficients of $A_r(\zeta)$ have the form $a\zeta^i$, where $a \in \mathbb{Z}$ and $i \geqslant 0$. Let t be an element of K and $\bar{A}_r(t)$ be the matrix obtained from $A_r(\zeta)$ by replacing each coefficient $a\zeta^i$ by $\bar{a}t^i \in K$, where $\bar{a}$ is the element of the prime field of K corresponding to $a \in \mathbb{Z}$. We now define $\bar{x}_r(t)$ to be the linear map of $\mathfrak{L}_K$ into itself represented by the matrix $\bar{A}_r(t)$ with respect to the basis $\{\bar{h}_r, r \in \Pi; \bar{e}_r, r \in \Phi\}$.

PROPOSITION 4.4.2. $\bar{x}_r(t)$ *is an automorphism of* $\mathfrak{L}_K$ *for each* $r \in \Phi$, $t \in K$.

PROOF. Observe first that $\bar{x}_r(t)$ is non-singular. Since $x_r(\zeta)x_r(-\zeta) = 1$, it follows immediately from the definitions that $\bar{x}_r(t)\bar{x}_r(-t) = 1$. Thus $\bar{x}_r(-t)$ is the inverse of $\bar{x}_r(t)$.

To show that $\bar{x}_r(t)$ is an automorphism of $\mathfrak{L}_K$ we consider the effect of $\bar{x}_r(t)$ on the basis of $\mathfrak{L}_K$. Let $v_1, v_2, \ldots$ be the Chevalley basis of $\mathfrak{L}$ and $\bar{v}_1, \bar{v}_2, \ldots$ be the corresponding basis of $\mathfrak{L}_K$. Suppose

$$[v_i v_j] = \sum_k \gamma_{ijk} v_k,$$

$$[\bar{v}_i \bar{v}_j] = \sum_k \bar{\gamma}_{ijk} \bar{v}_k,$$

where $\gamma_{ijk} \in \mathbb{Z}$ and $\bar{\gamma}_{ijk}$ are the corresponding elements of the prime field of K. Now we have

$$x_r(\zeta) . v_i = \sum_j A_r(\zeta)_{ij} v_j,$$

$$\bar{x}_r(t) . \bar{v}_i = \sum_j \bar{A}_r(t)_{ij} \bar{v}_j.$$

Thus $\bar{x}_r(t)$ is an automorphism of $\mathfrak{L}_K$ if and only if

$$\sum_{i',j'} \bar{A}_r(t)_{ii'}\bar{A}_r(t)_{jj'}\bar{\gamma}_{i'j'k} = \sum_{k'} \bar{\gamma}_{ijk'}\bar{A}_r(t)_{k'k}$$

for all i, j, k.

However, $x_r(\zeta)$ is an automorphism of $\mathfrak{L}$ for all $\zeta \in \mathbb{C}$. Thus

$$\sum_{i',j'} A_r(\zeta)_{ii'}A_r(\zeta)_{jj'}\gamma_{i'j'k} - \sum_{k'} \gamma_{ijk'}A_r(\zeta)_{k'k}$$

is a polynomial in $\mathbb{Z}[\zeta]$ which vanishes for all $\zeta \in \mathbb{C}$. It is therefore identically zero. It follows that the polynomial

$$\sum_{i',j'} \bar{A}_r(t)_{ii'}\bar{A}_r(t)_{jj'}\bar{\gamma}_{i'j'k} - \sum_{k'} \bar{\gamma}_{ijk'}\bar{A}_r(t)_{k'k}$$

in $K[t]$ is identically zero. Thus $\bar{x}_r(t)$ is an automorphism of $\mathfrak{L}_K$ for all $t \in K$. ∎

Now that we have established the results of 4.4.1 and 4.4.2 we shall simplify the notation. We shall write h_r for $\bar{h}_r$, e_r for $\bar{e}_r$, $x_r(t)$ for $\bar{x}_r(t)$, and $A_r(t)$ for $\bar{A}_r(t)$. This omission of the bars will not lead to confusion or inconsistency since the objects originally called h_r, e_r, $x_r(t)$, $A_r(t)$ are special cases of $\bar{h}_r$, $\bar{e}_r$, $\bar{x}_r(t)$, $\bar{A}_r(t)$ when $K = \mathbb{C}$.

We shall now define the Chevalley groups. The Chevalley group of type $\mathfrak{L}$ over the field K, denoted by $\mathfrak{L}(K)$, is defined to be the group of automorphisms of the Lie algebra $\mathfrak{L}_K$ generated by the $x_r(t)$ for all $r \in \Phi$, $t \in K$.

The generators of $\mathfrak{L}(K)$ operate on the elements of the Chevalley basis of $\mathfrak{L}_K$ according to the formulae:

$$x_r(t).e_r = e_r, \qquad\qquad r \in \Phi,$$

$$x_r(t).e_{-r} = e_{-r} + th_r - t^2 e_r, \qquad r \in \Phi,$$

$$x_r(t).h_s = h_s - A_{sr}te_r, \qquad r \in \Phi, s \in \Pi,$$

$$x_r(t).e_s = \sum_{i=0}^{q} M_{r,\,s,\,i}t^i e_{ir+s}$$

if $r, s \in \Phi$ are linearly independent.

PROPOSITION 4.4.3.   *The group $\mathfrak{L}(K)$ is determined up to isomorphism by the simple Lie algebra $\mathfrak{L}$ over $\mathbb{C}$ and the field K.*

PROOF.   We must show that $\mathfrak{L}(K)$ is independent of the choice of the

Chevalley basis of $\mathfrak{L}$. Let $\{h_r, r \in \Pi; e_r, r \in \Phi\}$ be a Chevalley basis of $\mathfrak{L}$. We must show that any Chevalley basis of $\mathfrak{L}$ can be transformed by an automorphism of $\mathfrak{L}$ into one of the form $\{h_r, r \in \Pi; \pm e_r, r \in \Phi\}$. Firstly, since any two Cartan subalgebras of $\mathfrak{L}$ can be transformed into one another by some automorphism of $\mathfrak{L}$ (3.5.2) we may restrict attention to Chevalley bases corresponding to a fixed Cartan subalgebra $\mathfrak{H}$. The isomorphism theorem 3.5.2 also shows that there is an automorphism of $\mathfrak{L}$ which transforms any set of fundamental roots into any other; and also that for a given system of fundamental roots there is an automorphism of $\mathfrak{L}$ which transforms any set of fundamental root vectors $e_r, r \in \Pi$, into any other. Thus we may restrict ourselves to Chevalley bases in which $\mathfrak{H}$, Π and the e_r for $r \in \Pi$ are fixed. Since the structure constants $N_{r,s}$ are all determined to within a sign we see, using 3.6.2, that each root vector e_r is determined to within a sign. Thus any Chevalley basis of $\mathfrak{L}$ can be transformed by an automorphism of $\mathfrak{L}$ into one of the form $\{h_r, r \in \Pi; \pm e_r, r \in \Phi\}$. It is now evident that the group generated by the elements $x_r(t)$ is independent of the Chevalley basis. For if $-e_r$ is chosen as a root vector instead of e_r, the generator $x_r(t)$ is simply replaced by $x_r(-t)$, its inverse. Hence the isomorphism type of the group $\mathfrak{L}(K)$ depends only upon $\mathfrak{L}$ and K. ∎

4.5 The Groups $A_1(K)$

The simplest examples of Chevalley groups are the groups $A_1(K)$. We shall show that these groups are isomorphic to the linear groups $PSL_2(K)$.

We note first that the simple Lie algebra A_1 over $\mathbb{C}$ can be represented as the algebra of 2×2 matrices of trace 0 under Lie multiplication $[xy] = xy - yx$. For if we define

$$h_r = \begin{pmatrix} 1 & 0 \\ 0 & -1 \end{pmatrix}, \qquad e_r = \begin{pmatrix} 0 & 1 \\ 0 & 0 \end{pmatrix}, \qquad e_{-r} = \begin{pmatrix} 0 & 0 \\ 1 & 0 \end{pmatrix},$$

we have

$$[h_r e_r] = 2e_r, \qquad [h_r e_{-r}] = -2e_{-r}, \qquad [e_r e_{-r}] = h_r.$$

These are the relations satisfied by a Chevalley basis of the simple algebra A_1. Note that the root vectors e_r, e_{-r} are represented by nilpotent matrices. We now require the following lemma.

LEMMA 4.5.1. *Let $\mathfrak{L}$ be a simple Lie algebra over $\mathbb{C}$ and suppose we have a representation of $\mathfrak{L}$ by matrices under Lie multiplication. Suppose*

$y \in \mathfrak{L}$ *is represented by a nilpotent matrix. Then* ad y *is a nilpotent derivation of* $\mathfrak{L}$ *and*

$$\exp (\text{ad } y).x = \exp y.x.(\exp y)^{-1}$$

for all $x \in \mathfrak{L}$. *Thus the image of* x *under the automorphism* exp (ad y) *is given by transforming by* exp y.

PROOF. We have

$$\text{ad } y.x = [yx] = yx - xy,$$

$$\frac{(\text{ad } y)^2}{2!}.x = \tfrac{1}{2}(y^2 x - 2yxy + xy^2).$$

We show

$$\frac{(\text{ad } y)^k}{k!}.x = \sum_{\substack{i,j \\ i+j=k}} \frac{y^i}{i!} x \frac{(-y)^j}{j!}.$$

This is true for $k = 1, 2$ and we prove it by induction. Assuming the above formula, by induction we have

$$\frac{(\text{ad } y)^{k+1}}{(k+1)!}.x = \frac{1}{k+1} \sum_{\substack{i,j \\ i+j=k}} \left(\frac{y^{i+1}}{i!} x \frac{(-y)^j}{j!} + \frac{y^i}{i!} x \frac{(-y)^{j+1}}{j!} \right)$$

$$= \sum_{\substack{m,n \\ m+n=k+1}} \frac{y^m}{m!} x \frac{(-y)^n}{n!} \binom{m+n}{k+1}$$

$$= \sum_{\substack{m,n \\ m+n=k+1}} \frac{y^m}{m!} x \frac{(-y)^n}{n!}.$$

Now y is nilpotent and so $((\text{ad } y)^k/k!)x = 0$ for sufficiently large values of k. Thus ad y is a nilpotent derivation of $\mathfrak{L}$. Also we have

$$\exp (\text{ad } y).x = \sum_{k=0}^{\infty} \frac{(\text{ad } y)^k}{k!} x$$

$$= \sum_{k=0}^{\infty} \sum_{\substack{i,j \\ i+j=k}} \frac{y^i}{i!} x \frac{(-y)^j}{j!}$$

$$= \sum_{i=0}^{\infty} \sum_{j=0}^{\infty} \frac{y^i}{i!} x \frac{(-y)^j}{j!}$$

$$= \exp y.x.(\exp y)^{-1}. \qquad \blacksquare$$

PROPOSITION 4.5.2. $A_1(K)$ *is isomorphic to* $PSL_2(K)$.

PROOF. We first apply 4.5.1 to the Lie algebra A_1 over $\mathbb{C}$. If $\zeta \in \mathbb{C}$ and e_r, e_{-r} are the matrices defined above we have

$$x_r(\zeta).x = \exp(\zeta e_r).x.\exp(\zeta e_r)^{-1},$$

$$x_{-r}(\zeta).x = \exp(\zeta e_{-r}).x.\exp(\zeta e_{-r})^{-1}.$$

We now pass to the Lie algebra $\mathfrak{L}_K$ for an arbitrary field K. $\mathfrak{L}_K$ is isomorphic to the Lie algebra of 2×2 matrices over K of trace 0. The Chevalley group $\mathfrak{L}_K$ is generated by the elements $x_r(t)$, $x_{-r}(t)$ as t runs through K. Now $x_r(t)$ is given by transformation by

$$\exp(te_r) = \begin{pmatrix} 1 & t \\ 0 & 1 \end{pmatrix}$$

and $x_{-r}(t)$ is given by transformation by

$$\exp(te_{-r}) = \begin{pmatrix} 1 & 0 \\ t & 1 \end{pmatrix}.$$

But the matrices

$$\begin{pmatrix} 1 & t \\ 0 & 1 \end{pmatrix}, \quad \begin{pmatrix} 1 & 0 \\ t & 1 \end{pmatrix}$$

generate $SL_2(K)$ as t runs through K. (A proof of this well-known fact will be given in section 6.1.) Thus there is a surjective homomorphism

$$SL_2(K) \rightarrow A_1(K)$$

under which the image of $m \in SL_2(K)$ is the automorphism $x \rightarrow mx\,m^{-1}$ of $\mathfrak{L}_K$. Under this homomorphism we have

$$\begin{pmatrix} 1 & t \\ 0 & 1 \end{pmatrix} \rightarrow x_r(t),$$

$$\begin{pmatrix} 1 & 0 \\ t & 1 \end{pmatrix} \rightarrow x_{-r}(t).$$

The kernel of the homomorphism consists of all $m \in SL_2(K)$ such that m commutes with all $x \in \mathfrak{L}_K$, and this is easily seen to be $\{\pm I_2\}$. Thus $A_1(K)$ is isomorphic to $PSL_2(K)$. ∎

CHAPTER 5

Unipotent Subgroups

5.1 The Subgroups U, V

Let $G = \mathfrak{L}(K)$ be the Chevalley group of type $\mathfrak{L}$ over K. G is generated by elements $x_r(t)$ for all $r \in \Phi$, $t \in K$. Now we have

$$x_r(t_1) . x_r(t_2) = \exp(t_1 \text{ ad } e_r) . \exp(t_2 \text{ ad } e_r)$$

$$= \exp(t_1 + t_2) \text{ ad } e_r$$

$$= x_r(t_1 + t_2).$$

Let X_r be the group generated by the elements $x_r(t)$ for all $t \in K$. X_r is a subgroup of G isomorphic to the additive group of K. For the map $t \rightarrow x_r(t)$ is an epimorphism from K to X_r; but if $x_r(t)$ is the identity, t must be 0, as can be seen from the operation of $x_r(t)$ on the basis elements of $\mathfrak{L}_K$. The subgroups X_r are called the *root subgroups* of G.

Let U be the subgroup of G generated by the elements $x_r(t)$ for $r \in \Phi^+$, $t \in K$; and V be generated by the $x_r(t)$ for $r \in \Phi^-$, $t \in K$. Then U and V clearly generate G, and U is generated by the root subgroups X_r corresponding to the positive roots, while V is generated by the negative root subgroups. We shall elucidate some of the properties of U, and analogous properties will clearly hold for V also.

We show first that the elements of U and V operate on $\mathfrak{L}_K$ as unipotent linear transformations. A linear transformation u of a vector space into itself is said to be unipotent if $u - 1$ is nilpotent. Let

$$\mathfrak{L}_K = \mathfrak{H} \oplus \sum_{r \in \Phi} \mathfrak{L}_r$$

be the decomposition of $\mathfrak{L}_K$ corresponding to the Cartan decomposition of $\mathfrak{L}$. We define $\mathfrak{L}_0 = \mathfrak{H}$ and

$$\mathfrak{L}_i = \sum_{h(r) = i} \mathfrak{L}_r$$

for each $i \neq 0$. Then

$$\mathfrak{L}_K = \bigoplus_i \mathfrak{L}_i.$$

Now if $r \in \Phi^+$ and $x \in \mathfrak{L}_i$ it is evident that

$$x_r(t) . x - x \in \sum_{j > i} \mathfrak{L}_j.$$

Since each element of U is a product of elements $x_r(t)$ for positive roots r, we have for each $u \in U$, $x \in \mathfrak{L}_i$:

$$u . x - x \in \sum_{j > i} \mathfrak{L}_j.$$

Thus $u - 1$ is a nilpotent linear transformation of $\mathfrak{L}_K$, whence u is unipotent. Similarly each element of V is a unipotent transformation of $\mathfrak{L}_K$.

The following lemma will be very useful in the subsequent development:

LEMMA 5.1.1. *Let $\mathfrak{L}$ be a simple Lie algebra over $\mathbb{C}$. Let y be an element of $\mathfrak{L}$ such that* ad y *is nilpotent and let θ be an automorphism of $\mathfrak{L}$. Then*

$$\theta . \exp (\text{ad } y) \theta^{-1} = \exp (\text{ad } \theta y).$$

PROOF. Let $x \in \mathfrak{L}$. Then we have

$$\theta . \exp (\text{ad } y) \, \theta^{-1}(x) = \theta \left(\sum_{i=0}^{\infty} \frac{1}{i!} \, [y, \ldots [y, \, \theta^{-1}(x)]] \right)$$

$$= \sum_{i=0}^{\infty} \frac{1}{i!} \, [\theta y, \ldots [\theta y, x]]$$

$$= \exp (\text{ad } \theta(y)) . x.$$

It follows that

$$\theta . \exp (\text{ad } y) \, \theta^{-1} = \exp (\text{ad } \theta y). \qquad \blacksquare$$

We now require a more technical lemma about certain nilpotent linear transformations of a vector space, which is a special case of the Campbell–Hausdorff formula.

LEMMA 5.1.2. *Let V be a vector space over a field of characteristic 0 and ξ, η be linear transformations of V into itself such that ξ, η and*

$$[\xi, \eta] = \xi \eta - \eta \xi$$

are nilpotent and $[\xi, \eta]$ commutes with ξ and η. Then $\xi + \eta$ is also nilpotent, and

$$\exp (\xi + \eta) = \exp \xi . \exp \eta . \exp (-\tfrac{1}{2}[\xi, \eta]).$$

PROOF. We shall show that

$$\frac{(\xi+\eta)^n}{n!} = \sum_{\substack{i,j,k \\ i+j+2k=n}} \frac{\xi^i}{i!}\frac{\eta^j}{j!}\frac{[\xi,\eta]^k}{2^k.k!}\cdot(-1)^k.$$

This is true for $n=1$ and 2, since

$$\tfrac{1}{2}(\xi+\eta)^2 = \tfrac{1}{2}\xi^2 + \tfrac{1}{2}\eta^2 + \xi\eta - \tfrac{1}{2}[\xi,\eta].$$

To prove it in general we require the formula

$$\eta\xi^i = \xi^i\eta - i\xi^{i-1}[\xi,\eta].$$

This is true for $i=1$ and its proof by induction is evident.

We assume the required result for $(\xi+\eta)^{n-1}/(n-1)!$. Then

$$\frac{(\xi+\eta)^n}{n!} = \frac{\xi+\eta}{n}\cdot\sum_{\substack{i,j,k \\ i+j+2k=n-1}} \frac{\xi^i}{i!}\frac{\eta^j}{j!}\frac{[\xi,\eta]^k}{2^k.k!}\cdot(-1)^k$$

$$= \frac{1}{n}\sum_{\substack{i,j,k \\ i+j+2k=n-1}} \frac{\xi^{i+1}}{i!}\frac{\eta^j}{j!}\frac{[\xi,\eta]^k}{2^k.k!}\cdot(-1)^k$$

$$+ \frac{1}{n}\sum_{\substack{i,j,k \\ i+j+2k=n-1}} \frac{\xi^i}{i!}\frac{\eta^{j+1}}{j!}\cdot\frac{[\xi,\eta]^k}{2^k.k!}\cdot(-1)^k$$

$$+ \frac{1}{n}\sum_{\substack{i,j,k \\ i+j+2k=n-1}} \frac{\xi^{i-1}}{(i-1)!}\frac{\eta^j}{j!}\cdot\frac{[\xi,\eta]^{k+1}}{2^k.k!}\cdot(-1)^{k+1}$$

$$= \sum_{\substack{i,j,k \\ i+j+2k=n}} \frac{\xi^i}{i!}\frac{\eta^j}{j!}\frac{[\xi,\eta]^k}{2^k.k!}\cdot(-1)^k\cdot\left(\frac{i}{n}+\frac{j}{n}+\frac{2k}{n}\right)$$

$$= \sum_{\substack{i,j,k \\ i+j+2k=n}} \frac{\xi^i}{i!}\frac{\eta^j}{j!}\frac{[\xi,\eta]^k}{2^k.k!}\cdot(-1)^k.$$

Since ξ, η and $[\xi,\eta]$ are all nilpotent,

$$\frac{(\xi+\eta)^n}{n!} = 0$$

for sufficiently large values of n. Also

$$\exp(\xi+\eta) = \exp\xi.\exp\eta.\exp(-\tfrac{1}{2}[\xi,\eta]),$$

as required. ∎

5.2 Chevalley's Commutator Formula

In the present section we shall derive a formula, due to Chevalley, which expresses the commutator of two generators of G as a product of generators. We work first over the complex field $\mathbb{C}$ and transfer at a later stage to an arbitrary field K.

Let $\mathfrak{L}$ be a simple Lie algebra over $\mathbb{C}$; r, s be linearly independent roots of $\mathfrak{L}$, and t, u be elements of $\mathbb{C}$. Consider the expression

$$x_r(t)x_s(u)x_r(t)^{-1}.$$

By 5.1.1 we have

$$x_r(t)x_s(u)x_r(t)^{-1} = x_r(t) \exp (\text{ad } ue_s)x_r(t)^{-1}$$

$$= \exp \text{ad } (x_r(t).ue_s)$$

$$= \exp \text{ad } \left(\sum_{i=0}^{q} M_{r,\,s,\,i} t^i u e_{ir+s} \right)$$

$$= \exp \left(\sum_{i=0}^{q} M_{r,\,s,\,i} t^i u \text{ ad } e_{ir+s} \right).$$

Now if r_1, r_2 are linearly independent roots such that r_1+r_2 is not a root, ad e_{r_1} and ad e_{r_2} commute. For

$$\text{ad } e_{r_1}.\text{ad } e_{r_2} - \text{ad } e_{r_2}.\text{ad } e_{r_1} = \text{ad}[e_{r_1}e_{r_2}] = 0.$$

Thus if $r+s$ is not a root ad (te_r) commutes with ad (ue_s) and hence $x_r(t)$ commutes with $x_s(u)$. On the other hand, if $r+s$ is a root, the integral combinations of r, s which are in Φ form a root system of type A_2, B_2 or G_2, by 3.6.3. We restrict attention to the roots of the form $ir+js$ with $i, j > 0$, and distinguish between a number of special cases.

(i) Suppose $r+2s$, $3r+2s$ are not roots. Then if $ir+js \in \Phi$ and $i > 0$, $j > 0$, we must have $j = 1$. This means that all the transformations ad e_{ir+s} commute. Thus

$$x_r(t)x_s(u)x_r(t)^{-1} = \prod_{i=0}^{q} \exp (M_{r,\,s,\,i} t^i u \text{ ad } e_{ir+s})$$

$$= \prod_{i=0}^{q} x_{ir+s}(M_{r,\,s,\,i} t^i u).$$

(ii) Suppose $r+2s$ is a root and the integral combinations of r, s in

Φ form a root system of type B_2. Then

$$x_r(t)x_s(u)x_r(t)^{-1} = \exp (u \text{ ad } e_s + N_{r, s}tu \text{ ad } e_{r+s}).$$

Let $\xi = u$ ad e_s and $\eta = N_{r, s}tu$ ad e_{r+s}. Then

$$[\xi, \eta] = N_{r, s}N_{s, r+s}tu^2 \text{ ad } e_{r+2s}.$$

Thus $[\xi, \eta]$ is nilpotent and commutes with ξ and η (because $r+3s$, $2r+3s$ are not in Φ). Applying 5.1.2 we obtain

$$x_r(t) \, x_s(u) \, x_r(t)^{-1}$$

$$= \exp (u \text{ ad } e_s).\exp (N_{r, s}tu \text{ ad } e_{r+s}).\exp (-\tfrac{1}{2}N_{r, s}N_{s, r+s}tu^2 \text{ ad } e_{r+2s})$$

$$= x_s(u) \, x_{r+s}(N_{r, s}tu) \, x_{r+2s}(M_{s, r, 2}tu^2).$$

(iii) Suppose the integral combinations of r, s in Φ form a root system of type G_2 in which $3r+2s$ is a root. Then the integral combinations of form $ir+js \in \Phi$ with $i > 0$, $j > 0$ are

$$r+s, \qquad 2r+s, \qquad 3r+s, \qquad 3r+2s.$$

Then

$$x_r(t) \, x_s(u) \, x_r(t)^{-1} = \exp \left(\sum_{i=0}^{3} M_{r, s, i}t^iu \text{ ad } e_{ir+s} \right).$$

Let

$$\xi = \sum_{i=0}^{2} M_{r, s, i}t^iu \text{ ad } e_{ir+s},$$

$$\eta = M_{r, s, 3}t^3u \text{ ad } e_{3r+s}.$$

Then

$$[\xi, \eta] = N_{s, 3r+s}M_{r, s, 3}t^3u^2 \text{ ad } e_{3r+2s}.$$

Thus $[\xi, \eta]$ is nilpotent and commutes with ξ, η. By 5.1.2 we have

$$x_r(t) \, x_s(u) \, x_r(t)^{-1} = \exp \left(\sum_{i=0}^{2} M_{r, s, i}t^iu \text{ ad } e_{ir+s} \right).\exp (M_{r, s, 3}t^3u \text{ ad } e_{3r+s})$$

$$\times \exp (-\tfrac{1}{2}N_{s, 3r+s}M_{r, s, 3}t^3u^2 \text{ ad } e_{3r+2s}).$$

We now make a second application of 5.1.2. Put

$$\xi = \sum_{i=0}^{1} M_{r, s, i}t^iu \text{ ad } e_{ir+s}, \qquad \eta = M_{r, s, 2}t^2u \text{ ad } e_{2r+s}.$$

Then

$$[\xi, \eta] = N_{r+s, 2r+s}M_{r, s, 1}M_{r, s, 2}t^3u^2 \text{ ad } e_{3r+2s}.$$

$[\xi, \eta]$ is again nilpotent and commutes with ξ, η. Thus

$$x_r(t)\, x_s(u)\, x_r(t)^{-1}$$

$$= \exp\,(u\ \mathrm{ad}\ e_s + M_{r,\ s,\ 1}M_{r,\ s,\ 2}t^3u^2\ \mathrm{ad}\ e_{3r+2s})$$

$$\times \exp\,(-\tfrac{1}{2}N_{r+s,\ 2r+s}M_{r,\ s,\ 1}M_{r,\ s,\ 2}t^3u^2\ \mathrm{ad}\ e_{3r+2s})$$

$$\times \exp\,(M_{r,\ s,\ 3}t^3u\ \mathrm{ad}\ e_{3r+s})$$

$$\times \exp\,(-\tfrac{1}{2}N_{s,\ 3r+s}M_{r,\ s,\ 3}t^3u^2\ \mathrm{ad}\ e_{3r+2s})$$

$$= x_s(u)\, x_{r+s}(M_{r,\ s,\ 1}tu)\, x_{2r+s}(M_{r,\ s,\ 2}t^2u)\, x_{3r+s}(M_{r,\ s,\ 3}t^3u)$$

$$\times x_{3r+2s}((-\tfrac{1}{2}N_{r+s,\ 2r+s}M_{r,\ s,\ 1}M_{r,\ s,\ 2} - \tfrac{1}{2}N_{s,\ 3r+s}M_{r,\ s,\ 3})t^3u^2).$$

Observe that the coefficient of x_{3r+2s} is

$$(-\tfrac{1}{2}N_{r+s,\ 2r+s}M_{r,\ s,\ 1}M_{r,\ s,\ 2} - \tfrac{1}{2}N_{s,\ 3r+s}M_{r,\ s,\ 3})\ t^3u^2$$

$$= (-\tfrac{1}{4}N_{r+s,\ 2r+s}N_{r,\ s}N_{r,\ r+s} - \tfrac{1}{2}N_{s,\ 3r+s}M_{r,\ s,\ 3})\ t^3u^2$$

$$= (\tfrac{1}{2}M_{r+s,\ r,\ 2} - \tfrac{1}{2}N_{s,\ 3r+s}M_{r,\ s,\ 3})\ t^3u^2,$$

since $N_{r,\ s} = \pm 1$.

This coefficient can be simplified by the following lemma:

LEMMA 5.2.1. $N_{s,\ 3r+s}M_{r,\ s,\ 3} = \tfrac{1}{3}M_{r+s,\ r,\ 2}.$

PROOF. We use the relation

$$s + (3r+s) + (-(r+s)) + (-(2r+s)) = 0.$$

By formula 4.1.2 (iv) relating the structure constants we have

$$\frac{N_{s,\ 3r+s}N_{-(r+s),\ -(2r+s)}}{3} + N_{-(r+s),\ s}N_{3r+s,\ -(2r+s)} = 0.$$

Using formulae 4.1.2 (ii), (iii) we obtain

$$\frac{-N_{s,\ 3r+s}N_{r+s,\ 2r+s}}{3} + \frac{N_{s,\ r}N_{-(2r+s),\ -r}}{3} = 0,$$

whence

$$N_{s,\ 3r+s}N_{r+s,\ 2r+s} = N_{s,\ r}N_{r,\ 2r+s}.$$

It follows that

$$N_{s,\ 3r+s}M_{r,\ s,\ 3} = \tfrac{1}{6}N_{s,\ 3r+s}N_{r,\ s}N_{r,\ r+s}N_{r,\ 2r+s}$$

$$= -\tfrac{1}{6}N_{r,\ r+s}N_{r+s,\ 2r+s} \qquad (\text{since } N_{s,\ 3r+s}^2 = 1)$$

$$= \tfrac{1}{6}N_{r+s,\ r}N_{r+s,\ 2r+s}$$

$$= \tfrac{1}{3}M_{r+s,\ r,\ 2}. \qquad \blacksquare$$

Applying 5.2.1, the coefficient of x_{3r+2s} in the expression for

$$x_r(t)\, x_s(u)\, x_r(t)^{-1}$$

is $\frac{1}{3}M_{r+s,\ r,\ 2}t^3u^2$. Thus

$$x_r(t)\, x_s(u)\, x_r(t)^{-1} = x_s(u)\, x_{r+s}(M_{r,\ s,\ 1}tu)\, x_{2r+s}(M_{r,\ s,\ 2}t^2u)$$

$$\times x_{3r+s}(M_{r,\ s,\ 3}t^3u)\, x_{3r+2s}(\tfrac{1}{3}M_{r+s,\ r,\ 2}t^3u^2).$$

(iv) Suppose the integral combinations of r, s in Φ form a root system of type G_2 in which $2r+3s$ is a root. Then the integral combinations of the form $ir+js \in \Phi$, with $i>0$, $j>0$ are

$$r+s, \qquad r+2s, \qquad r+3s, \qquad 2r+3s.$$

The formula of 5.1.2 can no longer be used to calculate

$$x_r(t)\, x_s(u)\, x_r(t)^{-1}.$$

We use instead an indirect method based on (iii). We have shown in (iii) (interchanging the rôles of r, s) that

$$x_s(u)\, x_r(t)\, x_s(u)^{-1} = x_r(t)\, x_{s+r}(M_{s,\ r,\ 1}ut)\, x_{2s+r}(M_{s,\ r,\ 2}u^2t)$$

$$\times x_{3s+r}(M_{s,\ r,\ 3}u^3t) . x_{3s+2r}(\tfrac{1}{3}M_{s+r,\ s,\ 2}u^3t^2)$$

$$= x_{s+r}(M_{s,\ r,\ 1}ut)\, x_{2s+r}(M_{s,\ r,\ 2}u^2t)\, x_r(t)$$

$$\times x_{3s+r}(M_{s,\ r,\ 3}u^3t) . x_r(t)^{-1}x_{3s+2r}(\tfrac{1}{3}M_{s+r,\ s,\ 2}u^3t^2)\, x_r(t)$$

$$= x_{s+r}(M_{s,\ r,\ 1}ut)\, x_{2s+r}(M_{s,\ r,\ 2}u^2t)\, x_{3s+r}(M_{s,\ r,\ 3}u^3t)$$

$$\times x_{3s+2r}((\tfrac{1}{3}M_{s+r,\ s,\ 2} + N_{r,\ 3s+r}M_{s,\ r,\ 3})u^3t^2) . x_r(t).$$

Now by 5.2.1,

$$N_{r,\ 3s+r}M_{s,\ r,\ 3} = \tfrac{1}{3}M_{s+r,\ s,\ 2}.$$

Thus

$$x_s(u)\, x_r(t)\, x_s(u)^{-1} = x_{s+r}(M_{s,\ r,\ 1}ut)\, x_{2s+r}(M_{s,\ r,\ 2}u^2t)\, x_{3s+r}(M_{s,\ r,\ 3}u^3t)$$

$$\times x_{3s+2r}(\tfrac{2}{3}M_{s+r,\ s,\ 2}u^3t^2) . x_r(t).$$

Replacing u by $-u$ in this formula gives

$$x_r(t)\, x_s(u)\, x_r(t)^{-1} = x_s(u)\, x_{s+r}(-M_{s,\ r,\ 1}ut)\, x_{2s+r}(M_{s,\ r,\ 2}u^2t)$$

$$\times x_{3s+r}(-M_{s,\ r,\ 3}u^3t) . x_{3s+2r}(-\tfrac{2}{3}M_{s+r,\ s,\ 2}u^3t^2).$$

(v) Suppose the integral combinations of r, s in Φ form a root system of type G_2 in which $r+2s \in \Phi$, $3r+2s \notin \Phi$, $3s+2r \notin \Phi$. Then the roots of

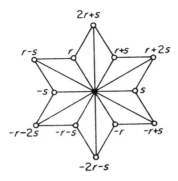

Figure 2

form $ir + js \in \Phi$ with $i > 0$, $j > 0$ are $r + s$, $2r + s$, $2s + r$. The root system in this case is shown in Figure 2. Then we have

$$x_r(t)\, x_s(u)\, x_r(t)^{-1}$$

$$= x_r(t)\, \exp\,(u\,\mathrm{ad}\,e_s)\, x_r(t)^{-1}$$

$$= \exp\,\mathrm{ad}\,\left(\sum_{i=0}^{q} M_{r,\,s,\,i} t^i u e_{ir+s} \right)$$

$$= \exp\,\mathrm{ad}\,(u e_s + M_{r,\,s,\,1} t u e_{r+s} + M_{r,\,s,\,2} t^2 u e_{2r+s})$$

$$= \exp\,\mathrm{ad}\,(u e_s + M_{r,\,s,\,1} t u e_{r+s})\, \exp\,\mathrm{ad}\,(M_{r,\,s,\,2} t^2 u e_{2r+s})$$

$$= \exp\,\mathrm{ad}\,(u e_s)\, \exp\,\mathrm{ad}\,(M_{r,\,s,\,1} t u e_{r+s})$$

$$\times \exp\,\mathrm{ad}\,(-\tfrac{1}{2} M_{r,\,s,\,1} N_{s,\,r+s} t u^2 e_{r+2s})\, \exp\,\mathrm{ad}\,(M_{r,\,s,\,2} t^2 u e_{2r+s})$$

$$= x_s(u)\, x_{r+s}(M_{r,\,s,\,1} t u)\, x_{2r+s}(M_{r,\,s,\,2} t^2 u)\, x_{r+2s}(M_{s,\,r,\,2} t u^2).$$

Now the cases (i), (ii), (iii), (iv), (v) considered above exhaust all possibilities. All the cases except G_2 are dealt with in (i) or (ii), and the three different ways in which r, s can be placed inside G_2 not satisfying (i) are dealt with in (iii), (iv), (v).

We may summarize the results of (i)–(v) in a single formula as follows:

$$x_r(t)\, x_s(u)\, x_r(t)^{-1} = x_s(u)\,.\,\prod_{i,\,j>0} x_{ir+js}(C_{ijrs} t^i u^j),$$

where

$$C_{i1rs} = M_{rsi},$$
$$C_{1jrs} = (-1)^j M_{srj},$$
$$C_{32rs} = \tfrac{1}{3} M_{r+s,\,r,\,2},$$
$$C_{23rs} = -\tfrac{2}{3} M_{s+r,\,s,\,2}.$$

The product is taken over all pairs $i > 0$, $j > 0$ such that $ir + js \in \Phi$, the order of the terms in the product being such that $i + j$ is increasing.‡

The numbers C_{ijrs} defined here are all rational integers. This is clear for C_{i1rs} and C_{1jrs}, while for C_{32rs} and C_{23rs} the absolute value of $M_{r+s,\,r,\,2}$, $M_{s+r,\,s,\,2}$ can be obtained by examining the root system of G_2. These both have value ± 3. Thus $C_{32rs} = \pm 1$ and $C_{23rs} = \pm 2$. Therefore in all cases we have $C_{ijrs} = \pm 1$, ± 2 or ± 3.

Now it follows from results established in chapter 4 that the validity of the equation

$$x_r(t)\, x_s(u)\, x_r(t)^{-1} = x_s(u) \cdot \prod_{i,\,j>0} x_{ir+js}(C_{ijrs} t^i u^j)$$

is equivalent to the vanishing of certain polynomials $f_k(t, u) = 0$, where $f_k(x, y) \in \mathbb{Z}[x, y]$. Let K be an arbitrary field and t, u be elements of K. Then the validity of the equation

$$x_r(t)\, x_s(u)\, x_r(t)^{-1} = x_s(u) \prod_{i,\,j>0} x_{ir+js}(C_{ijrs} t^i u^j)$$

in $G = \mathfrak{L}(K)$ is equivalent to the vanishing of the polynomials $\bar{f}_k(t, u) = 0$, where $\bar{f}_k(x, y) \in K[x, y]$ is the polynomial obtained from $f_k(x, y)$ by replacing each integral coefficient by its image in the prime subfield of K under the natural homomorphism. However, we have shown that $f_k(t, u) = 0$ for all t, $u \in \mathbb{C}$. Thus each f_k is identically zero. It follows that each $\bar{f}_k$ is identically zero. Thus $\bar{f}_k(t, u) = 0$ for all t, $u \in K$. We have therefore established the following theorem, known as Chevalley's commutator formula.

THEOREM 5.2.2. *Let $G = \mathfrak{L}(K)$ be a Chevalley group over an arbitrary field, r, s be linearly independent roots of $\mathfrak{L}$ and t, u be elements of K. Define the commutator*

$$[x_s(u),\, x_r(t)] = x_s(u)^{-1} x_r(t)^{-1} x_s(u)\, x_r(t).$$

Then we have

$$[x_s(u),\, x_r(t)] = \prod_{i,\,j>0} x_{ir+js}(C_{ijrs}(-t)^i u^j),$$

‡ This condition does not determine the order uniquely in the situation of case (v). However, the two terms with $i + j = 3$ commute, so both orders give the same result.

where the product is taken over all pairs of positive integers i, j for which $ir + js$ is a root, in order of increasing $i + j$. The constants C_{ijrs} are given by

$$C_{i1rs} = M_{rsi},$$
$$C_{1jrs} = (-1)^j M_{srj},$$
$$C_{32rs} = \tfrac{1}{3} M_{r+s, \, r, \, 2},$$
$$C_{23rs} = -\tfrac{2}{3} M_{s+r, \, s, \, 2}$$

Each C_{ijrs} is one of ± 1, ± 2, ± 3.

COROLLARY 5.2.3. *If r, $s \in \Phi^+$ are linearly independent then*

$$[x_s(u), x_r(t)] = \prod_{i,j > 0} x_{ir+js}(C_{ijrs}(-t)^i u^j),$$

where the product is taken over all positive roots of the form $ir + js$, $i > 0$, $j > 0$, in increasing order.

5.3 The Structure of U and V

We shall now use Chevalley's commutator formula to investigate the structure of U. We first require two lemmas.

LEMMA 5.3.1. *The ordering of the roots can be chosen compatible with the height function, i.e. so that $r \prec s$ implies $h(r) \leqslant h(s)$.*

PROOF. Let $\mathfrak{V}$ be the real vector space spanned by Φ and let $h : \mathfrak{V} \to \mathbb{R}$ be the linear map which coincides with the height function on Φ. Let $\mathfrak{V}_0$ be the null-space of h. Then dim $\mathfrak{V}_0 = l - 1$. Let v_1 be an element of $\mathfrak{V}$ with $h(v_1) = 1$ and $v_2, v_3, \ldots, v_l$ be a basis of $\mathfrak{V}_0$. Then

$$h\left(\sum_{i=1}^{l} \lambda_i v_i \right) = \lambda_1.$$

If we define $\mathfrak{V}^+$ to be the set of non-zero vectors whose first non-zero coefficient λ_i is positive we obtain an ordering compatible with the height function. ∎

LEMMA 5.3.2. *Let $r \in \Phi$. Then the element $h_r \in \mathfrak{L}_K$ is non-zero.*

PROOF. This is true by definition if $r \in \Pi$. The co-roots $h_r \in \mathfrak{L}$ form the dual root system Φ^* by 3.6.1, and Π^* is a fundamental system in Φ^*. Thus

$$h_r = n_1 h_{p_1} + \ldots + n_l h_{p_l},$$

where $\Pi = \{p_1, p_2, \ldots, p_l\}$ and $n_i \in \mathbb{Z}$.

Now it is only possible for $h_r = 0$ in $\mathfrak{L}_K$ if K has characteristic p and p divides each n_i, $i = 1, \ldots, l$. We suppose that this is so. Now r can be expressed in the form $w(p_i)$ for some $w \in W$ and some i. Then $w(h_{p_i}) = h_r$. The element w is represented with respect to the basis $h_{p_1}, \ldots, h_{p_l}$ by a matrix with coefficients in $\mathbb{Z}$, and one of the columns of this matrix is $(n_1, n_2, \ldots, n_l)$. Thus det w is divisible by p. Since det $w = \pm 1$ we have a contradiction. ∎

THEOREM 5.3.3. *Let* $G = \mathfrak{L}(K)$ *be a Chevalley group, U be the subgroup of G generated by the root subgroups X_r with $r \in \Phi^+$, and U_m be the subgroup generated by the X_r with $r \in \Phi^+$ and $h(r) \geqslant m$. Then:*

(i) *U is nilpotent and*

$$U = U_1 \supset U_2 \supset \ldots \supset U_h \supset 1$$

is a central series for U, where h is the greatest height of a root of $\mathfrak{L}$.

(ii) *Each element of U is uniquely expressible in the form*

$$\prod_{r_i \in \Phi^+} x_{r_i}(t_i),$$

where the product is taken over all positive roots in increasing order.

PROOF. Observe first that U_m is normal in U for each $m \geqslant 1$. For if $s \in \Phi^+$ and $h(s) \geqslant m$ we have

$$x_r(t)\, x_s(u)\, x_r(t)^{-1} \in U_m$$

for all $r \in \Phi^+$, by the commutator formula. Thus $x_r(t)\, U_m x_r(t)^{-1} \subseteq U_m$. Replacing t by $-t$ we see that $x_r(t)\, U_m x_r(t)^{-1} = U_m$. Since the $x_r(t)$ for $r \in \Phi^+$, $t \in K$ generate U, it follows that U_m is normal in U.

Let $[U_m, U_n]$ be the subgroup generated by all commutators $[a, b]$ for $a \in U_m$, $b \in U_n$. Let $r, s \in \Phi^+$ satisfy $h(r) \geqslant n$, $h(s) \geqslant m$. Then

$$h(ir + js) \geqslant m + n$$

for all $i, j > 0$. Thus

$$[x_s(u), x_r(t)] \in U_{m+n},$$

using the commutator formula. This means that the images of $x_r(t)$, $x_s(u)$ under the natural homomorphism commute in the factor group U/U_{m+n}.

Now elements of the form $x_s(u)$ generate U_m and elements of form $x_r(t)$ generate U_n. Thus each element of U_m commutes with each element of U_n in U/U_{m+n}. Hence $[a, b] \in U_{m+n}$ for all $a \in U_m$, $b \in U_n$. Thus

$$[U_m, U_n] \subseteq U_{m+n}.$$

In particular $[U, U_n] \subseteq U_{n+1}$ for each n. Thus each element of U_n/U_{n+1} commutes with each element of U/U_{n+1}, and so U_n/U_{n+1} is in the centre of U/U_{n+1}. Therefore we have a central series for U.

Now each element of U is a product of elements $x_r(t)$ with $r \in \Phi^+$. If in such a product we have a pair of consecutive terms $x_s(u) \, x_r(t)$ with $r \prec s$ we use the formula

$$x_s(u) \, x_r(t) = x_r(t) \, x_s(u) \prod_{i, j > 0} x_{ir+js}(C_{ijrs}(-t)^i u^j).$$

The terms $x_r(t) \, x_s(u)$ now occur in their natural order and all new terms introduced satisfy $h(ir + js) \geqslant h(r) + h(s)$. Thus the process of rearranging the terms in their natural order by means of the commutator formula terminates after a finite number of steps, and each element of U is expressible in the form

$$x_{r_1}(t_1) \, x_{r_2}(t_2) \ldots x_{r_N}(t_N),$$

where $0 \prec r_1 \prec r_2 \prec \ldots \prec r_N$ are the positive roots of $\mathfrak{L}$.

We now prove uniqueness. We shall prove by descending induction on m that

$$\prod_{h(s) \geqslant m} x_s(t_s) = \prod_{h(s) \geqslant m} x_s(t'_s)$$

implies $t_s = t'_s$, where the products are taken over all positive roots of height at least m in increasing order. The statement is trivially true for $m = h + 1$, since $U_{h+1} = 1$. We assume it true for $m + 1$. Let

$$u = \prod_{h(s) \geqslant m} x_s(t_s) = \prod_{h(s) \geqslant m} x_s(t'_s)$$

and let r be a positive root of height m. Consider the element $u . e_{-r}$ of $\mathfrak{L}_K$. If $h(s) > m$ and $-r + s \in \Phi$, then $-r + s \in \Phi^+$, since the ordering of the roots is compatible with the height. If $h(s) = m$ then $-r + s$ cannot be a root, since its height would be 0. Thus

$$u . e_{-r} = e_{-r} + t_r h_r + x,$$

where

$$x \in \sum_{r \in \Phi^+} \mathfrak{L}_r;$$

and similarly

$$u.e_{-r} = e_{-r} + t'_r h_r + y,$$

where

$$y \in \sum_{r \in \Phi^+} \mathfrak{L}_r.$$

But $\mathfrak{L}_K$ has the direct decomposition

$$\mathfrak{L}_K = \mathfrak{H} \oplus \sum_{r \in \Phi^+} \mathfrak{L}_r \oplus \sum_{r \in \Phi^-} \mathfrak{L}_r$$

and $h_r \neq 0$ by 5.3.2. Thus $t_r = t'_r$ and $x = y$. Hence

$$u = \prod_{h(s)=m} x_s(t_s) \cdot \prod_{h(s) \geqslant m+1} x_s(t_s) = \prod_{h(s)=m} x_s(t_s) \cdot \prod_{h(s) \geqslant m+1} x_s(t'_s).$$

It follows that

$$\left(\prod_{h(s)=m} x_s(t_s) \right)^{-1} . u = \prod_{h(s) \geqslant m+1} x_s(t_s) = \prod_{h(s) \geqslant m+1} x_s(t'_s),$$

whence $t_s = t'_s$ by induction, and the proof is complete. ∎

It is clear that similar results hold for V in terms of negative roots.

CHAPTER 6

The Subgroups $\langle X_r, X_{-r} \rangle$

In the preceding chapter we considered relations involving generators $x_r(t)$, $x_s(u)$ of a Chevalley group corresponding to linearly independent roots r, s, and showed in particular that two such elements generate a nilpotent subgroup. We shall now investigate relations involving generators of the form $x_r(t)$, $x_{-r}(u)$. We do this by determining the structure of the subgroup $\langle X_r, X_{-r} \rangle$ generated by two root subgroups corresponding to opposite roots. This subgroup turns out to be closely related to the group $SL_2(K)$; in fact there is a homomorphism from $SL_2(K)$ onto $\langle X_r, X_{-r} \rangle$. We prove the existence of this homomorphism first for the complex field. Regarding $\mathfrak{L}$ as a module for the group $\langle X_r, X_{-r} \rangle$, we show that the operation of $\langle X_r, X_{-r} \rangle$ on certain submodules of $\mathfrak{L}$ is the same as the operation of $SL_2(\mathbb{C})$ on the space of homogeneous polynomials in two variables of a suitable degree.‡

6.1 The Group $SL_2(K)$

LEMMA 6.1.1. *Let K be an arbitrary field. Then the group $SL_2(K)$ is generated by the elements*

$$\begin{pmatrix} 1 & t \\ 0 & 1 \end{pmatrix}, \quad \begin{pmatrix} 1 & 0 \\ t & 1 \end{pmatrix}$$

as t runs through K.

PROOF. Let

$$\begin{pmatrix} \alpha & \beta \\ \gamma & \delta \end{pmatrix}$$

be an element of $SL_2(K)$.
If $\gamma \neq 0$ we have

$$\begin{pmatrix} \alpha & \beta \\ \gamma & \delta \end{pmatrix} = \begin{pmatrix} 1 & (\alpha-1)\gamma^{-1} \\ 0 & 1 \end{pmatrix} \begin{pmatrix} 1 & 0 \\ \gamma & 1 \end{pmatrix} \begin{pmatrix} 1 & (\delta-1)\gamma^{-1} \\ 0 & 1 \end{pmatrix}.$$

‡ This method of establishing the existence of the homomorphism was described to the author by C. W. Curtis.

If $\beta \neq 0$ we have

$$\begin{pmatrix} \alpha & \beta \\ \gamma & \delta \end{pmatrix} = \begin{pmatrix} 1 & 0 \\ (\delta-1)\beta^{-1} & 1 \end{pmatrix} \begin{pmatrix} 1 & \beta \\ 0 & 1 \end{pmatrix} \begin{pmatrix} 1 & 0 \\ (\alpha-1)\beta^{-1} & 1 \end{pmatrix}.$$

If $\beta = \gamma = 0$ we have

$$\begin{pmatrix} \alpha & 0 \\ 0 & \alpha^{-1} \end{pmatrix} = \begin{pmatrix} 1 & 0 \\ \alpha^{-1}-1 & 1 \end{pmatrix} \begin{pmatrix} 1 & 1 \\ 0 & 1 \end{pmatrix} \begin{pmatrix} 1 & 0 \\ \alpha-1 & 1 \end{pmatrix} \begin{pmatrix} 1 & -\alpha^{-1} \\ 0 & 0 \end{pmatrix}.$$

It is clear from these relations that $SL_2(K)$ is generated in the manner described.   ∎

Let $R = \mathbb{C}[x, y]$ be the polynomial ring in two variables over $\mathbb{C}$ and R_q be the subspace of polynomials which are homogeneous of degree q. R_q has a basis

$$x^q, x^{q-1}y, x^{q-2}y^2, \dots, y^q$$

and dimension $q+1$. An element

$$\begin{pmatrix} \alpha & \beta \\ \gamma & \delta \end{pmatrix}$$

of $SL_2(K)$ operates on R by the transformation

$$x \to \alpha x + \gamma y,$$

$$y \to \beta x + \delta y,$$

and R_q is a representation space for $SL_2(\mathbb{C})$ for each q under this operation.

LEMMA 6.1.2. *Let $v_i = x^i y^{q-i}$, $i = 0, 1, \dots, q$. Then the images of v_i under transformation by*

$$\begin{pmatrix} 1 & t \\ 0 & 1 \end{pmatrix} \quad and \quad \begin{pmatrix} 1 & 0 \\ t & 1 \end{pmatrix}$$

are given by:

$$\begin{pmatrix} 1 & t \\ 0 & 1 \end{pmatrix} \cdot v_i = \sum_{j=0}^{q-i} \binom{q-i}{j} t^j v_{i+j},$$

$$\begin{pmatrix} 1 & 0 \\ t & 1 \end{pmatrix} \cdot v_i = \sum_{j=0}^{i} \binom{i}{j} t^j v_{i-j}.$$

PROOF. Since

$$\begin{pmatrix} 1 & t \\ 0 & 1 \end{pmatrix}$$

transforms x into itself and y into $tx+y$, and

$$\begin{pmatrix} 1 & 0 \\ t & 1 \end{pmatrix}$$

transforms y into itself and x into $x+ty$, the result is clear. ∎

6.2 The Homomorphism from $SL_2(\mathbb{C})$

We now consider the operation of the subgroup $\langle X_r, X_{-r} \rangle$ of $G = \mathfrak{L}(\mathbb{C})$ on the Lie algebra $\mathfrak{L}$. Each r-chain of roots gives rise to a submodule of $\mathfrak{L}$ under $\langle X_r, X_{-r} \rangle$ as follows. Let s be a root linearly independent of r which begins an r-chain. Let

$$s, r+s, 2r+s, \ldots, qr+s$$

be the r-chain of roots through s. Let $\mathfrak{M}_s$ be the subspace of $\mathfrak{L}$ defined by

$$\mathfrak{M}_s = \mathfrak{L}_s \oplus \mathfrak{L}_{r+s} \oplus \ldots \oplus \mathfrak{L}_{qr+s}.$$

Then $\mathfrak{M}_s$ is invariant under $\langle X_r, X_{-r} \rangle$, by the results of chapter 4. Let $\mathfrak{H}_r$ be the 1-dimensional subspace of $\mathfrak{H}$ spanned by the co-root h_r and $\mathfrak{H}_r^{\perp}$ be the subspace of $\mathfrak{H}$ of elements h with $(h, h_r)=0$. Then the subspaces $\mathfrak{L}_r \oplus \mathfrak{H}_r \oplus \mathfrak{L}_{-r}$ and $\mathfrak{H}_r^{\perp}$ are invariant under the action of $\langle X_r, X_{-r} \rangle$, and $\mathfrak{L}$ is the direct sum of these subspaces together with the spaces $\mathfrak{M}_s$ for the various r-chains of roots.

LEMMA 6.2.1. *Let r, s be linearly independent roots such that the r-chain through s is*

$$s, r+s, \ldots, qr+s.$$

Let $\mathfrak{M}_s = \mathfrak{L}_s \oplus \mathfrak{L}_{r+s} \oplus \ldots \oplus \mathfrak{L}_{qr+s}$ and write $f_i = e_{ir+s}$. Then $f_0, f_1, \ldots, f_q$ is a basis for $\mathfrak{M}_s$ and we have

$$x_r(t).f_i = \sum_{j=0}^{q-i} \epsilon_i \epsilon_{i+1} \ldots \epsilon_{i+j-1} \binom{i+j}{j} t^j f_{i+j},$$

$$x_{-r}(t).f_i = \sum_{j=0}^{i} \epsilon_{i-1} \epsilon_{i-2} \ldots \epsilon_{i-j} \binom{q-i+j}{j} t^j f_{i-j},$$

where the numbers ϵ_0, ϵ_1, ... *are all* ± 1 *and are defined by*

$$N_{r,\ ir+s} = \epsilon_i(i+1).$$

(Note that $N_{r,\ ir+s} = \pm(i+1)$ *by* 4.2.1.)

PROOF. We know from chapter 4 that

$$x_r(t).e_{ir+s} = \sum_{j=0}^{q-i} M_{r,\ ir+s,\ j}t^j e_{(i+j)r+s}$$

$$= \sum_{j=0}^{q-i} \epsilon_i\epsilon_{i+1}\ \ldots\ \epsilon_{i+j-1}\binom{i+j}{j}\ t^j\ e_{(i+j)r+s}.$$

Similarly we have

$$x_{-r}(t).e_{ir+s} = \sum_{j=0}^{i} M_{-r,\ ir+s,\ j}t^j\ e_{(i-j)r+s}.$$

However, using relations 4.1.2 (i), (ii), (iii) between the structure constants and also 4.1.1 we have

$$M_{-r,\ ir+s,\ j} = \frac{1}{j!}\ N_{-r,\ ir+s}N_{-r,\ (i-1)r+s}\ \ldots$$

$$= \frac{1}{j!}\ N_{r,\ (i-1)r+s}.\frac{q-i+1}{i}.N_{r,\ (i-2)r+s}.\frac{q-i+2}{i-1}\ \ldots$$

$$= \frac{1}{j!}\ \epsilon_{i-1}(q-i+1).\epsilon_{i-2}(q-i+2)\ \ldots$$

$$= \epsilon_{i-1}\epsilon_{i-2}\ \ldots\ \epsilon_{i-j}\binom{q-i+j}{j}.$$

Thus

$$x_{-r}(t).e_{ir+s} = \sum_{j=0}^{i} \epsilon_{i-1}\epsilon_{i-2}\ \ldots\ \epsilon_{i-j}\binom{q-i+j}{j}\ t^j\ e_{(i-j)r+s}.\ \blacksquare$$

We now show it is possible to choose a basis for R_q so that

$$\begin{pmatrix} 1 & t \\ 0 & 1 \end{pmatrix},\quad \begin{pmatrix} 1 & 0 \\ t & 1 \end{pmatrix}$$

operate on this basis in essentially the same way that $x_r(t)$, $x_{-r}(t)$ operate on the basis $f_0, f_1, \ldots, f_q$ of $\mathfrak{M}_s$.

LEMMA 6.2.2. *Let* $v_i = x^i y^{q-i} \in R_q$ *and* $\epsilon_0, \epsilon_1, \ldots, \epsilon_{q-1}$ *be integers equal to* ± 1. *Define*

$$u_i = \epsilon_0 \epsilon_1 \ldots \epsilon_{i-1} \binom{q}{i} v_i, \qquad i = 0, 1, \ldots, q.$$

Then

$$\begin{pmatrix} 1 & t \\ 0 & 1 \end{pmatrix} . u_i = \sum_{j=0}^{q-i} \epsilon_i \epsilon_{i+1} \ldots \epsilon_{i+j-1} \binom{i+j}{j} t^j u_{i+j},$$

$$\begin{pmatrix} 1 & 0 \\ t & 1 \end{pmatrix} . u_i = \sum_{j=0}^{i} \epsilon_{i-1} \epsilon_{i-2} \ldots \epsilon_{i-j} \binom{q-i+j}{j} t^j u_{i-j}.$$

PROOF. By 6.1.2 we have

$$\begin{pmatrix} 1 & t \\ 0 & 1 \end{pmatrix} . u_i = \sum_{j=0}^{q-i} \frac{\binom{q-i}{j}\binom{q}{i} \epsilon_0 \epsilon_1 \ldots \epsilon_{i-1}}{\binom{q}{i+j} \epsilon_0 \epsilon_1 \ldots \epsilon_{i+j-1}} t^j u_{i+j}$$

$$= \sum_{j=0}^{q-i} \epsilon_i \epsilon_{i+1} \ldots \epsilon_{i+j-1} \binom{i+j}{j} t^j u_{i+j}.$$

$$\begin{pmatrix} 1 & 0 \\ t & 1 \end{pmatrix} . u_i = \sum_{j=0}^{i} \frac{\binom{i}{j}\binom{q}{i} \epsilon_0 \epsilon_1 \ldots \epsilon_{i-1}}{\binom{q}{i-j} \epsilon_0 \epsilon_1 \ldots \epsilon_{i-j-1}} t^j u_{i-j}$$

$$= \sum_{j=0}^{i} \binom{q-i+j}{j} \epsilon_{i-1} \epsilon_{i-2} \ldots \epsilon_{i-j} t^j u_{i-j}. \qquad \blacksquare$$

We next compare the action of $\langle X_r, X_{-r}\rangle$ on $\mathfrak{L}_r \oplus \mathfrak{H}_r \oplus \mathfrak{L}_{-r}$ with the action of $SL_2(\mathbb{C})$ on R_2. We recall the relations:

$$x_r(t).e_r = e_r,$$

$$x_r(t).h_r = h_r - 2te_r,$$

$$x_r(t).e_{-r} = e_{-r} + th_r - t^2 e_r;$$

$$x_{-r}(t).e_r = e_r - th_r - t^2 e_{-r},$$

$$x_{-r}(t).h_r = h_r + 2te_{-r},$$

$$x_{-r}(t).e_{-r} = e_{-r}.$$

D

LEMMA 6.2.3.   *Consider the basis* $-x^2$, $2xy$, y^2 *of R_2.*

$$\begin{pmatrix} 1 & t \\ 0 & 1 \end{pmatrix} \text{ and } \begin{pmatrix} 1 & 0 \\ t & 1 \end{pmatrix}$$

operate on this basis as follows:

$$\begin{pmatrix} 1 & t \\ 0 & 1 \end{pmatrix}: \ -x^2 \rightarrow -x^2,$$

$$2xy \rightarrow 2xy - 2t.(-x^2),$$

$$y^2 \rightarrow y^2 + t.(2xy) - t^2.(-x^2);$$

$$\begin{pmatrix} 1 & 0 \\ t & 1 \end{pmatrix}: \ -x^2 \rightarrow -x^2 - t.(2xy) - t^2.y^2,$$

$$2xy \rightarrow 2xy + 2ty^2,$$

$$y^2 \rightarrow y^2.$$

PROOF.   This is clear. Thus

$$\begin{pmatrix} 1 & t \\ 0 & 1 \end{pmatrix}, \ \begin{pmatrix} 1 & 0 \\ t & 1 \end{pmatrix}$$

operate on the basis $-x^2$, $2xy$, y^2 of R_2 in the same way that $x_r(t)$, $x_{-r}(t)$ operate on the basis e_r, h_r, e_{-r} of $\mathfrak{L}_r \oplus \mathfrak{H}_r \oplus \mathfrak{L}_{-r}$.   ∎

We shall also require the following lemma.

LEMMA 6.2.4.   *Let*

$$\begin{pmatrix} \alpha & \beta \\ \gamma & \delta \end{pmatrix} \in SL_2(\mathbb{C}).$$

Then

$$\begin{pmatrix} \alpha & \beta \\ \gamma & \delta \end{pmatrix}.2xy = 2xy + 2(-\alpha\beta.(-x^2) + \beta\gamma.(2xy) + \gamma\delta.(y^2)).$$

PROOF.   This is evident since $\alpha\delta - \beta\gamma = 1$.   ∎

We can now prove the existence of the required homomorphism from $SL_2(\mathbb{C})$.

PROPOSITION 6.2.5. *There is a homomorphism from $SL_2(\mathbb{C})$ onto the subgroup $\langle X_r, X_{-r} \rangle$ of $G = \mathfrak{L}(\mathbb{C})$ under which*

$$\begin{pmatrix} 1 & t \\ 0 & 1 \end{pmatrix} \to x_r(t),$$

$$\begin{pmatrix} 1 & 0 \\ t & 1 \end{pmatrix} \to x_{-r}(t).$$

PROOF. Write $\mathfrak{M}_r = \mathfrak{L}_r \oplus \mathfrak{H}_r \oplus \mathfrak{L}_{-r}$. Then we have a decomposition of $\mathfrak{L}$

$$\mathfrak{L} = \mathfrak{H}_r^{\perp} \oplus \mathfrak{M}_r \oplus \sum_{s \in \Gamma} \mathfrak{M}_s$$

into a direct sum of submodules with respect to $\langle X_r, X_{-r} \rangle$, where Γ is the set of roots which are independent of r and begin an r-chain. Now $\langle X_r, X_{-r} \rangle$ operates as the identity on $\mathfrak{H}_r^{\perp}$. Lemmas 6.2.1, 6.2.2, and 6.2.3 show that there exists a module $\mathfrak{M}$ for $SL_2(\mathbb{C})$, which is a direct sum of spaces of homogeneous polynomials with a trivial module, such that

$$\begin{pmatrix} 1 & t \\ 0 & 1 \end{pmatrix}, \quad \begin{pmatrix} 1 & 0 \\ t & 1 \end{pmatrix}$$

operate on $\mathfrak{M}$ in the same way as $x_r(t)$, $x_{-r}(t)$ operate on $\mathfrak{L}$. Since the elements

$$\begin{pmatrix} 1 & t \\ 0 & 1 \end{pmatrix}, \quad \begin{pmatrix} 1 & 0 \\ t & 1 \end{pmatrix}$$

for $t \in \mathbb{C}$ generate $SL_2(\mathbb{C})$, there is a homomorphism from $SL_2(\mathbb{C})$ onto $\langle X_r, X_{-r} \rangle$ under which

$$\begin{pmatrix} 1 & t \\ 0 & 1 \end{pmatrix} \to x_r(t) \text{ and } \begin{pmatrix} 1 & 0 \\ t & 1 \end{pmatrix} \to x_{-r}(t). \qquad \blacksquare$$

6.3 The Homomorphism from $SL_2(K)$

We now transfer from the complex field to an arbitrary field K and establish the following result.

THEOREM 6.3.1. *Let* K *be any field. Then there is a homomorphism from* $SL_2(K)$ *onto the subgroup* $\langle X_r, X_{-r} \rangle$ *of* $G = \mathfrak{L}(K)$ *under which*

$$\begin{pmatrix} 1 & t \\ 0 & 1 \end{pmatrix} \to x_r(t),$$

$$\begin{pmatrix} 1 & 0 \\ t & 1 \end{pmatrix} \to x_{-r}(t).$$

PROOF. We already have the existence of such a homomorphism when $K = \mathbb{C}$. Consider the module $\mathfrak{M}$ for $SL_2(\mathbb{C})$ constructed in section 6.2. $\mathfrak{M}$ is a direct sum of modules R_q with a trivial module, and we obtain a basis for $\mathfrak{M}$ by combining the natural bases of the components R_q with a basis of the trivial module. With respect to this basis of $\mathfrak{M}$ an element

$$\begin{pmatrix} \alpha & \beta \\ \gamma & \delta \end{pmatrix}$$

of $SL_2(\mathbb{C})$ is represented by a matrix whose entries are polynomials in α, β, γ, δ with coefficients in $\mathbb{Z}$. This is clear because each component R_q gives such a polynomial representation. This basis for $\mathfrak{M}$ corresponds to a basis for $\mathfrak{L}$ as in section 6.2. This basis for $\mathfrak{L}$ is adapted to the decomposition

$$\mathfrak{L} = \mathfrak{H}_r^{\perp} \oplus \mathfrak{M}_r \oplus \sum_{s \in \Gamma} \mathfrak{M}_s$$

of 6.2.5, but is not necessarily a Chevalley basis of $\mathfrak{L}$. However, it differs from a Chevalley basis only on $\mathfrak{H}$. It will be necessary to show that when we change to a basis of $\mathfrak{M}$ corresponding to a Chevalley basis of $\mathfrak{L}$, the element

$$\begin{pmatrix} \alpha & \beta \\ \gamma & \delta \end{pmatrix}$$

of $SL_2(\mathbb{C})$ will still be represented by a matrix $F(\alpha, \beta, \gamma, \delta)$ whose entries are polynomials in α, β, γ, δ with coefficients in $\mathbb{Z}$.

With this in mind we first transform the basis elements of $\mathfrak{L}$ which lie in $\mathfrak{H}$ by an element of the Weyl group which transforms h_r into a fundamental co-root. Since the elements of the Weyl group operate on $\mathfrak{H}$ by integral unimodular transformations, this will not affect the property with which we are concerned. Thus we may assume that h_r is a fundamental co-root.

Now the remaining elements of the Chevalley basis which lie in $\mathfrak{H}$ are the other fundamental co-roots. Let h_s, $s \in \Pi$, be one of these. Then we have

$$h_s = \lambda . h_r + y,$$

where $y \in \mathfrak{H}_r^\perp$ and

$$\lambda = \frac{(h_r, h_s)}{(h_r, h_r)} = \frac{(s, r)}{(s, s)}.$$

In particular, 2λ is an integer. We note that if $\theta \in \langle X_r, X_{-r} \rangle$ then we have

$$\theta(h_s) = \lambda \theta(h_r) + y = h_s + \lambda(\theta(h_r) - h_r).$$

Now consider the action of $SL_2(\mathbb{C})$ on $\mathfrak{M}$. Let m_r, m_s be basis elements of $\mathfrak{M}$ corresponding to h_r, $h_s \in \mathfrak{L}$. Let

$$T = \begin{pmatrix} \alpha & \beta \\ \gamma & \delta \end{pmatrix} \in SL_2(\mathbb{C}).$$

Then we have

$$T(m_s) = m_s + \lambda . (T(m_r) - m_r).$$

By 6.2.4, $T(m_r) - m_r$ is a linear combination of basis elements of $\mathfrak{M}$ with coefficients which are polynomials in α, β, γ, δ with *even* integer coefficients. Since 2λ is an integer the entries in the matrix $F(\alpha, \beta, \gamma, \delta)$ are polynomials in α, β, γ, δ with coefficients in $\mathbb{Z}$.

We also observe from 6.2.4 that the polynomials in α, β, γ, δ occurring as coefficients in $T(m_r) - m_r$ are all homogeneous of degree 2. Since $\alpha\delta - \beta\gamma = 1$ we may write

$$T(m_s) = (\alpha\delta - \beta\gamma) m_s + \lambda . (T(m_r) - m_r).$$

Thus $T(m_s)$ is a linear combination of basis elements of $\mathfrak{M}$ with coefficients which may be taken as homogeneous polynomials in α, β, γ, δ of degree 2.

We now transfer to the field K and define a map

$$\begin{pmatrix} \alpha & \beta \\ \gamma & \delta \end{pmatrix} \to F(\alpha, \beta, \gamma, \delta),$$

where α, β, γ, $\delta \in K$. This maps $SL_2(K)$ into a group of matrices over K obtained by replacing the variables in the polynomial entries of $F(\alpha, \beta, \gamma, \delta)$

by elements of K. The condition for this map to be a homomorphism is the vanishing of the matrix

$$F(\alpha\alpha' + \beta\gamma', \, \alpha\beta' + \beta\delta', \, \gamma\alpha' + \delta\gamma', \, \gamma\beta' + \delta\delta') - F(\alpha, \, \beta, \, \gamma, \, \delta)F(\alpha', \, \beta', \, \gamma', \, \delta')$$

for all α, β, γ, δ, α', β', γ', $\delta' \in K$ satisfying $\alpha\delta - \beta\gamma = 1$, $\alpha'\delta' - \beta'\gamma' = 1$. This matrix certainly vanishes for all α, β, γ, δ, α', β', γ', $\delta' \in \mathbb{C}$ satisfying $\alpha\delta - \beta\gamma = 1$, $\alpha'\delta' - \beta'\gamma' = 1$, and we shall prove that the same is true for K.

We take a Chevalley basis of $\mathfrak{L}$ and decompose it into disjoint subsets consisting of:

    (i) the root vectors corresponding to the roots in a given r-chain.

    (ii) the remaining set e_r, e_{-r}, h_s for $s \in \Pi$. The basis obtained for $\mathfrak{M}$ corresponding to this Chevalley basis of $\mathfrak{L}$ decomposes in an analogous way, and the submodules spanned by the basis vectors in each individual subset are invariant under $SL_2(\mathbb{C})$. The matrix representing

$$\begin{pmatrix} \alpha & \beta \\ \gamma & \delta \end{pmatrix}$$

on a submodule corresponding to an r-chain of roots of length $q+1$ has as its coefficients homogeneous polynomials in α, β, γ, δ of degree q, and the matrix representing

$$\begin{pmatrix} \alpha & \beta \\ \gamma & \delta \end{pmatrix}$$

on the submodule corresponding to $\{e_r, \, e_{-r}, \, h_s; \, s \in \Pi\}$ has coefficients which are homogeneous polynomials in α, β, γ, δ of degree 2, as has been pointed out above. Thus the matrix $F(\alpha, \, \beta, \, \gamma, \, \delta)$ representing

$$\begin{pmatrix} \alpha & \beta \\ \gamma & \delta \end{pmatrix}$$

on $\mathfrak{M}$ is a diagonal sum of matrices $F_i(\alpha, \, \beta, \, \gamma, \, \delta)$ in which the coefficients are homogeneous polynomials in α, β, γ, δ of constant degree. Also the matrix

$$G(\alpha, \, \beta, \, \gamma, \, \delta, \, \alpha', \, \beta', \, \gamma', \, \delta') = F(\alpha\alpha' + \beta\gamma', \, \alpha\beta' + \beta\delta', \, \gamma\alpha' + \delta\gamma', \, \gamma\beta' + \delta\delta')$$
$$- F(\alpha, \, \beta, \, \gamma, \, \delta) \, F(\alpha', \, \beta', \, \gamma', \, \delta')$$

is a diagonal sum of matrices $G_i(\alpha, \, \beta, \, \gamma, \, \delta, \, \alpha', \, \beta', \, \gamma', \, \delta')$ in which the coefficients are homogeneous polynomials of constant degree in each of α, β, γ, δ and α', β', γ', δ' separately.

Since $\mathfrak{M}$ is a module for $SL_2(\mathbb{C})$, we have

$$G_i(\alpha, \beta, \gamma, \delta, \alpha', \beta', \gamma', \delta') = 0$$

whenever $\alpha\delta - \beta\gamma = 1$, $\alpha'\delta' - \beta'\gamma' = 1$. Since G_i is homogeneous in α, β, γ, δ and also in α', β', γ', δ', we have $G_i = 0$ whenever $\alpha\delta - \beta\gamma = \xi$, $\alpha'\delta' - \beta'\gamma' = \xi'$ for all ξ, $\xi' \neq 0$. Let $P(\alpha, \beta, \gamma, \delta, \alpha', \beta', \gamma', \delta')$ be a matrix coefficient of G_i. Then the polynomial

$$P(\alpha, \beta, \gamma, \delta, \alpha', \beta', \gamma', \delta') . (\alpha\delta - \beta\gamma)(\alpha'\delta' - \beta'\gamma')$$

vanishes for all α, β, γ, δ, α', β', γ', $\delta' \in \mathbb{C}$, so is identically zero. Since $\alpha\delta - \beta\gamma$ and $\alpha'\delta' - \beta'\gamma'$ are not identically zero, P is identically zero. Thus the matrix G is identically zero and we have

$$F(\alpha, \beta, \gamma, \delta) \, F(\alpha', \beta', \gamma', \delta') \equiv F(\alpha\alpha' + \beta\gamma', \alpha\beta' + \beta\delta', \gamma\alpha' + \delta\gamma', \gamma\beta' + \delta\delta')$$

in the sense that equating matrix coefficients gives a polynomial identity. This remains true when we transfer to the field K. Thus the map

$$\begin{pmatrix} \alpha & \beta \\ \gamma & \delta \end{pmatrix} \to F(\alpha, \beta, \gamma, \delta)$$

with α, β, γ, $\delta \in K$ is a homomorphism.

Let

$$A_r(\zeta) = F(1, \zeta, 0, 1),$$

$$A_{-r}(\zeta) = F(1, 0, \zeta, 1), \qquad \zeta \in \mathbb{C}.$$

Then $A_r(\zeta)$, $A_{-r}(\zeta)$ are the matrices representing $x_r(\zeta)$, $x_{-r}(\zeta)$ with respect to the Chevalley basis of $\mathfrak{L}$. For each $t \in K$ let $A_r(t)$, $A_{-r}(t)$ be the matrices obtained by replacing the complex variable ζ in $A_r(\zeta)$, $A_{-r}(\zeta)$ by the element t of K. Then $A_r(t)$, $A_{-r}(t)$ are the matrices representing the elements $x_r(t)$, $x_{-r}(t)$ of $\mathfrak{L}(K)$, by definition of the latter. We have

$$A_r(t) = F(1, t, 0, 1),$$

$$A_{-r}(t) = F(1, 0, t, 1), \qquad \text{for } t \in K.$$

Now $SL_2(K)$ is generated by the matrices

$$\begin{pmatrix} 1 & t \\ 0 & 1 \end{pmatrix}, \quad \begin{pmatrix} 1 & 0 \\ t & 1 \end{pmatrix}$$

for $t \in K$ and so the group of matrices $F(\alpha, \beta, \gamma, \delta)$ with α, β, γ, $\delta \in K$ and $\alpha\delta - \beta\gamma = 1$ is generated by $A_r(t)$, $A_{-r}(t)$ for $t \in K$. This group of matrices is therefore isomorphic to the subgroup $\langle X_r, X_{-r} \rangle$ of $\mathfrak{L}(K)$. Thus we have a homomorphism from $SL_2(K)$ onto $\langle X_r, X_{-r} \rangle$ under which

$$\begin{pmatrix} 1 & t \\ 0 & 1 \end{pmatrix} \to x_r(t) \quad \text{and} \quad \begin{pmatrix} 1 & 0 \\ t & 1 \end{pmatrix} \to x_{-r}(t).$$

This completes the proof.   ∎

6.4 The Elements $h_r(\lambda)$ and n_r

We shall denote by ϕ_r the homomorphism from $SL_2(K)$ into $\langle X_r, X_{-r} \rangle$ whose existence has just been proved. We consider next the images under ϕ_r of the diagonal matrices

$$\begin{pmatrix} \lambda & 0 \\ 0 & \lambda^{-1} \end{pmatrix}$$

in $SL_2(K)$ and the monomial matrix

$$\begin{pmatrix} 0 & 1 \\ -1 & 0 \end{pmatrix}.$$

The images of these matrices play an important rôle in the sequel.

PROPOSITION 6.4.1. *Let*

$$h_r(\lambda) = \phi_r \begin{pmatrix} \lambda & 0 \\ 0 & \lambda^{-1} \end{pmatrix}.$$

Then $h_r(\lambda)$ operates on the Chevalley basis of $\mathfrak{L}_K$ in the following manner:

$$h_r(\lambda) . h_s = h_s, \qquad s \in \Pi,$$
$$h_r(\lambda) . e_s = \lambda^{A_{rs}} e_s, \qquad s \in \Sigma.$$

PROOF. Suppose s is linearly independent of r. Let $\bar{s}$ be the root which begins the r-chain through s, and let $s = ir + \bar{s}$. Then $h_r(\lambda)$ operates on e_s in the same way that

$$\begin{pmatrix} \lambda & 0 \\ 0 & \lambda^{-1} \end{pmatrix}$$

operates on

$$u_i = \epsilon_0 \epsilon_1 \dots \epsilon_{i-1} \binom{q}{i} v_i,$$

where $v_i = x^i y^{q-i}$. Now

$$\begin{pmatrix} \lambda & 0 \\ 0 & \lambda^{-1} \end{pmatrix} . x^i y^{q-i} = \lambda^{2i-q} x^i y^{q-i}.$$

Thus we have

$$\begin{pmatrix} \lambda & 0 \\ 0 & \lambda^{-1} \end{pmatrix} . u_i = \lambda^{2i-q} u_i$$

and so

$$h_r(\lambda) . e_s = \lambda^{2i-q} e_s.$$

However, since the r-chain through s is

$$-ir+s, \ldots, s, \ldots, (q-i)\, r+s,$$

we have $A_{rs} = i - (q-i) = 2i - q$ by section 3.3. Hence

$$h_r(\lambda) . e_s = \lambda^{A_{rs}} e_s.$$

We know also that $h_r(\lambda)$ operates on e_r, h_r, e_{-r} in the same way that

$$\begin{pmatrix} \lambda & 0 \\ 0 & \lambda^{-1} \end{pmatrix}$$

operates on $-x^2$, $2xy$, y^2. Now

$$\begin{pmatrix} \lambda & 0 \\ 0 & \lambda^{-1} \end{pmatrix} : \quad -x^2 \to \lambda^2(-x^2)$$

$$2xy \to 2xy$$

$$y^2 \to \lambda^{-2} y^2.$$

Thus $h_r(\lambda) . e_r = \lambda^2 e_r$, $h_r(\lambda) . e_{-r} = \lambda^{-2} e_{-r}$. Finally, $h_r(\lambda)$ operates trivially on h_r and $\mathfrak{H}_r^{\perp}$, so operates trivially on the whole of $\mathfrak{H}$. ∎

PROPOSITION 6.4.2. *Let*

$$n_r = \phi_r \begin{pmatrix} 0 & 1 \\ -1 & 0 \end{pmatrix}.$$

Then n_r operates on the Chevalley basis of $\mathfrak{L}_K$ as follows:

$$n_r . h_s = h_{w_r(s)},$$

$$n_r . e_s = \eta_{r,\,s} e_{w_r(s)}, \quad \text{where } \eta_{r,\,s} = \pm 1.$$

PROOF. Suppose s is linearly independent of r and let $\bar{s} = -ir + s$ be the root beginning the r-chain through s. Then n_r operates on e_s in the same way that

$$\begin{pmatrix} 0 & 1 \\ -1 & 0 \end{pmatrix}$$

operates on

$$u_i = \epsilon_0 \epsilon_1 \ldots \epsilon_{i-1} \binom{q}{i} v_i,$$

where $v_i = x^i y^{q-i}$. Now we have

$$\begin{pmatrix} 0 & 1 \\ -1 & 0 \end{pmatrix} : \quad x^i y^{q-i} = (-1)^i x^{q-i} y^i.$$

Thus

$$\begin{pmatrix} 0 & 1 \\ -1 & 0 \end{pmatrix} : \quad v_i = (-1)^i v_{q-i}$$

and so

$$\begin{pmatrix} 0 & 1 \\ -1 & 0 \end{pmatrix} : \quad u_i = (-1)^i \; \frac{\epsilon_0 \epsilon_1 \ldots \epsilon_{i-1}}{\epsilon_0 \epsilon_1 \ldots \epsilon_{q-i-1}} \; u_{q-i}.$$

It follows that

$$n_r . e_s = \eta_{r,\, s} e_{(q-i)r + \bar{s}} = \eta_{r,\, s} e_{w_r(s)},$$

where

$$\eta_{r,\, s} = (-1)^i \; \frac{\epsilon_0 \epsilon_1 \ldots \epsilon_{i-1}}{\epsilon_0 \epsilon_1 \ldots \epsilon_{q-i-1}} = \pm 1.$$

Also n_r operates on e_r, h_r, e_{-r} in the same way that

$$\begin{pmatrix} 0 & 1 \\ -1 & 0 \end{pmatrix}$$

operates on $-x^2$, $2xy$, y^2.

Now

$$\begin{pmatrix} 0 & 1 \\ -1 & 0 \end{pmatrix} : \quad \begin{array}{l} -x^2 \to -y^2, \\ 2xy \to -2xy, \\ y^2 \to x^2. \end{array}$$

Thus $n_r . e_r = -e_{-r}$ and $n_r . e_{-r} = -e_r$. On $\mathfrak{H}$ we have $n_r . h_r = -h_r$ and $n_r . h = h$ if $(h_r, h) = 0$. Therefore $n_r . h_s = h_{w_r(s)}$ for all $s \in \Phi$. ∎

PROPOSITION 6.4.3. *The numbers $\eta_{r,\,s}$ defined in 6.4.2 satisfy the conditions:*

$$\eta_{r,\,r} = -1,$$

$$\eta_{r,\,-r} = -1,$$

$$\eta_{r,\,s}\,\eta_{r,\,w_r(s)} = (-1)^{A_{rs}},$$

$$\eta_{r,\,s}\,\eta_{r,\,-s} = 1.$$

PROOF. It has been shown in 6.4.2 that $\eta_{r,\,r} = -1$ and $\eta_{r,\,-r} = -1$. We therefore suppose that r, s are linearly independent. Let

$$-pr+s, \ldots, s, \ldots, qr+s$$

be the r-chain of roots through s. Then

$$\eta_{r,\,s} = (-1)^p \, \frac{\epsilon_0 \epsilon_1 \ldots \epsilon_{p-1}}{\epsilon_0 \epsilon_1 \ldots \epsilon_{q-1}},$$

where

$$N_{r,\,(i-p)r+s} = \epsilon_i(i+1).$$

The above chain is also the r-chain through $w_r(s) = (q-p)r+s$. Thus

$$\eta_{r,\,w_r(s)} = (-1)^q \, \frac{\epsilon_0 \epsilon_1 \ldots \epsilon_{q-1}}{\epsilon_0 \epsilon_1 \ldots \epsilon_{p-1}}.$$

Hence

$$\eta_{r,\,s}\,\eta_{r,\,w_r(s)} = (-1)^{p+q} = (-1)^{p-q} = (-1)^{A_{rs}}.$$

Now consider the r-chain through $-s$. This is

$$-qr-s, \ldots, -s, \ldots, pr-s.$$

Therefore we have

$$\eta_{r,\,-s} = (-1)^q \, \frac{\bar{\epsilon}_0 \bar{\epsilon}_1 \ldots \bar{\epsilon}_{q-1}}{\bar{\epsilon}_0 \bar{\epsilon}_1 \ldots \bar{\epsilon}_{p-1}},$$

where

$$N_{r,\,(i-q)r-s} = \bar{\epsilon}_i(i+1).$$

However, the relations between the structure constants proved in 4.1.2 show that

$$\frac{N_{r,\,jr+s}}{((j+1)r+s,\,(j+1)r+s)} = \frac{N_{-(j+1)r-s,\,r}}{(jr+s,\,jr+s)} = -\frac{N_{r,\,-(j+1)r-s}}{(jr+s,\,jr+s)}.$$

Since each ϵ, $\bar{\epsilon}$ is just the sign of the corresponding structure constant it follows that

$$\epsilon_{j+p}=-\bar{\epsilon}_{q-j-1}, \qquad j=0, 1, \ldots, (q-1).$$

Thus

$$\eta_{r, -s}=(-1)^q \frac{\bar{\epsilon}_0\bar{\epsilon}_1 \cdots \bar{\epsilon}_{q-1}}{\bar{\epsilon}_0\bar{\epsilon}_1 \cdots \bar{\epsilon}_{p-1}}=(-1)^p \frac{\epsilon_{p+q-1}\epsilon_{p+q-2} \cdots \epsilon_p}{\epsilon_{p+q-1}\epsilon_{p+q-2} \cdots \epsilon_q}.$$

It follows that

$$\eta_{r, s}\,\eta_{r, -s}=(-1)^p \frac{\epsilon_0\epsilon_1 \cdots \epsilon_{p-1}}{\epsilon_0\epsilon_1 \cdots \epsilon_{q-1}} \cdot (-1)^p \frac{\epsilon_{p+q-1} \cdots \epsilon_p}{\epsilon_{p+q-1} \cdots \epsilon_q}=1. \qquad \blacksquare$$

We shall also need some properties of the elements

$$n_r(t)=\phi_r\begin{pmatrix} 0 & t \\ -t^{-1} & 0 \end{pmatrix},$$

which are stated in the following lemma.

LEMMA 6.4.4.

(i) $n_r(1)=n_r,$

$n_r(-1)=n_r^{-1},$

(ii) $n_r(t)=x_r(t)x_{-r}(-t^{-1})x_r(t),$

(iii) $h_r(t)=n_r(t)n_r(-1).$

PROOF. These results follow from the corresponding relations in $SL_2(K)$ using the homomorphism ϕ_r. $\blacksquare$

We note that Proposition 6.4.2 shows that the element n_r of G operates on $\mathfrak{H}$ in the same way as the element w_r of W and that n_r also permutes the root spaces $\mathfrak{L}_s, s\in \Phi$, in the same way as w_r. This relationship between n_r and w_r has many implications which will be made use of in the development to follow.

CHAPTER 7

The Diagonal and Monomial Subgroups

7.1 Properties of the Subgroup H

We have introduced the elements $h_r(\lambda)$ of the Chevalley group $G = \mathfrak{L}(K)$ as the images of the diagonal matrices

$$\begin{pmatrix} \lambda & 0 \\ 0 & \lambda^{-1} \end{pmatrix}$$

under the homomorphism from $SL_2(K)$ into $\langle X_r, X_{-r} \rangle$. We define H to be the subgroup of G generated by the elements $h_r(\lambda)$ for all $r \in \Phi$, $\lambda \neq 0 \in K$.

Now $h_r(\lambda)$ operates on the Chevalley basis of $\mathfrak{L}_K$ by

$$h_r(\lambda).h_s = h_s, \qquad h_r(\lambda).e_s = \lambda^{A_{rs}} e_s.$$

Hence each element of H is an automorphism of $\mathfrak{L}_K$ which operates trivially on $\mathfrak{H}_K$ and transforms each root vector e_s into a multiple of itself. The coefficients arising in this way define naturally a character on the additive group generated by the roots, as we shall now describe.

Let $P = \mathbb{Z}\Phi$, the set of all linear combinations of elements of Φ with rational integer coefficients. P is the additive group generated by the roots of $\mathfrak{L}$. It is a free abelian group of rank l and has a basis consisting of the set of fundamental roots $\Pi = \{p_1, p_2, \ldots, p_l\}$. A homomorphism from the additive group P into the multiplicative group K^* of non-zero elements of K is called a K-character of P. Thus a K-character of P is a map from P into K^* satisfying the conditions

$$\chi(a+b) = \chi(a)\,\chi(b), \qquad a, b \in P,$$
$$\chi(-a) = \chi(a)^{-1}, \qquad a \in P.$$

Now a K-character of P is uniquely determined by its values on the fundamental roots. Thus if a map from Φ into K^* can be extended to a K-character of P, this extension is uniquely determined. We consider such maps of the form

$$s \to \lambda^{A_{rs}}, \qquad r \in \Phi, \lambda \in K^*.$$

97

This map from Φ into K^* can be extended to a K-character of P. For let $\chi_{r,\,\lambda}$ be the map from P into K^* given by

$$\chi_{r,\,\lambda}(a) = \lambda^{2(r,\,a)/(r,\,r)}.$$

Then $\chi_{r,\,\lambda}$ is a K-character of P which takes the value $\lambda^{A_{rs}}$ at the root s.

Now the K-characters of P form a multiplicative group. For if χ_1, χ_2 are K-characters so is $\chi_1\chi_2$, where

$$\chi_1\chi_2(a) = \chi_1(a)\,\chi_2(a), \qquad a \in P.$$

Moreover, each K-character χ of P gives rise to an automorphism $h(\chi)$ of $\mathfrak{L}_K$ defined by

$$h(\chi).h_s = h_s, \qquad h(\chi).e_s = \chi(s)\,e_s.$$

Observe that if $\chi = \chi_{r,\,\lambda}$ then $h(\chi) = h_r(\lambda)$. The automorphisms of $\mathfrak{L}_K$ of the form $h(\chi)$ form a subgroup $\hat{H}$ of the full automorphism group of $\mathfrak{L}_K$. For

$$h(\chi_1).h(\chi_2) = h(\chi_1\chi_2).$$

Also the map $\chi \to h(\chi)$ is an isomorphism from the group of K-characters of P onto the group $\hat{H}$.

Now, as H is the subgroup of G generated by the automorphisms $h_r(\lambda)$ for all $r \in \Phi$, $\lambda \in K^*$, H is a subgroup of $\hat{H}$. Each element of H therefore has the form $h(\chi)$ for some K-character χ. We shall now consider which K-characters χ give rise to automorphisms $h(\chi)$ which are in H.

Let $q_1, q_2, \ldots, q_l$ be the basis of $\mathfrak{H}_{\mathbb{R}}$ dual to the basis $h_{p_1}, h_{p_2}, \ldots, h_{p_l}$ of fundamental co-roots. Thus $q_1, \ldots, q_l$ are defined by

$$(h_{p_i}, q_j) = \begin{cases} 1 & \text{if } i=j, \\ 0 & \text{if } i \neq j. \end{cases}$$

The non-singularity of the Killing form shows that $q_1, \ldots, q_l$ form a basis for $\mathfrak{H}$. They are called the *fundamental weights* of $\mathfrak{L}$. In particular, each fundamental root p_i is a linear combination of $q_1, \ldots, q_l$. Let

$$p_i = \mu_{i1}q_1 + \ldots + \mu_{il}q_l.$$

Then $(h_{p_j}, p_i) = \mu_{ij}$. Thus we have

$$\mu_{ij} = 2\,\frac{(p_j, p_i)}{(p_j, p_j)} = A_{p_j\,p_i} = A_{ji}$$

in the notation of 3.5.2. Thus

$$p_i = \sum_{j=1}^{l} A_{ji}q_j$$

and p_i is actually an integral combination of $q_1, \ldots, q_l$.

Let Q be the set of linear combinations of $q_1, \ldots, q_l$ with rational integer coefficients. Q is the additive group generated by the fundamental weights and is free abelian of rank l. Since each p_i is in Q, P is a subgroup of Q. The index $|Q : P|$ of P in Q is finite. In fact, since

$$p_i = \sum_{j=1}^{l} A_{ji} q_j,$$

we have $|Q : P| = \det (A_{ji})$. We shall write $|Q : P| = \Delta$. The values of Δ for the different types of algebra may be calculated from the Cartan matrices shown in Section 3.6, and are as follows.

$\mathfrak{L}$:	A_l	B_l	C_l	D_l	G_2	F_4	E_6	E_7	E_8,
Δ:	$l+1$	2	2	4	1	1	3	2	1.

(Δ is equal to the order of the fundamental group of the Lie group of the given type. See Adams[1].)

Now every K-character of Q gives rise to a K-character of P by restriction. However, not every K-character of P need be the restriction of some K-character of Q.

We can now describe which K-characters of P give rise to elements of H.

THEOREM 7.1.1. *H is the subgroup of G consisting of all automorphisms $h(\chi)$ of $\mathfrak{L}_K$ for which χ is a K-character of P which can be extended to a K-character of Q.*

PROOF. H is generated by the elements $h(\chi_r, \lambda)$ for all $r \in \Phi$, $\lambda \in K^*$. Now $\chi_{r, \lambda}$ can be extended to the K-character of Q given by

$$\chi_{r, \lambda}(a) = \lambda^{2(r, a)/(r, r)}, \qquad a \in Q.$$

(Note that $2(r, a)/(r, r) \in \mathbb{Z}$ since $2r/(r, r)$ is a co-root and $a \in Q$.) Thus each element of H has the form $h(\chi)$, where χ is a K-character of Q.

Conversely, let χ be any K-character of Q. Let

$$\chi(q_1) = \lambda_1, \quad \chi(q_2) = \lambda_2, \ldots, \chi(q_l) = \lambda_l.$$

Now χ_{p_i, λ_i} takes value λ_i at q_i and value 1 at q_j for all $j \neq i$. For

$$\chi_{p_i, \lambda_i}(q_j) = \lambda_i^{2(p_i, q_j)/(p_i, p_i)} = \lambda_i^{\delta_{ji}}.$$

Thus we have

$$\chi = \chi_{p_1, \lambda_1} \chi_{p_2, \lambda_2} \cdots \chi_{p_l, \lambda_l}$$

and so

$$h(\chi) = h(\chi_{p_1, \, \lambda_1}) \, h(\chi_{p_2, \, \lambda_2}) \ldots h(\chi_{p_l, \, \lambda_l}).$$

Therefore $h(\chi) \in H$ and the theorem is proved.   ∎

We now consider the relation between H and the subgroups U and V of G. Note first that H normalizes each root subgroup X_r. For

$$h(\chi) \, x_r(t) \, h(\chi)^{-1} = h(\chi) \exp \mathrm{ad} \, (te_r) \, h(\chi)^{-1}$$
$$= \exp \mathrm{ad} \, (h(\chi) \cdot te_r)$$
$$= \exp \mathrm{ad} \, (\chi(r) \, te_r)$$

by 5.1.1. Thus

$$h(\chi) \, x_r(t) \, h(\chi)^{-1} = x_r(\chi(r)t),$$

whence

$$h(\chi) \, X_r h(\chi)^{-1} = X_r.$$

It follows that H normalizes U and V, which are generated by root subgroups. In particular, UH and VH are both subgroups of G.

Note. The same argument shows that the group $\hat{H}$ also normalizes U and V. Although $\hat{H}$ is a group of automorphisms of $\mathfrak{L}_K$, we have not shown that $\hat{H}$ is contained in G, and in fact this is not in general so. However it is evident now that $\hat{H}$ normalizes G in the group of all automorphisms of $\mathfrak{L}_K$, since G is generated by U and V.

LEMMA 7.1.2. $UH \cap V = 1$.

PROOF.   Let $\theta \in UH \cap V$ and consider the effect of θ on the Chevalley basis of $\mathfrak{L}_K$. Since $\theta \in V$ we have

$$\theta \cdot h_s = h_s + x, \qquad \text{where } x \in \sum_{r \in \Phi^-} \mathfrak{L}_r.$$

Since $\theta \in UH$ we have

$$x \in \mathfrak{H} \oplus \sum_{r \in \Phi^+} \mathfrak{L}_r.$$

However

$$\mathfrak{L} = \mathfrak{H} \oplus \sum_{r \in \Phi^+} \mathfrak{L}_r \oplus \sum_{r \in \Phi^-} \mathfrak{L}_r$$

and so $x = 0$. Thus $\theta \cdot h_s = h_s$.

Now let $\theta . e_s = e_s + y$. Suppose $s \in \Phi^+$. Then

$$y \in \mathfrak{H} \oplus \sum_{r \prec s} \mathfrak{L}_r$$

since $\theta \in V$. Also

$$y \in \sum_{r \succeq s} \mathfrak{L}_r$$

since $\theta \in UH$. But we have

$$\mathfrak{L} = \mathfrak{H} \oplus \sum_{r \prec s} \mathfrak{L}_r \oplus \sum_{r \succeq s} \mathfrak{L}_r;$$

hence $y = 0$. Thus $\theta . e_s = e_s$.

A similar argument applies if $s \in \Phi^-$ and so $\theta = 1$. ∎

Note. A similar argument shows that $VH \cap U = 1$. In particular we have $H \cap V = 1$ and $H \cap U = 1$. These results clearly apply also when H is replaced by $\hat{H}$.

COROLLARY 7.1.3. $UH \cap VH = H$.

PROOF. $UH \cap VH = (UH \cap V) H = H$.

7.2 Properties of the Subgroup N

Let N be the subgroup of G generated by H and the elements n_r for all $r \in \Phi$. In investigating the properties of N we begin with a lemma describing the way in which the elements n_r transform the root subgroups.

LEMMA 7.2.1. *Let* $r, s \in \Phi$. *Then*

(i) $n_r . x_s(t) . n_r^{-1} = x_{w_r(s)}(\eta_r, {}_s t)$,

(ii) $\quad n_r X_s n_r^{-1} = X_{w_r(s)}$.

PROOF. $n_r x_s(t) n_r^{-1} = n_r \exp \operatorname{ad} (t e_s) n_r^-$

$\qquad = \exp \operatorname{ad} (n_r . t e_s)$

$\qquad = \exp \operatorname{ad} (\eta_r, {}_s t e_{w_r(s)})$

$\qquad = x_{w_r(s)}(\eta_r, {}_s t),$

using 5.1.1 and 6.4.2. ∎

The most important property of the subgroup N is the one we shall now prove.

THEOREM 7.2.2. *There is a homomorphism from N onto W with kernel H under which $n_r \to w_r$ for all $r \in \Phi$. Thus H is a normal subgroup of N and N/H is isomorphic to W. If $n \in N$, $h(\chi) \in H$, we have*

$$nh(\chi)\, n^{-1} = h(\chi'),$$

where $\chi'(r) = \chi(w^{-1}(r))$, w being the image of n under the above homomorphism.

PROOF. By 6.4.2 n_r operates on the Chevalley basis of $\mathfrak{L}_K$ by

$$n_r . h_s = h_{w_r(s)}, \qquad n_r . e_s = \eta_{r,\, s} e_{w_r(s)}.$$

Thus n_r transforms the root space $\mathfrak{L}_s$ into $\mathfrak{L}_{w_r(s)}$, and so n_r permutes the root spaces in the same way that the element w_r of W permutes the roots. Furthermore, each element of H transforms each root space into itself. Thus there is a homomorphism of N onto W which maps n_r into w_r and which contains H in its kernel.

We show next that H is a normal subgroup of N. Let $n \in N$ and $h(\chi) \in H$. Then we have

$$nh(\chi)\, n^{-1} . h_s = nh(\chi) . (n^{-1} h_s) = n . (n^{-1} h_s) = h_s,$$
$$nh(\chi)\, n^{-1} . e_s = nh(\chi) . (\eta e_{w^{-1}\, (s)}),$$

where w is the image of n under the above homomorphism and $\eta = \pm 1$. Thus

$$\begin{aligned} nh(\chi)\, n^{-1} . e_s &= \eta \chi(w^{-1}(s))\, n . e_{w^{-1}\, (s)} \\ &= \eta \chi(w^{-1}(s))\, \eta^{-1} e_s \\ &= \chi'(s)\, e_s, \end{aligned}$$

where $\chi'(s) = \chi(w^{-1}(s))$. Hence we have

$$nh(\chi)\, n^{-1} = h(\chi')$$

and so H is normal in N.

We thus have a homomorphism from N/H onto W under which $n_r H$ is mapped to w_r. We show this is an isomorphism. Let $r,\ s \in \Phi$. Then

$$\begin{aligned} n_r n_s n_r^{-1} &= n_r x_s(1)\, x_{-s}(-1)\, x_s(1)\, n_r^{-1}, \quad \text{by } 6.4.4, \\ &= x_{w_r(s)}(\eta_r,\ s)\, x_{w_r(-s)}(-\eta_r,\ -s)\, x_{w_r(s)}(\eta_r,\ s), \quad \text{by } 7.2.1, \\ &= x_{w_r(s)}(\eta_r,\ s)\, x_{-w_r(s)}(-\eta_{r,\ s}^{-1})\, x_{w_r(s)}(\eta_r,\ s), \quad \text{by } 6.4.3, \\ &= n_{w_r(s)}(\eta_r,\ s) \\ &= h_{w_r(s)}(\eta_r,\ s)\, n_{w_r(s)}, \quad \text{by } 6.4.4. \end{aligned}$$

Since $h_{w_r(s)}(\eta_r s) \in H$ we have

$$n_r H . n_s H . (n_r H)^{-1} = n_{w_r(s)} H.$$

Since $n_r^2 = h(-1) \in H$ we also have

$$(n_r H)^2 = H.$$

Now W is defined as an abstract group by generators w_r, $r \in \Phi$, subject to relations $w_r^2 = 1$, $w_r w_s w_r^{-1} = w_{w_r(s)}$ (2.4.3).

Hence there is a homomorphism from W onto N/H under which w_r is mapped to $n_r H$. Since this is the inverse map of the one obtained above, it is an isomorphism. Thus N/H is isomorphic to W. ■

COROLLARY 7.2.3. $N \cap \hat{H} = H.$

PROOF.   Let $n \in N \cap \hat{H}$. Since $n \in \hat{H}$ we have $n . \mathfrak{L}_s = \mathfrak{L}_s$ for each $s \in \Phi$. Thus n is in the kernel of the homomorphism mapping N to W, and so $n \in H$. ■

COROLLARY 7.2.4.   (i) $U \cap N = 1$, (ii) $UH \cap N = H.$

PROOF.   Let $\theta \in U \cap N$. Then we have

$$\theta . e_s = \lambda e_{w(s)},$$

where $\lambda \in K^*$ and $w \in W$ is the image of θ under the homomorphism from N to W. However

$$\theta . e_s = e_s + x,$$

where

$$x \in \mathfrak{H} \oplus \sum_{r \succ s} \mathfrak{L}_r,$$

since $\theta \in U$. It follows that $x = 0$, $\lambda = 1$ and $w(s) = s$. This is valid for all $s \in \Phi$, and so $w = 1$. This means that $\theta \in H$. But $U \cap H = 1$ and therefore $\theta = 1$. Hence $U \cap N = 1$.

Finally we have

$$UH \cap N = (U \cap N) H = H. \qquad ■$$

Note.   A similar argument shows that $U\hat{H} \cap N = H.$

CHAPTER 8

The Bruhat Decomposition

8.1 Bruhat's Lemma

Let B be the subgroup UH of the Chevalley group G. In the case when the field K is algebraically closed the subgroup B and its conjugates in G are usually called Borel subgroups of G. In this case G may be interpreted as an algebraic group and the Borel subgroups are precisely the maximal soluble connected subgroups of G. It was proved by Borel that any two such subgroups are conjugate (cf. Borel [3]). We shall investigate the decomposition of G into double cosets BgB with respect to B, following an investigation by Bruhat of the double coset decomposition of a semisimple Lie group with respect to a maximal soluble connected subgroup (cf. Bruhat [1]).

We shall first require some additional information about the subgroup U of G. Let $r \in \Pi$ be a fundamental root and define

$$U_r = \prod_{s \in \Phi^+ - \{r\}} X_s,$$

where the product is taken over all the positive roots other than r in increasing order. Chevalley's commutator formula shows that U_r is a subgroup of U and that the root subgroups in the product can in fact be taken in any order.

LEMMA 8.1.1. *Let $r \in \Pi$. Then X_r and X_{-r} normalize U_r.*

PROOF. The commutator formula 5.2.2 shows that $[X_r, U_r] \subseteq U_r$, hence X_r normalizes U_r. Now consider the subgroup $[X_{-r}, U_r]$. Suppose s is a positive root distinct from r and suppose $-ir + js \in \Phi$, where $i > 0$, $j > 0$. Then $-ir + js \in \Phi^+$. For when this root is expressed as an integral combination of fundamental roots, at least one coefficient is positive. Thus each coefficient must be non-negative and we have a positive root. The root r itself cannot occur in this form. Applying the commutator formula once more we have $[X_{-r}, U_r] \subseteq U_r$. Thus X_{-r} normalizes U_r also. ∎

COROLLARY 8.1.2. (i) $U = U_r X_r = X_r U_r$, (ii) $U_r X_{-r} = X_{-r} U_r$.

PROOF. U_r is permutable with X_r and X_{-r} since both normalize U_r. Since U_r and X_r are permutable subgroups of U, $U_r X_r$ is a subgroup of U. However, it contains all the root subgroups of U, so is the whole of U. ■

COROLLARY 8.1.3. *n_r normalizes U_r.*

PROOF. The normalizer of U_r contains X_r and X_{-r} so contains n_r, since $n_r \in \langle X_r, X_{-r} \rangle$. ■

We define similarly, for each $r \in \Pi$,

$$V_{-r} = \prod_{s \in \Phi^- - \{-r\}} X_s,$$

where the product is taken over the negative roots other than $-r$ in decreasing order. Then X_r and X_{-r} normalize V_{-r} and we have

$$V = V_{-r} X_{-r} = X_{-r} V_{-r} \text{ and } V_{-r} X_r = X_r V_{-r}.$$

Furthermore n_r normalizes V_{-r}.

We observe that the above results depend essentially upon the fact that r is a fundamental root. They are not valid if r is an arbitrary positive root. We shall use them to prove the following propositions.

PROPOSITION 8.1.4. *Let $r \in \Pi$. Then the subset $B \cup Bn_r B$ is a subgroup of G.*

PROOF. We must show that this subset is closed under multiplication and inversion. Inversion is clear, since

$$B^{-1} = B,$$

$$(Bn_r B)^{-1} = B^{-1} n_r^{-1} B^{-1} = Bn_r^{-1} B = Bn_r B.$$

In order to prove that the subset is closed under multiplication it is sufficient to show that

$$n_r Bn_r \subseteq B \cup Bn_r B.$$

Now $B = UH = X_r U_r H$, as in 8.1.2. Then we have

$$n_r Bn_r = n_r Bn_r^{-1}$$

$$= n_r X_r U_r Hn_r^{-1}$$

$$= X_{-r} U_r H \subseteq X_{-r} B,$$

since n_r normalizes U_r. Thus it is sufficient to show that $X_{-r} \subseteq B \cup Bn_rB$, i.e. that $x_{-r}(t) \in B \cup Bn_rB$ for all $t \in K$. If $t = 0$, $x_{-r}(t) \in B$. If $t \neq 0$ we have

$$x_{-r}(t) = x_r(t^{-1}) \, n_r(-t^{-1}) \, x_r(t^{-1}), \text{ by } 6.4.4 \text{ (ii)},$$

$$= x_r(t^{-1}) \, h_r(-t^{-1}) \, n_r x_r(t^{-1}), \text{ by } 6.4.4 \text{ (iii)}.$$

Thus $x_{-r}(t) \in Bn_rB$ and the proof is complete.   ∎

The following proposition is of crucial importance in the later development.

PROPOSITION 8.1.5. *Let* $r \in \Pi$, $n \in N$ *and* w *be the image of* n *under the natural homomorphism from* N *to* W. *Then*

$$BnB \,.\, Bn_rB \subseteq Bnn_rB \cup BnB.$$

In particular, if $w(r) \in \Phi^+$ *we have*

$$BnB \,.\, Bn_rB = Bnn_rB,$$

whereas if $w(r) \in \Phi^-$ *the set* $BnB \,.\, Bn_rB$ *contains elements from both double cosets* Bnn_rB, BnB.

PROOF. Suppose first that $w(r) \in \Phi^+$. Then we have

$$BnB \,.\, Bn_rB = BnX_rU_rHn_rB$$

$$= BnX_rn^{-1} \,.\, nn_r \,.\, n_r^{-1}U_rHn_rB.$$

Now

$$nX_rn^{-1} = X_{w(r)} \subseteq B,$$

since $w(r) \in \Phi^+$. Also

$$n_r^{-1}U_rHn_r = U_rH \subseteq B,$$

by 8.1.3. Hence $BnB \,.\, Bn_rB \subseteq Bnn_rB$. The reverse inclusion is clear.

Now suppose that $w(r) \in \Phi^-$. Let $w' = ww_r$. Then $w'(r) \in \Phi^+$. Let n' be an element of N mapping to w' under the natural homomorphism. Then we have

$$Bn'B \,.\, Bn_rB = Bn'n_rB,$$

as above. Thus

$$BnB \cdot Bn_r B = Bn'n_r B \cdot Bn_r B$$
$$= Bn'B \cdot Bn_r B \cdot Bn_r B$$
$$\subseteq Bn'B \cdot (B \cup Bn_r B), \text{ by } 8.1.4,$$
$$= Bn'B \cup Bn'B \cdot Bn_r B$$
$$= Bn'B \cup Bn'n_r B$$
$$= Bnn_r B \cup BnB.$$

Thus

$$BnB \cdot Bn_r B \subseteq Bnn_r B \cup BnB.$$

If $w(r) \in \Phi^-$, the set $BnB \cdot Bn_r B$ contains elements from both $Bnn_r B$ and BnB. For nn_r is an element in this set from $Bnn_r B$, and $nx_r(1)n_r$ is an element from BnB, since

$$nx_r(1) \, n_r = nn_r \cdot n_r^{-1} x_r(1) \, n_r = nn_r \cdot x_{-r}(-1)$$
$$= nn_r \cdot x_r(-1) \, n_r x_r(-1) \in Bn'B \cdot Bn_r B = BnB. \qquad \blacksquare$$

COROLLARY 8.1.6. *Let $r \in \Pi$, $n \in N$ and w be the image of n in W. Then*

$$Bn_r B \cdot BnB \subseteq Bn_r nB \cup BnB.$$

PROOF. Let $g \in Bn_r B \cdot BnB$. Then $g^{-1} \in Bn^{-1}B \cdot Bn_r B$. Hence, by 8.1.5, $g^{-1} \in Bn^{-1}n_r B$ or $g^{-1} \in Bn^{-1}B$. It follows that $g \in Bn_r nB$ or $g \in BnB$. $\qquad \blacksquare$

8.2 Groups with a (B, N)-Pair

It is now useful to introduce the general concept of a (B, N)-pair in a group. This concept, originally due to Tits [11], is useful not merely for deriving further properties of the Chevalley groups, but also in connection with the 'twisted groups' discussed in chapters 13 and 14, and with the associated geometrical structures described in chapter 15.

A pair of subgroups B, N of a group G is called a (B, N)-pair if the following axioms are satisfied:

BN 1. G is generated by B and N.

BN 2. $B \cap N$ is a normal subgroup of N.

BN 3. The group $W = N/B \cap N$ is generated by a set of elements w_i, $i \in I$, such that $w_i^2 = 1$.

BN 4. If $n_i \in N$ maps to w_i under the natural homomorphism of N into W, and if n is any element of N, then

$$Bn_i B . BnB \subseteq Bn_i nB \cup BnB.$$

BN 5. If n_i is as above, then $n_i Bn_i \neq B$.

PROPOSITION 8.2.1.   *The Chevalley group* $G = \mathfrak{L}(K)$ *has a* (B, N)-*pair*.

PROOF.   We show that the subgroups of G previously denoted by B, N satisfy the above axioms. By 7.2.4 we have $B \cap N = H$, and by 7.2.2 we know that H is normal in N and that N/H is isomorphic to the Weyl group W. The Weyl group is generated by the fundamental reflections w_r, $r \in \Pi$, and we take these as the generating involutions in *BN* 3. By 8.1.6 we have, for each $r \in \Pi$,

$$Bn_r B . BnB \subseteq Bn_r nB \cup BnB.$$

If $r \in \Pi$ we have $X_r \subseteq B$, but $n_r X_r n_r = X_{-r}$ is not in B. Thus $n_r Bn_r \neq B$. Finally, G is generated by B and N. For G is generated by the root subgroups X_r, and each root r has the form $w(r_i)$ for some $w \in W$ and $r_i \in \Pi$. Let n be an element of N mapping to w under the natural homomorphism. Then

$$X_r = X_{w(r_i)} = n X_{r_i} n^{-1}$$

by 7.2.1. Thus X_r is in the subgroup generated by B and N, whence $\langle B, N \rangle = G$. Therefore G has a (B, N)-pair. ■

We now derive some consequences of the axioms for a (B, N)-pair.

PROPOSITION 8.2.2.   *Let* G *be a group with a* (B, N)-*pair. Then*
(i) $G = BNB$.
(ii) *For each subset* J *of* I, *let* W_J *be the subgroup of* W *generated by the elements* w_i *for* $i \in J$, *and* N_J *be the subgroup of* N *mapping to* W_J *under the natural homomorphism. Then* $P_J = BN_J B$ *is a subgroup of* G.

PROOF.   We prove (ii) first. $BN_J B$ is certainly closed under inversion, and so we must prove it closed under multiplication. Let $n \in N_J$. Then n may be written in the form $n = n_1 n_2 \ldots n_k$, where each n_i maps under the natural homomorphism to a generator w_i of W with $i \in J$. Thus

$$nBN_J B = n_1 n_2 \ldots n_k BN_J B$$

$$\subseteq n_1 n_2 \ldots n_{k-1} BN_J B \subseteq \ldots \subseteq BN_J B$$

by repeated application of $BN\,4$. Hence

$$N_J BN_J B \subseteq BN_J B$$

and it follows that

$$BN_J B \cdot BN_J B \subseteq BN_J B.$$

Thus $BN_J B$ is a subgroup of G.

We now take $J = I$. Then BNB is a subgroup of G. This subgroup certainly contains B and N, and must be G since B and N generate G. ∎

It is evident from 8.2.2 that every double coset BgB contains an element of N. We now consider the question of when two elements of N lie in the same double coset.

PROPOSITION 8.2.3. *Let G be a group with a (B, N)-pair. Let n, n' be elements of N. Then $BnB = Bn'B$ if and only if n, n' map to the same element of W under the natural homomorphism from N into W. Thus there is a natural 1–1 correspondence between double cosets of B in G and elements of W.*

(We note that in the case where G is a Chevalley group and B a Borel subgroup this result implies that the number of double cosets of B in G is finite, and equal to the order of the Weyl group. For a general group with a (B, N)-pair, the group W need not be finite.)

PROOF. Each element of W is a product of generators w_i, $i \in I$. We denote by $l(w)$ the shortest length of any expression for w as a product of such generators. Suppose $BnB = Bn'B$, where n, n' correspond to w, w' respectively, and suppose that $l(w) \leqslant l(w')$. We show that $w = w'$ by induction on $l(w)$. If $l(w) = 0$ we have $w = 1$ and $BnB = B$. Thus $Bn'B = B$ and so $n' \in B \cap N$. Hence $w' = 1$.

Now suppose $l(w) > 0$. Then $w = w_i w''$, where $i \in I$ and $l(w'') = l(w) - 1$. Let n_i, n'' be elements of N corresponding to w_i, $w'' \in W$. Then

$$n_i n'' B \subseteq Bn'B.$$

Thus by $BN\,4$ we have

$$n'' B \subseteq n_i Bn'B \subseteq Bn_i n'B \cup Bn'B.$$

Hence $Bn''B = Bn_i n'B$ or $Bn''B = Bn'B$. By induction we have $w'' = w_i w'$ or $w'' = w'$. The latter is impossible since $l(w'') < l(w')$. Thus $w'' = w_i w'$ and so $w' = w_i w'' = w$.

Since every double coset of B in G has the form BnB for some $n \in N$, there is a 1–1 correspondence

$$BnB \leftrightarrow (B \cap N)n$$

between double cosets of B in G and elements of W. ∎

The next two propositions establish analogues of the more detailed results concerning double coset multiplication proved for Chevalley groups in 8.1.5.

PROPOSITION 8.2.4. *Let G be a group with a (B, N)-pair. Let w_i, $i \in I$, be one of the distinguished generators of W and w be an element of W such that $l(w_i w) \geqslant l(w)$. Let n_i, n be elements of N mapped by the natural homomorphism into w_i, $w \in W$. Then $Bn_iB . BnB \subseteq Bn_inB$.*

PROOF. We again use induction on $l(w)$. If $l(w) = 0$ then $w = 1$ and $n \in B$, so the result is clear. Suppose $l(w) > 0$. Then $w = w'w_j$ for some $j \in I$, where $i(w') = l(w) - 1$.

Suppose by way of contradiction that the result is false. Then $Bn_iB . BnB$ intersects BnB, by BN 4. Let n', n_j be elements of N mapped by the natural homomorphism to w', $w_j \in W$. Then

$$n_i Bn \cap BnB \neq \phi$$

and therefore

$$n_i Bn' \cap BnBn_j \neq \phi.$$

Now $l(w_i w') \geqslant l(w')$ and so, by induction, we have

$$n_i Bn' \subseteq Bn_i n'B.$$

Thus

$$Bn_i n'B \cap BnBn_j \neq \phi.$$

Now it follows easily from BN 4 that

$$BnB . Bn_jB \subseteq Bnn_jB \cup BnB.$$

Hence $Bn_in'B$ intersects either Bnn_jB or BnB. Thus, by 8.2.3, $w_iw' = ww_j$ or $w_iw' = w$. The former equation gives $w_iw' = w'$, whence $w_i = 1$, which contradicts BN 5. The latter gives $w_iw = w'$, which implies $l(w_iw) < l(w)$, a contradiction. ∎

PROPOSITION 8.2.5. *Suppose the notation is as in 8.2.4. Then, if $l(w_iw) \leqslant l(w)$, $Bn_iB . BnB$ has non-empty intersection with BnB.*

PROOF. By BN 4 we have

$$n_i Bn_i \subseteq B \cup Bn_i B.$$

Since $n_i Bn_i \neq B$ we have

$$n_i Bn_i \cap Bn_i B \neq \phi.$$

It follows that

$$n_i B \cap Bn_i Bn_i \neq \phi$$

$$n_i Bn \cap Bn_i Bn_i n \neq \phi.$$

Now $l(w_i . w_i w) \geqslant l(w_i w)$, thus by 8.2.4

$$n_i Bn_i n \subseteq BnB.$$

Hence $n_i Bn$ intersects BnB and the result follows. ∎

Note. It follows from 8.2.4 and 8.2.5 that $l(w_i w) \neq l(w)$.

COROLLARY 8.2.6.   *Suppose the notation is as in* 8.2.4. *If* $l(w_i w) < l(w)$ *then* $n_i \in BnBn^{-1}B$.

PROOF.   By 8.2.5 we have

$$n_i Bn \cap BnB \neq \phi.$$

Therefore $n_i \in BnBn^{-1}B$, as required. ∎

8.3 Parabolic Subgroups

Let G be a group with a (B, N)-pair. A parabolic subgroup of G is one which contains some conjugate gBg^{-1} of B. For example, the subgroup P_J of 8.2.2 is a parabolic subgroup. We shall show that the subgroups P_J for the various subsets J of I are the only subgroups of G containing B, and therefore that every parabolic subgroup is conjugate to some P_J. Furthermore, the subgroups P_J are all distinct and non-conjugate in G, and each of them is equal to its normalizer.

We first prove a proposition describing the subgroup generated by B and a single element of N.

PROPOSITION 8.3.1.   *Let G be a group with a (B, N)-pair and let n be an element of N. Let $w \in W$ be the image of n under the natural* *homomorphism and let*

$$w = w_{i_1} w_{i_2} \ldots w_{i_k}, \qquad i_1, i_2, \ldots \in I,$$

where $l(w) = k$. Let J be the subset $\{i_1, i_2, \ldots, i_k\}$ of I. Then the following three subgroups of G are equal:

   (i) $\langle B, n \rangle$,

   (ii) $\langle B, nBn^{-1} \rangle$,

   (iii) $P_J = BN_J B$.

PROOF. We have $n = n_{i_1} n_{i_2} \ldots n_{i_k}$, where $n_{i_\alpha} \in N$ corresponds to $w_{i_\alpha} \in W$. Each n_{i_α} is in N_J, so $n \in N_J$. Thus we have the inclusions

$$\langle B, nBn^{-1} \rangle \subseteq \langle B, n \rangle \subseteq P_J.$$

Now P_J is the subgroup generated by B and the elements $n_{i_1}, n_{i_2}, \ldots, n_{i_k}$. Since $l(w_{i_1} w) < l(w)$ we have, by 8.2.6, $n_{i_1} \in \langle B, nBn^{-1} \rangle$. Since

$$l(w_{i_2} w_{i_1} w) < l(w_{i_1} w)$$

we have

$$n_{i_2} \in \langle B, n_{i_1} n B n^{-1} n_{i_1}^{-1} \rangle \subseteq \langle B, nBn^{-1} \rangle$$

and arguing similarly we see that each of $n_{i_1}, n_{i_2}, \ldots, n_{i_k}$ lies in $\langle B, nBn^{-1} \rangle$. Hence P_J is contained in $\langle B, nBn^{-1} \rangle$ and the result follows. ∎

THEOREM 8.3.2. *Let G be a group with a (B, N)-pair. Then the subgroups P_J are the only subgroups of G containing B.*

PROOF. Let M be a subgroup of G containing B. M is a union of double cosets of B in G, and each such double coset contains an element of N. Thus M is generated by B and a certain set of elements of N. Let n_α be any element of N. Then, by 8.3.1, the subgroup generated by B and n_α is P_{J_α}, for a suitable subset J_α of I. Thus M is generated by subgroups P_{J_α} for a family of subsets J_α of I, and so $M = P_J$, where

$$J = \bigcup_\alpha J_\alpha.$$ ∎

THEOREM 8.3.3. *Let G be a group with a (B, N)-pair. Then each subgroup P_J of G is equal to its normalizer. Furthermore distinct subgroups P_J, P_K cannot be conjugate in G.*

PROOF. The normalizer $\mathfrak{N}(P_J)$ is generated by B and elements of N.

Let $n \in N \cap \mathfrak{N}(P_J)$. Then we have

$$P_J \supseteq \langle B, nBn^{-1} \rangle = \langle B, n \rangle$$

by 8.3.1. Therefore $n \in P_J$. It follows that $P_J = \mathfrak{N}(P_J)$.

Suppose now that P_J, P_K are conjugate subgroups of G. Let, $gP_Jg^{-1} = P_K$ where $g = bnb'$ with b, $b' \in B$ and $n \in N$. Then $nP_Jn^{-1} = P_K$. Thus

$$P_K \supseteq \langle B, nBn^{-1} \rangle = \langle B, n \rangle$$

by 8.3.1. Therefore $n \in P_K$ and so $P_K = P_J$. Thus distinct subgroups P_J, P_K are non-conjugate. ∎

THEOREM 8.3.4. *Let G be a group with a (B, N)-pair. Then the subgroups P_J for distinct subsets J of I are all distinct. Furthermore we have $P_J \cap P_K = P_{J \cap K}$. Thus the subgroups P_J form a lattice isomorphic to the lattice of subsets of I.*

PROOF. It is necessary to show first that the elements w_i of W for $i \in I$ form a minimal set of generators of W. Suppose we remove one element j from I, and assume by way of contradiction that W is generated by the elements w_i for $i \in I - \{j\}$. Then G has a (B, N)-pair with I replaced by $I - \{j\}$, since the axioms BN 1–5 are still satisfied. Let

$$w_j = w_{i_1} w_{i_2} \ldots w_{i_k}, \qquad i_\alpha \in I - \{j\},$$

be an expression of minimal length of w_j in terms of the remaining generators. Let n_j be an element of N corresponding to $w_j \in W$, and let $J = \{i_1, i_2, \ldots, i_k\}$. Then, by 8.3.1, we have $\langle B, n_j \rangle = BN_JB$. However, $\langle B, n_j \rangle = B \cup Bn_jB$ by 8.3.1 applied to the full set of generators of W. Thus $BN_JB = B \cup Bn_jB$. By 8.2.3 this implies that $W_J = \{1, w_j\}$, which gives a contradiction. Thus the elements w_i, $i \in I$, form a minimal set of generators.

Now let J, K be any two subsets of I. $P_J \cap P_K$ is a subgroup of G containing B, thus $P_J \cap P_K = P_L$ for some subset L of I by 8.3.2. Now $P_{J \cap K} \subseteq P_L$. We suppose by way of contradiction that $P_{J \cap K} \neq P_L$. Then certainly L is not contained in $J \cap K$ and we may assume without loss of generality that L is not contained in J. However, $P_L \subseteq P_J$ and so $N_L \subseteq N_J$ by 8.2.3. Hence $W_L \subseteq W_J$. Let $i \in L - J$. Then $w_i \in W_L$ and so $w_i \in W_J$. But this means that w_i is expressible in terms of the remaining generators of W, which is impossible. Thus $P_J \cap P_K = P_{J \cap K}$.

We show finally that the subgroups P_J are all distinct. Suppose $P_J = P_K$. If $J \neq K$ we may assume that $J - K$ is non-empty. Since

$$P_J = P_J \cap P_K = P_{J \cap K},$$

we have $W_J = W_{J \cap K}$, as before. This again means that some generator w_i is expressible in terms of the remainder, which gives a contradiction. Thus $P_J = P_K$ implies that $J = K$. ∎

We emphasize here that all the results of sections 8.2 and 8.3 have been obtained using only the axioms for a (B, N)-pair.

8.4 A Canonical Form

We now return to the situation in which G is a Chevalley group $\mathfrak{L}(K)$. The results just proved on the double coset decomposition are valid for G since G has a (B, N)-pair. Thus every element of G can be written in the form $b_1 n b_2$, where $b_1, b_2 \in B$ and $n \in N$. However, it may be possible to express an element of G in such a form in a number of different ways. We seek a canonical form for elements of G, i.e. a way of decomposing an element of G so that each element has a unique expression in the given form.

A subset Ψ of Φ is called a closed set of roots if, whenever $r, s \in \Psi$ and $ir + js \in \Phi$, where i, j are positive integers, the root $ir + js$ is in Ψ also. We consider the situation in which Φ^+ is expressed as the disjoint union of two closed subsets. Suppose $\Phi^+ = \Psi_1 \cup \Psi_2$, where $\Psi_1 \cap \Psi_2$ is empty and Ψ_1, Ψ_2 are closed. Let

$$U^1 = \prod_{r \in \Psi_1} X_r \text{ and } U^2 = \prod_{r \in \Psi_2} X_r,$$

where the products are taken over the roots in increasing order. The commutator formula shows that U^1, U^2 are subgroups of U and that the same subgroups are obtained if the factors are taken in a different order.

LEMMA 8.4.1. $U = U^1 U^2$, $U^1 \cap U^2 = 1$.

PROOF. Let $u \in U^1 \cap U^2$. Then

$$u = \prod_{r_i \in \Phi^+} x_{r_i}(t_i).$$

Since $u \in U^1$ we have $t_i = 0$ whenever $r_i \in \Psi_2$. Since $u \in U^2$ we have $t_i = 0$ whenever $r_i \in \Psi_1$. Thus $t_i = 0$ for all i and $u = 1$.

We define

$$U_m = \prod_{h(r) \geqslant m} X_r$$

for $m = 1, 2, \ldots$ and also define $U_m^1 = U_m \cap U^1$, $U_m^2 = U_m \cap U^2$. We show $U_m = U_m^1 U_m^2$ for all m. This is clear if m is sufficiently large, and we prove it by descending induction. Assume inductively that $U_{m+1} = U_{m+1}^1 U_{m+1}^2$. Now we have

$$U_m = \prod_{h(r) = m} X_r \cdot U_{m+1}$$

and U_m/U_{m+1} is abelian. Thus

$$U_m = U_m^1 U_m^2 U_{m+1}.$$

Since U_{m+1} is normal in U_m we have

$$U_m = U_m^1 U_m^2 U_{m+1} = U_m^1 U_{m+1} U_m^2$$
$$= U_m^1 U_{m+1}^1 U_{m+1}^2 U_m^2 = U_m^1 U_m^2.$$

Finally, putting $m = 1$, we have $U = U^1 U^2$ as required. ∎

COROLLARY 8.4.2. *Each element $u \in U$ is uniquely expressible in the form $u = u_1 u_2$ with $u_1 \in U^1$, $u_2 \in U^2$.*

A decomposition of Φ^+ of the type we have been considering is determined by each element of the Weyl group W. Let $w \in W$ and define

$$\Psi_1 = \{ r \in \Phi^+; \ w(r) \in \Phi^+ \},$$

$$\Psi_2 = \{ r \in \Phi^+; \ w(r) \in \Phi^- \}.$$

Then Ψ_1, Ψ_2 are disjoint closed subsets whose union is Φ^+. We define

$$U_w^+ = \prod_{r \in \Psi_1} X_r,$$

$$U_w^- = \prod_{r \in \Psi_2} X_r.$$

Then $U = U_w^+ U_w^-$ and $U_w^+ \cap U_w^- = 1$.

We can now describe the required canonical form for elements of G.

THEOREM 8.4.3. *For each $w \in W$ choose a coset representative $n_w \in N$ which maps to w under the natural homomorphism. Then each element* *of G is expressible in just one way in the form*

$$g = bn_w u,$$

where $b \in B$ and $u \in U_w^-$.

PROOF. We shall show that $n_w U_w^+ n_w^{-1} \subseteq U$. By 7.2.1 we have

$$n_r X_s n_r^{-1} = X_{w_r(s)},$$

for all r, $s \in \Phi$. Let $w = w_{r_1} w_{r_2} \ldots w_{r_k}$ be an expression of w as a product of reflections. Then n_w and $n_{r_1} n_{r_2} \ldots n_{r_k}$ both have image w under the natural homomorphism, hence

$$n_w = h n_{r_1} \ldots n_{r_k}$$

for some $h \in H$. Thus

$$n_w X_s n_w^{-1} = h n_{r_1} \ldots n_{r_k} X_s n_{r_1}^{-1} \ldots n_{r_k}^{-1} h^{-1}$$

$$= h X_{w(s)} h^{-1} = X_{w(s)}$$

by 7.2.1. It follows that

$$n_w U_w^+ n_w^{-1} = n_w \cdot \prod_{s \in \Psi_1} X_s \cdot n_w^{-1} \subseteq U.$$

(One can prove similarly that $n_w U_w^- n_w^{-1} \subseteq V$.)

Now consider the double coset $Bn_w B$. We have

$$Bn_w B = Bn_w HU = Bn_w HU_w^+ U_w^-$$

$$= BHn_w U_w^+ U_w^- \subseteq BUn_w U_w^- = Bn_w U_w^-.$$

But clearly $Bn_w U_w^- \subseteq Bn_w B$, so we have equality. Since each element $g \in G$ is in some double coset $Bn_w B$, g may be expressed in the form $g = bn_w u$, where $b \in B$, $u \in U_w^-$.

To show uniqueness, suppose

$$b_1 n_w u_1 = b_2 n_{w'} u_2,$$

where b_1, $b_2 \in B$, w, $w' \in W$, $u_1 \in U_w^-$, $u_2 \in U_{w'}^-$. By 8.2.3 we have $w = w'$, thus $n_w = n_{w'}$. It follows that

$$b_2^{-1} b_1 = n_w u_2 u_1^{-1} n_w^{-1}.$$

Since $n_w U_w^- n_w^{-1} \subseteq V$, this element is in $B \cap V$, so is 1 by 7.1.2. Thus $b_1 = b_2$ and $u_1 = u_2$. ∎

COROLLARY 8.4.4. *Each element of G has a unique expression in the form $g = u_1 h n_w u$, where $u_1 \in U$, $h \in H$, $w \in W$, $u \in U_w^-$.*

One consequence of the canonical form we have established is the following useful result.

PROPOSITION 8.4.5. *Two elements of H which are conjugate in G are conjugate in N.*

PROOF. Let h_1, $h_2 \in H$ satisfy $h_2 = g h_1 g^{-1}$. Let $g = u_1 h n_w u$ as in 8.4.4. Then

$$u_1 h n_w u h_1 = h_2 u_1 h n_w u.$$

It follows that

$$u_1 . h n_w h_1 n_w^{-1} . n_w h_1^{-1} u h_1 = h_2 u_1 h_2^{-1} . h_2 h . n_w . u.$$

By the uniqueness of expression of elements of G in canonical form we have

$$u_1 = h_2 u_1 h_2^{-1}, \qquad h n_w h_1 n_w^{-1} = h_2 h, \qquad h_1^{-1} u h_1 = u.$$

In particular, $h_2 = n_w h_1 n_w^{-1}$, so h_1, h_2 are conjugate in N. ∎

We conclude this section with a result concerning the group $\hat{H}$ defined in chapter 7.

PROPOSITION 8.4.6. $G \cap \hat{H} = H$.

PROOF. Let $g \in G \cap \hat{H}$. Then $g = b_1 n b_2$, where b_1, $b_2 \in B$ and $n \in N$. Thus

$$n = b_1^{-1} g b_2^{-1} \in B \hat{H} B.$$

Now we have

$$B \hat{H} B = U H \hat{H} H U = U \hat{H} U = U \hat{H}.$$

Thus $n \in U \hat{H} \cap N$. But $U \hat{H} \cap N = H$ by 7.2.4. Hence $n \in H$ and $g \in B$. But now $g \in B \cap \hat{H}$ and

$$B \cap \hat{H} = U H \cap \hat{H} = (U \cap \hat{H}) H = H,$$

since $U \cap \hat{H} = 1$ by 7.1.2. ∎

E

COROLLARY 8.4.7. *Let $\hat{G}$ be the group of automorphisms of $\mathfrak{L}_K$ generated by G and $\hat{H}$. Then G is a normal subgroup of $\hat{G}$ and $\hat{G}/G$ is isomorphic to $\hat{H}/H$.*

PROOF. $\hat{H}$ normalizes G by the remark following 7.1.1. Thus G is a normal subgroup of $\hat{G}$ and $\hat{G} = G\hat{H}$. Also

$$\hat{G}/G = G\hat{H}/G \cong \hat{H}/G \cap \hat{H} = \hat{H}/H$$

by 8.4.6.

8.5 The Levi Decomposition

We shall now derive further information about the parabolic subgroups of the Chevalley group G. We have seen in sections 8.2 and 8.3 that every parabolic subgroup of G is conjugate to one of the subgroups $P_J = BN_JB$ for some subset J of the set Π of fundamental roots. Furthermore distinct subsets of Π give rise to non-conjugate subgroups P_J. Thus G has 2^l conjugacy classes of parabolic subgroups.

We first give an alternative way of describing the subgroup P_J. Let Φ_J be the set of roots which are integral combinations of roots in J and $\bar{\Phi}_J$ be the set of roots which do not belong to Φ_J.

PROPOSITION 8.5.1. *P_J is the subgroup of G generated by H and the root subgroups X_r for $r \in \Phi^+ \cup \Phi_J$.*

PROOF. $P_J = BN_JB$ is the subgroup of G generated by B and the elements n_r for $r \in J$. Since $n_r \in \langle X_r, X_{-r} \rangle$, it is clear that

$$P_J \subseteq \langle H, X_r; r \in \Phi^+ \cup \Phi_J \rangle.$$

Conversely, it is clear that $H \subseteq P_J$ and $X_r \subseteq P_J$ if $r \in \Phi^+$. Thus we consider the subgroups X_r with $r \in \Phi^- \cap \Phi_J$. By 2.5.1 each root in Φ_J is the image of a root in J under an element of W_J. Thus

$$r = w_{r_1} w_{r_2} \ldots w_{r_k}(s),$$

where $r_1, \ldots, r_k, s \in J$. Hence

$$n_{r_1} n_{r_2} \ldots n_{r_k} X_s n_{r_k}^{-1} \ldots n_2^{-1} n_{r_1}^{-1} = X_r$$

by 7.2.1. Therefore $X_r \subseteq P_J$ as required.   ∎

If J is the empty set, then $P_J = B$. Now B admits a decomposition into the semi-direct product of U and H. This semi-direct decomposition generalizes in fact to any parabolic subgroup, as we shall now show. For any subset J of Π we define U_J to be the subgroup of G generated by the root subgroups X_r for which $r \in \Phi^+ \cap \overline{\Phi}_J$. $\Phi^+ \cap \overline{\Phi}_J$ is evidently a closed set of roots, therefore

$$U_J = \prod_{r \in \Phi^+ \cap \overline{\Phi}_J} X_r,$$

by Chevalley's commutator formula. (The factors may be taken in any order.) We also define L_J to be the subgroup of G generated by H and the root subgroups X_r for all $r \in \Phi_J$.

THEOREM 8.5.2. (i) U_J is a normal subgroup of P_J.
(ii) $P_J = U_J L_J$ and $U_J \cap L_J = 1$.
(iii) P_J is the normalizer of U_J in G.
The decomposition of P_J into the semi-direct product of U_J and L_J is called the Levi decomposition, and L_J and its conjugates in U_J are called the Levi subgroups of P_J.

PROOF. We show that the subgroups generating P_J all normalize U_J. It is clear that H normalizes U_J. Let r be a positive root. If $s \in \Phi^+ \cap \overline{\Phi}_J$ all roots of form $ir + js$, where i, j are positive integers, are also in $\Phi^+ \cap \overline{\Phi}_J$. Thus the commutator formula shows that X_r normalizes U_J. Now suppose $r \in \Phi^- \cap \overline{\Phi}_J$. Then $-r$ is not in $\Phi^+ \cap \overline{\Phi}_J$, and if s is any root in $\Phi^+ \cap \overline{\Phi}_J$, all roots of form $ir + js$, where i, j are positive integers, are in $\Phi^+ \cap \overline{\Phi}_J$. For $ir + js$ involves some fundamental root not in J with a positive coefficient. Hence X_r normalizes U_J in this case also, and so U_J is normal in P_J by 8.5.1.

Now 8.5.1 shows that P_J is generated by U_J and L_J. Since U_J is normal in P_J we have $P_J = U_J L_J$.

We now consider $U_J \cap L_J$. Let $\theta \in U_J \cap L_J$ and consider the effect of θ on the Chevalley basis of L_K. We have

$$\theta . h_s = h_s + x,$$

where

$$x \in \sum_{r \in \Phi^+ \cap \overline{\Phi}_J} \mathfrak{L}_r.$$

since $\theta \in U_J$. But

$$\theta . h_s \in H \oplus \sum_{r \in \Phi_J} \mathfrak{L}_r,$$

since $\theta \in L_J$. However

$$\mathfrak{L} = \mathfrak{H} \oplus \sum_{r \in \Phi_J} \mathfrak{L}_r \oplus \sum_{r \in \overline{\Phi}_J} \mathfrak{L}_r$$

and so $\theta . h_s = h_s$.

Now consider $\theta . e_s$. Suppose first that $s \in \Phi_J$. Then

$$\theta . e_s = e_s + x,$$

where

$$x \in \sum_{r \in \overline{\Phi}_J} \mathfrak{L}_r,$$

since $\theta \in U_J$. Also

$$\theta . e_s \in \mathfrak{H} \oplus \sum_{r \in \Phi_J} \mathfrak{L}_r.$$

since $\theta \in L_J$. It follows that $\theta . e_s = e_s$.

Now suppose $s \in \overline{\Phi}_J$. Then

$$\theta . e_s = e_s + x,$$

where

$$x \in \mathfrak{H} \oplus \sum_{r - s \in \overline{\Phi}_J} \mathfrak{L}_r,$$

since $\theta \in U_J$. But

$$\theta . e_s \in \mathfrak{L}_s \oplus \sum_{r - s \in \Phi_J} \mathfrak{L}_r,$$

since $\theta \in L_J$. Thus $\theta . e_s = e_s$ in this case also. Hence θ operates trivially on the Chevalley basis, so $\theta = 1$.

Finally consider the subgroup $\mathfrak{N}_G(U_J)$. This subgroup contains P_J, so must be one of the subgroups P_{J_1} for some subset $J_1 \supseteq J$. If $J_1 \supset J$, choose a root $r \in J_1$ with $r \notin J$. Then $r \in \Phi^+ \cap \overline{\Phi}_J$ and so $X_r \subseteq U_J$. Also $n_r \in \mathfrak{N}_G(U_J)$. However $n_r X_r n_r^{-1} = X_{-r}$ is not contained in U_J and we have a contradiction. Therefore $J_1 = J$ and $\mathfrak{N}_G(U_J) = P_J$. ∎

8.6 The Finite Chevalley Groups

We now consider the special case in which the base field K is the finite field $GF(q)$ with q elements, where q is an arbitrary prime power. G is then a group of non-singular linear transformations of a space over a finite field, so is a finite group. The Chevalley group of type $\mathfrak{L}$ over $GF(q)$ will be denoted by $\mathfrak{L}(q)$. We shall obtain a formula for the orders of the groups $\mathfrak{L}(q)$.

We begin with the canonical form for elements of G established in 8.4.4. Each element of G has a unique expression in the form $u_1 h n_w u$, where $u_1 \in U$, $h \in H$, $w \in W$, $u \in U_w^-$. Thus we have

$$| G | = \sum_{w \in W} | B n_w B | = \sum_{w \in W} | U H n_w U_w^- | = | U | . | H | . \sum_{w \in W} | U_w^- | .$$

Now each element of U is uniquely expressible in the form

$$\prod_{r \in \Phi^+} x_{r_i}(t_i)$$

with $t_i \in GF(q)$. Thus $| U | = q^N$, where $N = | \Phi^+ |$. Also each element of U_w^- is uniquely expressible in the form

$$\prod_{\substack{r_i \in \Phi^+ \\ w(r_i) \in \Phi^-}} x_{r_i}(t_i),$$

and so

$$| U_w^- | = q^{l(w)}.$$

We now consider the order of H. By 7.1.1 H is the set of automorphisms $h(\chi)$ of $\mathfrak{L}_K$ where χ is a K-character of P which can be extended to a K-character of Q. H is a subgroup of $\hat{H}$, the group of automorphisms $h(\chi)$ as χ runs through all K-characters of P. Since P is isomorphic to Q (both are free abelian groups of rank l), $\hat{H}$ is isomorphic to the group of automorphisms $h(\chi)$ for all K-characters χ of Q. Thus there is an epimorphism $\hat{H} \to H$ obtained by restricting χ from Q to P. The kernel consists of the automorphisms $h(\chi)$, where χ is a K-character of Q which is the identity on P. The kernel is therefore isomorphic to the group of K-characters of the factor group Q/P.

Now $| \hat{H} | = (q-1)^l$, since the image of each generator under a K-character can be chosen in $q-1$ ways. Thus

$$| H | = \frac{1}{d}(q-1)^l,$$

where d is the order of the group of K-characters of Q/P. Now it has been shown in section 7.1 that the bases $p_1, \ldots, p_l; q_1, \ldots, q_l$ of fundamental roots and fundamental weights for P, Q respectively are related by

$$p_i = \sum_{j=1}^{l} A_{ji} q_j.$$

Thus the factor group Q/P is generated by elements $\bar{q}_1, \ldots, \bar{q}_l$ subject to relations

$$\sum_{j=1}^{l} A_{ji}\bar{q}_j = 0.$$

A consideration of the Cartan integers A_{ij} (listed in section 3.6 for the different types of simple algebra) shows that Q/P has the following structure ($\mathbb{Z}_i$ denotes a cyclic group of order i):

$\mathfrak{L}$:	A_l	B_l	C_l	D_{2k+1}	D_{2k}	G_2	F_4	E_6	E_7	E_8
Q/P:	$\mathbb{Z}_{l+1}$	$\mathbb{Z}_2$	$\mathbb{Z}_2$	$\mathbb{Z}_4$	$\mathbb{Z}_2 \times \mathbb{Z}_2$	1	1	$\mathbb{Z}_3$	$\mathbb{Z}_2$	1

Now the number of K-characters of $\mathbb{Z}_i$ is the number of ith roots of 1 in $GF(q)$, which is $(i, q-1)$. Thus the number d of K-characters of Q/P has the following values:

$\mathfrak{L}$:	A_l	B_l	C_l	D_{2k+1}	D_{2k}
d:	$(l+1, q-1)$	$(2, q-1)$	$(2, q-1)$	$(4, q-1)$	$(2, q-1)^2$

$\mathfrak{L}$:	G_2	F_4	E_6	E_7	E_8
d:	1	1	$(3, q-1)$	$(2, q-1)$	1

The two cases for type D may be summarized by writing $d = (4, q^l - 1)$.

We now collect together the information obtained above and obtain the following formula.

PROPOSITION 8.6.1. *Let $G = \mathfrak{L}(q)$. Then*

$$|G| = \frac{1}{d} q^N (q-1)^l \sum_{w \in W} q^{l(w)}.$$

Although this is a simple theoretical formula for the order of G, it is not the best formula in practice as the expression

$$\sum_{w \in W} q^{l(w)}$$

is very cumbersome. We shall show in the following chapter how this expression can be simplified.

CHAPTER 9

Polynomial Invariants of the Weyl Group

9.1 The Algebra of Polynomial Invariants

We have seen how the Bruhat decomposition enables us to obtain a formula for the orders of the finite Chevalley groups which involves the polynomial

$$\sum_{w \in W} t^{l(w)}.$$

We shall show in the present chapter that this polynomial factorizes into a product of terms of form

$$1 + t + t^2 + \ldots + t^{d_i - 1} = \frac{t^{d_i} - 1}{t - 1},$$

where $d_1, \ldots, d_l$ are the degrees of certain basic polynomial invariants of W. The numbers $d_1, \ldots, d_l$ have other interpretations also, as we shall see in the following chapter, but we describe them first in terms of polynomial invariants as this appears to be the approach which generalizes most readily to the twisted groups to be considered in chapters 13, 14.

Let W be a Weyl group operating on the Euclidean space $\mathcal{V}$, and let $e_1, e_2, \ldots, e_l$ be an orthonormal basis of $\mathcal{V}$. Then each $x \in \mathcal{V}$ can be written in the form

$$x = x_1 e_1 + \ldots + x_l e_l, \qquad x_i \in \mathbb{R}.$$

Given any polynomial $P(x_1, \ldots, x_l)$ in $x_1, \ldots, x_l$, P may be regarded as a map from $\mathcal{V}$ to $\mathbb{R}$. Let $\mathcal{S}$ be the algebra of all such polynomial functions on $\mathcal{V}$. $\mathcal{S}$ is independent of the basis chosen for $\mathcal{V}$. For the homogeneous polynomials of degree one in $\mathcal{S}$ form the dual space $\hat{\mathcal{V}}$ of $\mathcal{V}$, and $\mathcal{S}$ is then the symmetric algebra of $\hat{\mathcal{V}}$; viz., the algebra of symmetric element in the tensor space

$$\mathbb{R}1 \oplus \hat{\mathcal{V}} \oplus (\hat{\mathcal{V}} \otimes \hat{\mathcal{V}}) \oplus (\hat{\mathcal{V}} \otimes \hat{\mathcal{V}} \otimes \hat{\mathcal{V}}) \oplus \ldots.$$

Now the action of W on $\mathcal{V}$ may be transferred in a natural way to an action on $\hat{\mathcal{V}}$ by defining $w(f)$, $w \in W, f \in \hat{\mathcal{V}}$ by

$$w(f)(w(x)) = f(x), \qquad x \in \mathcal{V}.$$

123

The action of W on $\check{\mathfrak{P}}$ may then be extended to an action on $\mathfrak{S}$ by defining

$$w(P)(x) = P(w^{-1}(x)), \qquad P \in \mathfrak{S}, \quad x \in \mathfrak{P}.$$

A polynomial function in $\mathfrak{S}$ is called an invariant of W if $w(P) = P$ for all $w \in W$. The invariants form a subring $\mathfrak{I}$ of $\mathfrak{S}$. Now $\mathfrak{S}$ is the polynomial ring $\mathbb{R}[x_1, x_2, \ldots, x_l]$, and we shall show that its subring of invariants is also a polynomial ring, i.e.

$$\mathfrak{I} = \mathbb{R}[I_1, I_2, \ldots, I_l],$$

where $I_1, I_2, \ldots, I_l$ are certain elements of $\mathfrak{I}$.

Example 9.1.1. The Weyl group of type A_l is isomorphic to the symmetric group S_{l+1}. It is best described operating on the subspace of an $(l+1)$-dimensional Euclidean space with orthonormal basis $e_0, e_1, \ldots, e_l$ whose elements satisfy $x_0 + x_1 + \ldots + x_l = 0$. The elements of W operate on $\mathfrak{P}$ by permuting the coordinates $x_0, x_1, \ldots, x_l$ in all possible ways. Thus the polynomial invariants are the symmetric polynomials in $x_0, x_1, \ldots, x_l$. These are all polynomials in the elementary symmetric polynomials. Since $x_0 + x_1 + \ldots + x_l = 0$ we have $\mathfrak{I} = \mathbb{R}[I_1, I_2, \ldots, I_l]$, where I_i is the elementary symmetric polynomial of degree $i+1$.

Example 9.1.2. The Weyl group of type B_l operates on a Euclidean space with orthonormal basis $e_1, \ldots, e_l$ by permuting the coordinates $x_1, \ldots, x_l$ in all possible ways and by changing the signs arbitrarily. Thus the polynomial invariants are the symmetric polynomials in $x_1^2, x_2^2, \ldots, x_l^2$. These are all polynomials in the elementary symmetric polynomials in $x_1^2, \ldots, x_l^2$. Thus $\mathfrak{I} = \mathbb{R}[I_1, I_2, \ldots, I_l]$, where I_i is the ith elementary symmetric polynomial in $x_1^2, \ldots, x_l^2$.

In general, a polynomial of form $\lambda x_1^{k_1} x_2^{k_2} \ldots x_m^{k_m}$, $\lambda \in \mathbb{R}$, is called a monomial of degree $k_1 + k_2 + \ldots + k_m$. The degree of an arbitrary polynomial P is the greatest degree of any monomial constituent and will be denoted by deg P. A polynomial is called homogeneous if all its monomial constituents have the same degree.

For each $P \in \mathfrak{S}$, the average of P under W is defined by

$$\text{Av } P = \frac{1}{|W|} \sum_{w \in W} w(P).$$

LEMMA 9.1.3. *If* $P \in \mathfrak{S}$ *then* Av $P \in \mathfrak{I}$.

PROOF. Let $w \in W$. Then we have

$$w(\text{Av } P) = \frac{1}{|W|} \sum_{w' \in W} ww'(P) = \frac{1}{|W|} \sum_{w' \in W} w'(P) = \text{Av } P. \qquad \blacksquare$$

9.2 A Theorem of Chevalley

The fact that $\mathfrak{I}$ is a polynomial ring was originally proved by Chevalley [5]. In order to prove Chevalley's theorem we first need a preliminary lemma. Let $\mathfrak{S}^+$ be the set of polynomials in $\mathfrak{S}$ with constant term 0, and let $\mathfrak{I}^+ = \mathfrak{I} \cap \mathfrak{S}^+$. Let $\mathfrak{S}\mathfrak{I}^+$ be the ideal of $\mathfrak{S}$ generated by $\mathfrak{I}^+$. The elements of $\mathfrak{S}\mathfrak{I}^+$ therefore have form $P_1 J_1 + \ldots + P_k J_k$, where $P_i \in \mathfrak{S}$, $J_i \in \mathfrak{I}^+$ for each i.

LEMMA 9.2.1. *Suppose J_1, J_2, ..., J_k are elements of $\mathfrak{I}$ such that J_1 is not in the ideal of $\mathfrak{I}$ generated by J_2, ..., J_k. Let P_1, P_2, ..., P_k be homogeneous polynomials in $\mathfrak{S}$ such that $P_1 J_1 + \ldots + P_k J_k = 0$. Then $P_1 \in \mathfrak{S}\mathfrak{I}^+$.*

PROOF. We show first that J_1 is not in the ideal of $\mathfrak{S}$ generated by J_2, ..., J_k. Suppose this were false. Then

$$J_1 = Q_2 J_2 + \ldots + Q_k J_k, \qquad Q_i \in \mathfrak{S}.$$

Now for each $w \in W$ we have

$$w(Q_i J_i) = w(Q_i) \, w(J_i) = w(Q_i) J_i.$$

Thus $\text{Av } (Q_i J_i) = (\text{Av } Q_i) J_i$. It follows that

$$J_1 = \text{Av } J_1 = (\text{Av } Q_2) J_2 + \ldots + (\text{Av } Q_k) J_k.$$

Since $\text{Av } Q_i \in \mathfrak{I}$, this means that J_1 is in the ideal of $\mathfrak{I}$ generated by J_2, ..., J_k, a contradiction.

We prove that $P_1 \in \mathfrak{S}\mathfrak{I}^+$ by induction on deg P_1. If deg $P_1 = 0$, P_1 is constant. Since J_1 is not in the ideal of $\mathfrak{S}$ generated by J_2, ..., J_k, we must have $P_1 = 0$. Thus $P_1 \in \mathfrak{S}\mathfrak{I}^+$ in this case.

We now assume deg $P_1 > 0$. The root system Φ is a finite subset of $\mathfrak{P}$ and, for each $r \in \Phi$, the hyperplane orthogonal to r is given by an equation $H_r = 0$, where H_r is a homogeneous polynomial of degree 1. Consider the polynomial $w_r(P_i) - P_i$. This polynomial vanishes at all points for which H_r is zero. Since H_r is an irreducible polynomial, H_r divides

$w_r(P_i) - P_i$, thus

$$w_r(P_i) - P_i = H_r \cdot \bar{P}_i, \qquad \bar{P}_i \in \mathfrak{S}.$$

Now P_i is homogeneous, hence $w_r(P_i)$ is homogeneous of the same degree. Thus $w_r(P_i) - P_i$ is also homogeneous, and it follows that $\bar{P}_i$ is homogeneous. Moreover $\deg \bar{P}_i < \deg P_i$. Now we have

$$P_1 J_1 + \ldots + P_k J_k = 0.$$

Thus

$$w_r(P_1) J_1 + \ldots + w_r(P_k) J_k = 0$$

and so

$$H_r(\bar{P}_1 J_1 + \ldots + \bar{P}_k J_k) = 0.$$

Since H_r is not identically zero we have

$$\bar{P}_1 J_1 + \ldots + \bar{P}_k J_k = 0.$$

But $\deg \bar{P}_1 < \deg P_1$ and so $\bar{P}_1 \in \mathfrak{S}\mathfrak{J}^+$ by induction. Thus

$$w_r(P_1) - P_1 \in \mathfrak{S}\mathfrak{J}^+.$$

Now $w(\mathfrak{J}^+) = \mathfrak{J}^+$ for each $w \in W$, hence $w(\mathfrak{S}\mathfrak{J}^+) = \mathfrak{S}\mathfrak{J}^+$. Thus each $w \in W$ operates naturally on the quotient ring $\mathfrak{S}/\mathfrak{S}\mathfrak{J}^+$. We have seen that w_r operates trivially on P_1 in this quotient ring for each $r \in \Phi$. Since the w_r generate W, each $w \in W$ operates trivially on P_1 in the quotient ring. Thus

$$w(P_1) - P_1 \in \mathfrak{S}\mathfrak{J}^+.$$

It follows that $\text{Av } P_1 - P_1 \in \mathfrak{S}\mathfrak{J}^+$. But P_1 is homogeneous with $\deg P_1 > 0$, hence $\text{Av } P_1 \in \mathfrak{J}^+$. In particular $\text{Av } P_1 \in \mathfrak{S}\mathfrak{J}^+$ and so $P_1 \in \mathfrak{S}\mathfrak{J}^+$ as required. ∎

Now the ideal $\mathfrak{S}\mathfrak{J}^+$ of $\mathfrak{S}$ is generated by the homogeneous elements of $\mathfrak{J}$ of positive degree. By Hilbert's basis theorem there is a finite subset of this generating set which generates $\mathfrak{S}\mathfrak{J}^+$. Thus there is a set $I_1, I_2, \ldots, I_n$ of homogeneous polynomials in $\mathfrak{J}$ such that $I_1, \ldots, I_n$ generates $\mathfrak{S}\mathfrak{J}^+$ but no proper subset has this property.

THEOREM 9.2.2. *There is no polynomial $P \neq 0$ such that*

$$P(I_1, \ldots, I_n) = 0.$$

Thus $I_1, \ldots, I_n$ are algebraically independent.

PROOF. Suppose $P(I_1, \ldots, I_n) = 0$ with $P \neq 0$. We may assume, by comparing terms of a given degree, that all monomials in $I_1, \ldots, I_n$ which occur in P have the same degree d in $x_1, \ldots, x_l$. Let $P_i = \partial P / \partial I_i$. Then $P_i(I_1, \ldots, I_n)$, $i = 1, \ldots, n$, are elements of $\mathfrak{J}$ and not all the P_i are zero. Let $\mathfrak{K}$ be the ideal of $\mathfrak{J}$ generated by $P_1, P_2, \ldots, P_n$. We may choose the notation so that $P_1, \ldots, P_m$ but no proper subset generate $\mathfrak{K}$ as an ideal of $\mathfrak{J}$. Then there exist polynomials $Q_{i,j} \in \mathfrak{J}$ such that

$$P_i = \sum_{j=1}^{m} Q_{i,j} P_j, \qquad i > m.$$

Now each P_i is homogeneous in $x_1, \ldots, x_l$ of degree $d - \deg I_i$. Thus, by comparing terms of the same degree in $x_1, \ldots, x_l$ on both sides, we may assume each $Q_{i,j}$ is homogeneous of degree $\deg P_i - \deg P_j$. Thus $\deg Q_{i,j} = \deg I_j - \deg I_i$.

Now $P(I_1, \ldots, I_n) = 0$, thus $\partial P / \partial x_k = 0$ for $k = 1, \ldots l$. Hence

$$\sum_{i=1}^{n} \frac{\partial P}{\partial I_i} \frac{\partial I_i}{\partial x_k} = 0,$$

$$\sum_{i=1}^{n} P_i \frac{\partial I_i}{\partial x_k} = 0.$$

It follows that

$$\sum_{i=1}^{m} P_i \frac{\partial I_i}{\partial x_k} + \sum_{i=m+1}^{n} \left(\sum_{j=1}^{m} Q_{i,j} P_j \frac{\partial I_i}{\partial x_k} \right) = 0,$$

$$\sum_{i=1}^{m} P_i \left(\frac{\partial I_i}{\partial x_k} + \sum_{j=m+1}^{n} Q_{j,i} \frac{\partial I_j}{\partial x_k} \right) = 0.$$

We now apply 9.2.1. $P_1, \ldots, P_m$ are in $\mathfrak{J}$ and P_1 is not in the ideal of $\mathfrak{J}$ generated by $P_2, \ldots, P_m$. Each of the polynomials

$$\frac{\partial I_i}{\partial x_k} + \sum_{j=m+1}^{n} Q_{j,i} \frac{\partial I_j}{\partial x_k}, \qquad i = 1, \ldots, m,$$

is homogeneous in $x_1, \ldots, x_l$ of degree $\deg I_i - 1$. Thus by 9.2.1 we have

$$\frac{\partial I_1}{\partial x_k} + \sum_{j=m+1}^{n} Q_{j,1} \frac{\partial I_j}{\partial x_k} \in \mathfrak{S}\mathfrak{J}^{+}.$$

We now multiply this polynomial by x_k and sum over k. For a homogeneous polynomial I_j in $x_1, \ldots, x_l$ we have, by Euler's formula,

$$\sum_{k=1}^{l} x_k \frac{\partial I_j}{\partial x_k} = \deg I_j . I_j.$$

Therefore

$$\deg I_1 . I_1 + \sum_{j=m+1}^{n} \deg I_j . Q_{j,\,1} I_j = \sum_{i=1}^{n} I_i R_i,$$

where each $R_i \in \mathfrak{S}^+$. All the terms on the left-hand side are homogeneous of degree $\deg I_1$. Comparing terms of this degree on both sides we obtain

$$\deg I_1 . I_1 + \sum_{j=m+1}^{n} \deg I_j . Q_{j,\,1} I_j = \sum_{i} I_i S_i,$$

where the sum on the right extends over some subset of $1, \ldots, n$ not including $i=1$ (since the monomials in $I_1 R_1$ have too large a degree). It follows now that I_1 is in the ideal of $\mathfrak{S}$ generated by $I_2, \ldots, I_n$ and we have a contradiction. This completes the proof. ∎

THEOREM 9.2.3. *Every element of $\mathfrak{I}$ is a polynomial in $I_1, \ldots, I_n$.*

PROOF. It is sufficient to prove this for homogeneous polynomials in $\mathfrak{I}$. Let $J \in \mathfrak{I}$ be homogeneous. We use induction on $\deg J$, the result being clear if $\deg J = 0$. Suppose $\deg J > 0$. Then $J \in \mathfrak{I}^+$ and in particular $J \in \mathfrak{S}\mathfrak{I}^+$. Thus we have

$$J = P_1 I_1 + \ldots + P_n I_n$$

for certain $P_1, \ldots, P_n \in \mathfrak{S}$. Since $J, I_1, \ldots, I_n$ are all homogeneous we may assume each P_i is homogeneous also, with $\deg P_i = \deg J - \deg I_i$. Then

$$J = \mathrm{Av}\, J = \mathrm{Av}\, P_1 . I_1 + \ldots + \mathrm{Av}\, P_n . I_n.$$

$\mathrm{Av}\, P_1, \ldots, \mathrm{Av}\, P_n$ are homogeneous polynomials in $\mathfrak{I}$ of degree less than $\deg J$. Thus they are polynomials in $I_1, \ldots, I_n$ by induction, and so J is also. ∎

COROLLARY 9.2.4. $\mathfrak{I} = \mathbb{R}[I_1, \ldots, I_n]$ *is isomorphic to the polynomial ring in n generators over $\mathbb{R}$.*

PROOF. This follows from 9.2.2 and 9.2.3. ∎

$I_1, \ldots, I_n$ is called a set of *basic polynomial invariants* of W. We now determine the number of invariants in a basic set.

THEOREM 9.2.5. *The number of invariants in a basic set is equal to the rank of the Weyl group.*

PROOF. In the above notation we must show that $n = l$. Let

$$K = \mathbb{R}(x_1, \ldots, x_l)$$

be the field of rational functions in $x_1, \ldots, x_l$ over $\mathbb{R}$. Similarly let $k = \mathbb{R}(I_1, \ldots, I_n)$ be the field of rational functions in $I_1, \ldots, I_n$ over $\mathbb{R}$. Then we have

$$\mathbb{R} \subset k \subset K.$$

Since $x_1, \ldots, x_l$ are algebraically independent over $\mathbb{R}$, the transcendence degree of K over $\mathbb{R}$ is given by

$$\text{tr. deg. } K/\mathbb{R} = l.$$

Also by 9.2.2 we have

$$\text{tr. deg. } k/\mathbb{R} = n.$$

Since

$$\text{tr. deg. } K/\mathbb{R} = \text{tr. deg. } k/\mathbb{R} + \text{tr. deg. } K/k$$

by Galois theory (see, for example, Jacobson [2]), we consider tr. deg. K/k. Now K is generated over k by $x_1, \ldots, x_l$. However, each x_i is an algebraic element over k. For the polynomial

$$\prod_{w \in W} (t - w(x_i))$$

has x_i as a root, and its coefficients are the elementary symmetric polynomials in $w(x_i)$ for all $w \in W$. These coefficients are invariant under each element of W so are in $\mathfrak{I}$ and therefore in k. Thus K is generated over k by a finite number of algebraic elements over k, and so

$$\text{tr. deg. } K/k = 0.$$

It follows that $n = l$.   ∎

9.3 The Degrees of the Basic Invariants

Now the set $I_1, \ldots, I_l$ of basic polynomial invariants of W is not uniquely determined. It is not difficult to show, however, that the degrees of the polynomials in a basic set are uniquely determined.

PROPOSITION 9.3.1. *Let $I_1, \ldots, I_l$; $I_1', \ldots, I_l'$ be two sets of basic* *polynomial invariants. Then we may arrange the numbering so that* deg $I_i = $ deg I_i' *for* $i = 1, \ldots, l$.

PROOF. Each of $I_1', \ldots, I_l'$ is expressible as a polynomial in $I_1, \ldots, I_l$ and conversely. Consider the matrices

$$\left(\frac{\partial I_i}{\partial I_j'}\right), \quad \left(\frac{\partial I_i'}{\partial I_j}\right).$$

These are inverse matrices since

$$\sum_{k=1}^{l} \frac{\partial I_i}{\partial I_k'} \frac{\partial I_k'}{\partial I_j} = \frac{\partial I_i}{\partial I_j} = \begin{cases} 1 & \text{if } i = j, \\ 0 & \text{if } i \neq j. \end{cases}$$

Thus the determinant $|\partial I_i / \partial I_j'|$ is non-zero. It follows that for some permutation ρ of $1, 2, \ldots, l$,

$$\prod_{i=1}^{l} \frac{\partial I_i}{\partial I_{\rho(i)}'} \neq 0.$$

By renumbering $I_1', \ldots, I_l'$ we may assume ρ is the identity. Thus $\partial I_i / \partial I_i' \neq 0$ for each i. This means that I_i, as a polynomial in $I_1', \ldots, I_l'$, involves I_i' and so deg $I_i \geqslant$ deg I_i'. In particular

$$\sum_{i=1}^{l} \deg I_i \geqslant \sum_{i=1}^{l} \deg I_i'.$$

By symmetry we must have equality. Thus deg $I_i =$ deg I_i' for each i. ∎

We shall denote the degrees of the basic invariants by $d_1, d_2, \ldots, d_l$, and shall derive some properties of this set of integers.

LEMMA 9.3.2. *Let M be any finite-dimensional W-module over a field of characteristic 0. Then the dimension of the subspace formed by the W-invariants of M (i.e. the elements $x \in M$ such that $w(x) = x$ for all $w \in W$) is equal to the trace of the linear transformation*

$$\frac{1}{|W|} \sum_{w \in W} w$$

of M.

PROOF. Let

$$T = \frac{1}{|W|} \sum_{w \in W} w.$$

Then $T^2 = T$. Let M_0 be the set of $x \in M$ with $T(x) = 0$ and M_1 be the set of $x \in M$ with $T(x) = x$. Then $M = M_0 \oplus M_1$ since T is idempotent. The trace of T is the dimension of M_1. Also the W-invariants of M are just the elements which lie in M_1. For if x is W-invariant,

$$T(x) = \frac{1}{|W|} \sum_{w \in W} x = x,$$

whereas conversely if $T(x) = x$ we have

$$w(x) = wT(x) = T(x) = x$$

for all $w \in W$. This completes the proof. ∎

We now return to the natural representation of W on $\mathfrak{P}$. Let $w \in W$ be an element with eigenvalues $\lambda_1, \ldots, \lambda_l$ on $\mathfrak{P}$. The λ_i are in the complex field. Let $\mathfrak{P}_{\mathbb{C}}$ be the complexification of $\mathfrak{P}$. Then w may be represented by the diagonal matrix diag $(\lambda_1, \ldots, \lambda_l)$ with respect to a suitable basis of $\mathfrak{P}_{\mathbb{C}}$. Thus, for each $t \in \mathbb{C}$,

$$\det (1 - tw) = (1 - \lambda_1 t)(1 - \lambda_2 t) \ldots (1 - \lambda_l t).$$

If we consider t now as an indeterminate, $1/\det (1 - tw)$ may be expressed as a power series in t. In fact

$$\frac{1}{\det (1 - tw)} = (1 + \lambda_1 t + \lambda_1^2 t^2 + \ldots)(1 + \lambda_2 t + \lambda_2^2 t^2 + \ldots)$$

$$\ldots (1 + \lambda_l t + \lambda_l^2 t^2 + \ldots)$$

$$= \sum_{n \geqslant 0} \left(\sum_{k_1 + \ldots + k_l = n} \lambda_1^{k_1} \lambda_2^{k_2} \ldots \lambda_l^{k_l} \right) t^n.$$

We write Av $1/\det (1 - tw)$ to denote the power series

$$\frac{1}{|W|} \sum_{w \in W} \frac{1}{\det (1 - tw)}$$

in t. The next result shows that this power series can be expressed simply in terms of the degrees $d_1, \ldots, d_l$.

PROPOSITION 9.3.3.

$$\text{Av } \frac{1}{\det (1 - tw)} = \prod_{i=1}^{l} \frac{1}{(1 - t^{d_i})}$$

as power series in t.

PROOF. We have earlier defined the action of W on the dual space $\hat{\mathfrak{P}}$ of $\mathfrak{P}$ and on the ring $\mathfrak{S}$ of polynomial functions on $\mathfrak{P}$. In the present proof we take $\mathfrak{P}$, $\hat{\mathfrak{P}}$, $\mathfrak{S}$ over the complex field instead of the real field as before.

$\lambda_1, \ldots, \lambda_l$ are the eigenvalues of w on $\mathfrak{P}$, thus $\lambda_1^{-1}, \ldots, \lambda_l^{-1}$ are the eigenvalues of w on $\hat{\mathfrak{P}}$. However these eigenvalues are roots of unity and w is a real transformation, hence the eigenvalues of w on $\hat{\mathfrak{P}}$ are $\lambda_1, \ldots, \lambda_l$. Let $y_1, \ldots, y_l$ be corresponding eigenvectors spanning $\hat{\mathfrak{P}}$. $y_1, \ldots, y_l$ are linear combinations (possibly complex) of the original basis $x_1, \ldots, x_l$ of $\hat{\mathfrak{P}}$. The polynomial functions $y_1^{k_1} y_2^{k_2} \ldots y_l^{k_l}$ for all sets of non-negative integers $k_1, \ldots, k_l$ with $k_1 + \ldots + k_l = n$ form a basis for $\mathfrak{S}_n$, the space of polynomial functions which are homogeneous of degree n. Since $w(y_i) = \lambda_i y_i$ we have

$$w(y_1^{k_1} \ldots y_l^{k_l}) = \lambda_1^{k_1} \ldots \lambda_l^{k_l} y_1^{k_1} \ldots y_l^{k_l}.$$

Thus the eigenvalues of w on $\mathfrak{S}_n$ are the complex numbers $\lambda_1^{k_1} \ldots \lambda_l^{k_l}$. Hence the coefficient of t^n in the power series $1/\det(1 - tw)$ is the sum of the eigenvalues of w on $\mathfrak{S}_n$, i.e. the trace of w on $\mathfrak{S}_n$. It follows that the coefficient of t^n in the power series Av $1/\det(1 - tw)$ is the trace of the linear transformation

$$\frac{1}{|W|} \sum_{w \in W} w$$

on $\mathfrak{S}_n$. By 9.3.2 this is the dimension of the space $\mathfrak{S}_n \cap \mathfrak{I} = \mathfrak{I}_n$. Now $\mathfrak{I}_n$ has as a basis the set of all polynomials $I_1^{e_1} I_2^{e_2} \ldots I_l^{e_l}$ of degree n, where $I_1, \ldots, I_l$ are a set of basic invariants. Since deg $I_i = d_i$ the number of such polynomials is the number of solutions of the equation

$$d_1 e_1 + d_2 e_2 + \ldots + d_l e_l = n,$$

which is in turn the coefficient of t^n in the power series

$$(1 + t^{d_1} + t^{2d_1} + \ldots)(1 + t^{d_2} + t^{2d_2} + \ldots) \ldots (1 + t^{d_l} + t^{2d_l} + \ldots).$$

This is the coefficient of t^n in

$$\prod_{i=1}^{l} \frac{1}{(1 - t^{d_i})}$$

and so the result is proved. ∎

By applying this result we can determine the sum and the product of the degrees $d_1, \ldots, d_l$.

THEOREM 9.3.4.   (i) $d_1 d_2 \ldots d_l = |W|$,
(ii) $d_1 + d_2 + \ldots + d_l = N + l$.

PROOF.   Consider det $(1 - tw)$. This is $(1-t)^l$ if $w = 1$, $(1-t)^{l-1}(1+t)$
if w is a reflection, and a polynomial not divisible by $(1-t)^{l-1}$ otherwise.

Now the only elements of W which are reflections are the elements
w_r with $r \in \Phi$. For suppose W contains a reflection w_H in a hyperplane
H not orthogonal to any root. H contains a vector v not orthogonal to
any root, and $w_H(v) = v$. v lies in a chamber C (see section 2.3), and
since $w_H(C)$ is also a chamber containing v we have $w_H(C) = C$. Hence
$w_H = 1$ by 2.3.2 and we have a contradiction. Therefore W contains
exactly N reflections, where $N = |\Phi^+|$.

We now apply 9.3.3. Multiplying both sides by $(1-t)^l$ we have

$$\prod_{i=1}^{l} \frac{1}{1 + t + \ldots + t^{d_i - 1}} = \frac{1}{|W|} \left(1 + \frac{N(1-t)}{(1+t)} + (1-t)^2 F(t) \right),$$

where $F(t)$ is some rational function whose denominator is not divisible
by $1 - t$. (The first term on the right comes from the unit element, the
second from the N reflections, and the third from the remaining elements
of W.)

Putting $t = 1$ we obtain

$$\frac{1}{d_1 d_2 \ldots d_l} = \frac{1}{|W|},$$

hence $|W| = d_1 d_2 \ldots d_l$.

To obtain the second result we first differentiate and then put $t = 1$.
We have

$$\prod_{i=1}^{l} \frac{1}{1 + t + \ldots + t^{d_i - 1}} \left(\sum_{i=1}^{l} - \frac{1 + 2t + \ldots + (d_i - 1) t^{d_i - 2}}{1 + t + \ldots + t^{d_i - 1}} \right)$$

$$= - \frac{N}{|W|} \cdot \frac{1}{1 + t} + G(t),$$

where $G(t)$ is a rational function whose numerator is divisible by $1 - t$.
Putting $t = 1$ we obtain

$$- \frac{1}{2} \cdot \frac{1}{d_1 d_2 \ldots d_l} \cdot \sum_{i=1}^{l} (d_i - 1) = - \frac{N}{2|W|}.$$

It follows that

$$d_1 + d_2 + \ldots + d_l = N + l. \qquad \blacksquare$$

Let $I_1, \ldots, I_l$ be a set of basic polynomial invariants and

$$\partial(I_1, \ldots, I_l)/\partial(x_1, \ldots, x_l)$$

be the Jacobian of the map

$$(x_1, \ldots, x_l) \rightarrow (I_1(x_1, \ldots, x_l), \ldots, I_l(x_1, \ldots, x_l)).$$

The Jacobian is the matrix whose (i, j)-coefficient is $\partial I_i/\partial x_j$. Let

$$J = \left| \frac{\partial(I_1, \ldots, I_l)}{\partial(x_1, \ldots, x_l)} \right|$$

be the determinant of this matrix. J is a homogeneous polynomial in $x_1, \ldots, x_l$ of degree

$$\sum_{i=1}^{l} (d_i - 1).$$

By 9.3.4 J has degree N. We now show that J has a factorization into linear factors.

THEOREM 9.3.5.   *For each root $r \in \Phi^+$ let $H_r = 0$ be the equation of the hyperplane orthogonal to r. Then*

$$J = \lambda . \prod_{r \in \Phi^+} H_r$$

for some $\lambda \in \mathbb{R}$.

PROOF.   Consider the map T of $\mathcal{V}$ into itself given by

$$T(x_1, \ldots, x_l) = (I_1(x_1, \ldots, x_l), \ldots, I_l(x_1, \ldots, x_l)).$$

For each point $x = (x_1, \ldots, x_l)$ of $\mathcal{V}$ at which $J \neq 0$ there exist open neighbourhoods of x, $T(x)$ in $1-1$ correspondence under T. (See, for example Loomis and Sternberg [1].) Now suppose x lies in a reflecting hyperplane $H_r, r \in \Phi$. Then any open neighbourhood of x contains points a, b such that $a \neq b$ but $w_r(a) = b$. Then we have

$$I_i(b) = I_i(w_r(a)) = w_r(I_i(a)) = I_i(a)$$

for $i = 1, \ldots, l$. Thus $T(b) = T(a)$ and so $J = 0$ at x. Hence the polynomial J vanishes at each point at which the linear polynomial H_r vanishes. Since H_r is an irreducible polynomial, H_r divides J. This is true for all $r \in \Phi^+$, thus

$$\prod_{r \in \Phi^+} H_r$$

divides J. But

$$\deg J = N = \deg \left(\prod_{r \in \Phi^+} H_r \right).$$

Hence

$$J = \lambda . \prod_{r \in \Phi^+} H_r$$

for some $\lambda \in \mathbb{R}$. ∎

9.4 A Theorem of Solomon

We now turn to the proof of the identity

$$\sum_{w \in W} t^{l(w)} = \prod_{i=1}^{l} \left(\frac{t^{d_i} - 1}{t - 1} \right).$$

This factorization of the polynomial $\sum t^{l(w)}$ was first proved by Bott [1], with a suitable interpretation of the integers $d_1, \ldots, d_l$, by considerations involving the topology of Lie groups. We give here a proof due to Solomon [4] in a slightly modified form due to Steinberg [16], where $d_1, \ldots, d_l$ are interpreted as the degrees of the basic polynomial invariants of W.

We define

$$P_W(t) = \sum_{w \in W} t^{l(w)},$$

$$\bar{P}_W(t) = \prod_{i=1}^{l} \frac{t^{d_i} - 1}{t - 1},$$

and $P_{W_J}(t)$, $\bar{P}_{W_J}(t)$ denote the corresponding polynomials for the Weyl groups W_J, where J is any subset of Π. The idea of Solomon's proof is to show that $P_W(t)$, $\bar{P}_W(t)$ satisfy the following identities:

$$\sum_J (-1)^{|J|} \frac{P_W(t)}{P_{W_J}(t)} = t^N,$$

$$\sum_J (-1)^{|J|} \frac{\bar{P}_W(t)}{\bar{P}_{W_J}(t)} = t^N.$$

Assuming by induction that $P_{W_J}(t) = \bar{P}_{W_J}(t)$ whenever J is a proper subset of Π, it follows from these identities that $P_W(t) = \bar{P}_W(t)$.

We concentrate first on the polynomial $P_W(t)$.

LEMMA 9.4.1.   *Let w be an element of W_J. Then $l(w)$ is the same whether w is regarded as an element of the Weyl group W or the Weyl group W_J.*

PROOF.   Suppose r is a positive root not in Φ_J. Then r is a positive combination of roots in Π involving some root in $\Pi - J$. If $r_i \in J$, $w_{r_i}(r)$ still involves this root in $\Pi - J$ with a positive coefficient. Thus $w_{r_i}(r)$ is still a positive root not in Φ_J. Repeating this argument we see that $w(r)$ is a positive root not in Φ_J for all $w \in W_J$. Thus all positive roots transformed by w into negative roots are in Φ_J. The result follows by 2.2.2.   ∎

Lemma 9.4.1 shows that we may write $l(w)$ for $w \in W_J$ without ambiguity.

We now consider the operation of W on the Coxeter complex, as described in section 2.6. By 2.6.3 each element of the Coxeter complex can be transformed under W into just one element of the form

$$C_J = \left\{ v; \; \begin{array}{l} (v, r) = 0 \text{ for } r \in J \\ (v, r) > 0 \text{ for } r \in \Pi - J \end{array} \right\}.$$

Consider the orbit of the Coxeter complex under W containing C_J. For $w \in W$ we define $n_J(w)$ to be the number of elements in this orbit which are fixed by w. By 2.6.1 $n_J(w)$ is equal to the number of left cosets xW_J fixed by w under left multiplication.

Now $n_J(w)$ has a useful interpretation in terms of the theory of characters. If G is any finite group and H a subgroup of G we define the operations of restriction and induction of class functions in the usual way. If χ is a class function on G we denote by χ_H the restriction of χ to H, and if ϕ is a class function on H we denote by ϕ^G the induced class function on G. ϕ^G is defined by

$$\phi^G(g) = \frac{1}{|H|} \sum_{x \in G} \phi(xgx^{-1})$$

summed over those elements $x \in G$ for which $xgx^{-1} \in H$.

LEMMA 9.4.2.   *Let 1_{W_J} be the unit character of W_J. Thus $1_{W_J}(w) = 1$ for all $w \in W_J$. Then $1_{W_J}^W(w) = n_J(w)$ for all $w \in W$.*

PROOF.

$$1_{W_J}^W(w) = \frac{1}{|W_J|} \sum_x 1$$

summed over the elements $x \in W$ for which $xwx^{-1} \in W_J$. Now $xwx^{-1} \in W_J$ if and only if $xwx^{-1}(C_J) = C_J$ by 2.6.1, and this holds if and only if w fixes $x^{-1}(C_J)$. Now there are $n_J(w)$ such elements $x^{-1}(C_J)$ fixed by w and for each one of them the element x may be chosen in $| W_J |$ ways. Thus

$$1_{W_J}^W(w) = \frac{1}{| W_J |} \cdot | W_J | n_J(w) = n_J(w). \qquad \blacksquare$$

PROPOSITION 9.4.3.

$$\sum_J (-1)^{| J |} n_J(w) = \det w.$$

To prove this we again consider the operation of w on the Coxeter complex. However, we first require a preliminary lemma. In this we consider not the Coxeter complex defined in terms of the reflecting hyperplanes, but a complex $\mathcal{K}$ defined similarly by any finite set of hyperplanes of $\mathfrak{P}$. The dimension of an element $\mathfrak{K} \in \mathcal{K}$ is defined as the dimension of the smallest subspace of $\mathfrak{P}$ containing $\mathfrak{K}$.

LEMMA 9.4.4. *Let n_i be the number of elements of $\mathcal{K}$ of dimension i. Then*

$$\sum_i (-1)^i n_i = (-1)^{\dim \mathfrak{P}}.$$

PROOF. We use induction on the number of hyperplanes. Suppose the result proved for a system of n hyperplanes and that an additional hyperplane H is then added. Each element of $\mathcal{K}$ of dimension i which is cut in two by H has corresponding to it an element of dimension $i-1$ in H separating the two parts. Thus the sum

$$\sum_i (-1)^i n_i$$

remains unchanged. $\blacksquare$

PROOF of 9.4.3. Let U be the subspace of $\mathfrak{P}$ of elements fixed by w. The elements of the Coxeter complex which are fixed by w are just the ones which lie in U, by 2.6.2. We now apply 9.4.4 to U. n_i is the number of elements of the Coxeter complex which have dimension i and lie in U. Since $\dim C_J = l - | J |$ we have

$$n_i = \sum_{|J| = l - i} n_J(w).$$

Thus, by 9.4.4,

$$\sum_i (-1)^i n_i = (-1)^l \sum_J (-1)^{|J|} n_J(w) = (-1)^{\dim U}.$$

Now w, being an orthogonal transformation of $\mathfrak{V}$, has eigenvalues which are 1, -1 or pairs of complex conjugates of modulus 1. Thus

$$\det w = (-1)^{l-\dim U}.$$

It follows that

$$\sum_J (-1)^{|J|} n_J(w) = \det w. \qquad \blacksquare$$

THEOREM 9.4.5.

$$\sum_J (-1)^{|J|} \frac{P_W(t)}{P_{W_J}(t)} = t^N.$$

PROOF. Let D_J be the set of distinguished coset representatives of W_J in W defined in 2.5.8. Define

$$P_{D_J}(t) = \sum_{w \in D_J} t^{l(w)}.$$

Then, by 2.5.8, we have

$$P_W(t) = P_{W_J}(t) . P_{D_J}(t).$$

Therefore

$$\sum_J (-1)^{|J|} \frac{P_W(t)}{P_{W_J}(t)} = \sum_J (-1)^{|J|} P_{D_J}(t)$$

$$= \sum_J (-1)^{|J|} \left(\sum_{\substack{w \\ w(J) \subseteq \Phi^+}} t^{l(w)} \right)$$

$$= \sum_{w \in W} \left(\sum_{\substack{J \\ w(J) \subseteq \Phi^+}} (-1)^{|J|} \right) t^{l(w)}.$$

Let J_w be the set of roots $r \in \Pi$ such that $w(r) \in \Phi^+$. Then the coefficient of $t^{l(w)}$ in the above sum is

$$\sum_{J \subseteq J_w} (-1)^{|J|} = (1-1)^{|J_w|}.$$

This is 0 unless J_w is the empty set, when it is 1. However if J_w is empty w transforms every positive root into a negative root, so $w = w_0$ by 2.2.6. Hence the above sum is $t^{l(w_0)} = t^N$ and the result is proved. $\qquad \blacksquare$

We now turn to consider the polynomial

$$\bar{P}_W(t) = \prod_{i=1}^{l} \left(\frac{t^{d_i} - 1}{t - 1} \right).$$

In order to show that $\bar{P}_W(t)$ satisfies the required identity it is necessary to look more closely at the operation of W on the ring $\mathfrak{S}$ of polynomial functions on $\mathcal{V}$. As before, $\mathfrak{I}$ denotes the set of invariant polynomials under W. A polynomial $P \in \mathfrak{S}$ is said to be alternating if $w(P) = \det w . P$ for all $w \in W$, and the set of alternating polynomials is denoted by $\hat{\mathfrak{I}}$. As before, H_r denotes the linear form representing the hyperplane orthogonal to the root r, defined by $H_r(x) = (r, x)$.

LEMMA 9.4.6. *A polynomial $P \in \mathfrak{S}$ is alternating if and only if it is the product of an invariant polynomial with*

$$\prod_{r \in \Phi^+} H_r.$$

PROOF. We show first that

$$\prod_{r \in \Phi^+} H_r$$

is alternating. If s is any root we have $w_s(H_r) = H_{w_s(r)}$. For

$$w_s(H_r(x)) = H_r(w_s(x)) = (r, w_s(x))$$
$$= (w_s(r), x) = H_{w_s(r)}(x).$$

Let $r_i \in \Pi$ be any fundamental root. Then

$$w_{r_i}\left(\prod_{r \in \Phi^+} H_r \right) = -\left(\prod_{r \in \Phi^+} H_r \right),$$

since w_{r_i} transforms r_i into $-r_i$ and permutes the remaining positive roots amongst themselves. Hence

$$w\left(\prod_{r \in \Phi^+} H_r \right) = \det w . \prod_{r \in \Phi^+} H_r$$

and so

$$\prod_{r \in \Phi^+} H_r$$

is alternating. Furthermore if $Q \in \mathfrak{I}$ we have

$$w\left(\prod_{r \in \Phi^+} H_r . Q \right) = \det w . \prod_{r \in \Phi^+} H_r . Q.$$

and so
$$\prod_{r \in \Phi^+} H_r \cdot Q$$
is also alternating.

Now let P be any alternating polynomial. Then $w_r(P) = -P$ for each $r \in \Phi$. Let x be an element in the hyperplane orthogonal to r. Then

$$-P(x) = w_r(P(x)) = P(w_r(x)) = P(x).$$

Hence $P(x) = 0$. Thus P vanishes at all points for which H_r vanishes, therefore H_r divides P. In fact

$$\prod_{r \in \Phi^+} H_r$$

divides P. Thus

$$P = \prod_{r \in \Phi^+} H_r \cdot Q$$

for some $Q \in \mathfrak{S}$. But now

$$\det w \cdot P = w(P) = w\left(\prod_{r \in \Phi^+} H_r \right) \cdot w(Q) = \det w \cdot \prod_{r \in \Phi^+} H_r \cdot w(Q)$$

Therefore $w(Q) = Q$ for all $w \in W$ and so $Q \in \mathfrak{I}$. Thus every alternating element is the product of

$$\prod_{r \in \Phi^+} H_r$$

with an invariant element.   ∎

Let $\mathfrak{S}_n$ be the set of homogeneous polynomial functions on $\mathfrak{P}$ of degree n, let $\mathfrak{I}_n = \mathfrak{I} \cap \mathfrak{S}_n$ and $\hat{\mathfrak{I}}_n = \hat{\mathfrak{I}} \cap \mathfrak{S}_n$. Let $\mathfrak{I}_J$ be the set of polynomial functions in $\mathfrak{S}$ invariant under W_J and $(\mathfrak{I}_J)_n = \mathfrak{I}_J \cap \mathfrak{S}_n . \mathfrak{I}_n$, $(\mathfrak{I}_J)_n$ and $\hat{\mathfrak{I}}_n$ are all finite dimensional vector spaces over $\mathbb{R}$. Also we have

$$\dim \hat{\mathfrak{I}}_n = \dim \mathfrak{I}_{n-N}$$

for all $n \geqslant N$ by 9.4.6, since

$$\prod_{r \in \Phi^+} H_r$$

is homogeneous of degree N. If $n < N$ then $\dim \hat{\mathfrak{I}}_n = 0$.

PROPOSITION 9.4.7.

$$\sum_J (-1)^{|J|} \dim (\mathfrak{I}_J)_n = \dim \hat{\mathfrak{I}}_n.$$

PROOF. Let $\chi(w) = \mathrm{tr}_{\mathfrak{H}_n} w$ be the trace of w on $\mathfrak{H}_n$. Let χ_{W_J} be the restriction of χ to W_J and $\chi^W_{W_J}$ be the induced character of W. Then $\chi^W_{W_J} = \chi \cdot 1^W_{W_J}$.

For

$$\chi^W_{W_J}(w) = \frac{1}{|W_J|} \sum_{\substack{x \in W \\ xwx^{-1} \in W_J}} \chi_{W_J}(xwx^{-1}) = \frac{1}{|W_J|} \sum_{\substack{x \in W \\ xwx^{-1} \in W_J}} \chi(w)$$

$$= \chi(w) \cdot \frac{1}{|W_J|} \sum_{\substack{x \in W \\ xwx^{-1} \in W_J}} 1 = \chi(w) \cdot 1^W_{W_J}(w).$$

Now by 9.4.2 $1^W_{W_J}(w) = n_J(w)$ and by 9.4.3 we have

$$\sum_J (-1)^{|J|} n_J(w) = \det w.$$

It follows that

$$\sum_J (-1)^{|J|} \chi^W_{W_J}(w) = \det w \cdot \chi(w).$$

We now average over W and obtain

$$\sum_J (-1)^{|J|} \chi^W_{W_J}\left(\frac{1}{|W|} \sum_{w \in W} w \right) = \frac{1}{|W|} \sum_{w \in W} \det w \cdot \chi(w).$$

The left-hand side can be simplified since

$$\chi^W_{W_J}\left(\frac{1}{|W|} \sum_{w \in W} w \right) = \chi_{W_J}\left(\frac{1}{|W_J|} \sum_{w \in W_J} w \right).$$

For

$$\chi^W_{W_J}\left(\frac{1}{|W|} \sum_{w \in W} w \right) = \frac{1}{|W_J|} \cdot \frac{1}{|W|} \sum_{\substack{x, w \\ xwx^{-1} \in W_J}} \chi_{W_J}(xwx^{-1})$$

$$= \frac{1}{|W_J|} \sum_{y \in W_J} \chi_{W_J}(y).$$

Thus we obtain

$$\sum_J (-1)^{|J|} \chi_{W_J}\left(\frac{1}{|W_J|} \sum_{w \in W_J} w \right) = \frac{1}{|W|} \sum_{w \in W} \det w \cdot \chi(w).$$

We now apply 9.3.2.

$$\chi_{W_J}\left(\frac{1}{|W_J|} \sum_{w \in W_J} w \right)$$

is the dimension of the subspace of invariants of $\mathfrak{S}_n$ under W_J, viz., dim $(\mathfrak{I}_J)_n$. Also

$$\frac{1}{|W|} \sum_{w \in W} \det w \cdot \chi(w)$$

is the dimension of the subspace of invariants of $\mathfrak{S}_n$ under the W-action

$$P \overset{w}{\to} \det w \cdot w(P)$$

However $\det w \cdot w(P) = P$ if and only if $w(P) = \det w \cdot P$; thus we require the dimension of the space of alternating elements of $\mathfrak{S}_n$, viz., dim $\hat{\mathfrak{I}}_n$. Hence

$$\sum_J (-1)^{|J|} \dim (\mathfrak{I}_J)_n = \dim \hat{\mathfrak{I}}_n. \qquad \blacksquare$$

We are now able to show that the polynomial $\bar{P}_W(t)$ satisfies the analogue of 9.4.5.

PROPOSITION 9.4.8.

$$\sum_J (-1)^{|J|} \frac{\bar{P}_W(t)}{\bar{P}_{W_J}(t)} = t^N.$$

PROOF. We prove this result by comparing the coefficients of t^n in the power series expansion of the rational functions

$$\sum_J \frac{(-1)^{|J|}}{(1-t)^l \bar{P}_{W_J}(t)} \quad \text{and} \quad \frac{t^N}{(1-t)^l \bar{P}_W(t)}.$$

Now

$$\frac{1}{(1-t)^l \bar{P}_W(t)} = \prod_{i=1}^{l} \frac{1}{(1-t^{d_i})}$$

and the coefficient of t^n in the expansion of this function is dim $\mathfrak{I}_n$ as in 9.3.3. Thus the coefficient of t^n in

$$\frac{t^N}{(1-t)^l \bar{P}_W(t)}$$

is dim $\mathfrak{I}_{n-N}$, and this is equal to dim $\hat{\mathfrak{I}}_n$ by 9.4.6.

We now consider the operation of W_J on $\mathfrak{P}$. We have

$$\bar{P}_{W_J}(t) = \prod_{i=1}^{|J|} \left(\frac{t^{d_i'} - 1}{t - 1} \right),$$

where $d'_1, \ldots, d'_{|J|}$ are the degrees of the basic invariants of W_J on $\mathfrak{P}_J$, the subspace of $\mathfrak{P}$ spanned by roots in J. Since W_J operates as the identity on the subspace of $\mathfrak{P}$ orthogonal to $\mathfrak{P}_J$, the degrees of the basic polynomial invariants of W_J on $\mathfrak{P}$ are $d'_1, \ldots, d'_{|J|}, 1, \ldots, 1$. Thus the coefficient of t^n in

$$\left(\prod_{i=1}^{|J|} \frac{1}{(1-t^{d'_i})} \right) \cdot \frac{1}{(1-t^{l-|J|})} = \frac{1}{(1-t)^l \bar{P}_{W_J}(t)}$$

is $\dim (\mathfrak{J}_J)_n$, also as in 9.3.3. However

$$\sum_J (-1)^{|J|} \dim (\mathfrak{J}_J)_n = \dim \hat{\mathfrak{J}}_n$$

by 9.4.7. It follows that the rational functions

$$\sum_J \frac{(-1)^{|J|}}{(1-t)^l \bar{P}_{W_J}(t)}, \qquad \frac{t^N}{(1-t)^l \bar{P}_W(t)}$$

have the same power series expansion, and so are equal.   ∎

We now complete the proof that $P_W(t) = \bar{P}_W(t)$.

THEOREM 9.4.9.

$$\sum_{w \in W} t^{l(w)} = \prod_{i=1}^l \left(\frac{t^{d_i} - 1}{t - 1} \right).$$

PROOF. We use induction on $l = \text{rank } W$. We may assume inductively that $P_{W_J}(t) = \bar{P}_{W_J}(t)$ for all proper subsets J of Π. By 9.4.5 and 9.4.8 we have

$$\frac{t^N - (-1)^{|\Pi|}}{P_W(t)} = \sum_{J \subset \Pi} \frac{(-1)^{|J|}}{P_{W_J}(t)},$$

$$\frac{t^N - (-1)^{|\Pi|}}{\bar{P}_W(t)} = \sum_{J \subset \Pi} \frac{(-1)^{|J|}}{\bar{P}_{W_J}(t)}.$$

Since the right-hand sides are equal by induction it follows that

$$P_W(t) = \bar{P}_W(t).$$ ∎

We have now been able to derive a multiplicative formula for the orders of the finite Chevalley groups.

THEOREM 9.4.10.   *Let $G = \mathfrak{L}(q)$ be a finite Chevalley group. Then*

$$|G| = \frac{1}{d} q^N (q^{d_1} - 1)(q^{d_2} - 1) \ldots (q^{d_l} - 1),$$

where d is defined as in section 8.6, N is the number of positive roots of $\mathfrak{L}$, and $d_1, \ldots, d_l$ are the degrees of the basic polynomial invariants of W.

PROOF. This follows from 8.6.1 and 9.4.9.   ■

CHAPTER 10

The Exponents of the Weyl Group

In the present chapter we shall show that the set of integers $d_1, \ldots, d_l$ occurring in the multiplicative formula for the order of the finite Chevalley groups can be obtained in three essentially different ways. They have been defined as the degrees of the basic polynomial invariants of the Weyl group, but we shall show that they can also be defined in terms of the eigenvalues of the Coxeter elements of the Weyl group and also in terms of the partition of the positive roots into roots of a given height. This latter definition gives a particularly simple way of calculating the integers in the individual cases.

The present chapter may be regarded as a digression from the main theme of this book, and the reader primarily interested in the properties of the Chevalley groups and twisted groups may prefer to omit it at a first reading, referring to 10.2.4 and 10.2.5 for a knowledge of the numbers $d_1, \ldots, d_l$ in the individual groups. The equivalent definitions of these integers are nevertheless of considerable interest, and it is probable that their significance is not yet fully understood.

10.1 A Theorem of Weyl

We prove first a well known factorization theorem of Weyl. In order to do this we derive some properties of the vector

$$S = \tfrac{1}{2} \sum_{r \in \Phi^+} r.$$

LEMMA 10.1.1. *If $p_i \in \Pi$ then $w_{p_i}(S) = S - p_i$.*

PROOF. This is evident from the fact that w_{p_i} transforms p_i into $-p_i$ and permutes the other positive roots. ∎

LEMMA 10.1.2. *S lies in the fundamental chamber.*

PROOF. It follows from 10.1.1 that

$$\frac{2(p_i, S)}{(p_i, p_i)} = 1$$

for all $p_i \in \Pi$. Thus $(p_i, S) > 0$ for all p_i.   ∎

LEMMA 10.1.3. *Let* $q_1, \ldots, q_l$ *be the fundamental weights. Then* $S = q_1 + \ldots + q_l$.

PROOF. S is certainly a linear combination

$$\sum_{i=1}^{l} \lambda_i q_i$$

of $q_1, \ldots, q_l$. However

$$\left(\frac{2p_i}{(p_i, p_i)}, q_j \right) = \begin{cases} 1 & \text{if } i = j, \\ 0 & \text{if } i \neq j, \end{cases}$$

as in section 7.1. Thus

$$\left(\frac{2p_i}{(p_i, p_i)}, S \right) = \lambda_i$$

and therefore $\lambda_i = 1$ for $i = 1, \ldots, l$.   ∎

LEMMA 10.1.4. *Let* $w \in W$. *Then*

$$w(S) = S - \sum_{r \in \Omega} r,$$

where Ω *is a subset of* Φ^+ *with* $|\Omega| = l(w)$.

PROOF. Let Ω be the set of roots in Φ^+ which are not in $w(\Phi^+)$. Then

$$S - w(S) = \tfrac{1}{2} \sum_{r \in \Omega} r - \tfrac{1}{2} \sum_{-w(r) \in \Omega} w(r) = \sum_{r \in \Omega} r.$$

Moreover, $|\Omega| = l(w^{-1}) = l(w)$, using 2.2.2.   ∎

LEMMA 10.1.5. *Let* Ω *be any subset of* Φ^+. *Then if*

$$S - \sum_{r \in \Omega} r$$

lies in the fundamental chamber, Ω *is empty.*

PROOF. Let

$$x = \sum_{r \in \Omega} r$$

and $p_i \in \Pi$. Then

$$\left(S - x, \frac{2p_i}{(p_i, p_i)} \right) = 1 - \frac{2(x, p_i)}{(p_i, p_i)} > 0.$$

Since $2(x, p_i)/(p_i, p_i)$ is an integer, this means that $2(x, p_i)/(p_i, p_i) \leqslant 0$. Therefore $(x, p_i) \leqslant 0$ for all $p_i \in \Pi$. However, x is a sum of positive roots and so

$$x = \sum_{i=1}^{l} \lambda_i p_i,$$

with each $\lambda_i \geqslant 0$. Hence we have

$$(x, x) = \sum_{i=1}^{l} \lambda_i (x, p_i) \leqslant 0.$$

It follows that $(x, x) = 0$, whence $x = 0$ and Ω is empty. ∎

LEMMA 10.1.6. *Let Ω be a subset of Φ^+. Then*

$$S - \sum_{r \in \Omega} r$$

is either in one of the reflecting hyperplanes or is a transform $w(S)$ of S by some element of the Weyl group.

PROOF. Suppose

$$S - \sum_{r \in \Omega} r$$

is not in any reflecting hyperplane. Then

$$S - \sum_{r \in \Omega} r$$

lies in some chamber. By 2.3.2 there exists $w \in W$ such that

$$w \left(S - \sum_{r \in \Omega} r \right)$$

is in the fundamental chamber. However

$$w \left(S - \sum_{r \in \Omega} r \right) = S - \sum_{r \in \Omega_1} r$$

for some other subset Ω_1 of Φ^+. (This follows from the definition of S and the fact that w permutes the roots.) Thus by 10.1.5 Ω_1 is empty and so

$$S - \sum_{r \in \Omega} r$$

is a transform of S.   ■

As usual we denote by Q the set of integral combinations

$$\sum_{i=1}^{l} n_i q_i$$

of the fundamental weights. Q is an additive abelian group. It will be more convenient in the following discussion to regard Q as a multiplicative group instead, and so we define $e(Q)$ to be a multiplicative group isomorphic to Q. The elements of $e(Q)$ have form

$$\prod_{i=1}^{l} e(q_i)^{n_i}$$

and the map e from Q to $e(Q)$ satisfies

$$e(a+b) = e(a).e(b), \qquad a, b \in Q,$$
$$e(-a) = e(a)^{-1}.$$

Let A be the rational group algebra of $e(Q)$. The elements of A are finite sums

$$\sum_x \lambda_x e(x),$$

where $x \in Q$ and $\lambda_x \in \mathbb{Q}$. The natural operation of W on Q can be transferred in the obvious way to an operation on $e(Q)$ and then extended by linearity to give an operation of W on A, thus making A into a W-module. We consider the alternating elements of A, viz., those which satisfy

$$w(a) = \det w . a$$

for all $w \in W$. We denote by θ the linear map of A into itself given by

$$\theta = \sum_{w \in W} \det w . w.$$

LEMMA 10.1.7. *The image of A under θ is the set of all alternating elements of A.*

PROOF.   Let $w \in W$ and $a \in A$. Then we have

$$w . \theta(a) = w . \sum_{w' \in W} \det w' . w'(a)$$

$$= \det w . \sum_{w' \in W} \det (ww') . ww'(a)$$

$$= \det w . \theta(a).$$

Thus $\theta(a)$ is alternating.

Now let a be any alternating element of A. Then

$$\theta(a) = \sum_{w \in W} \det w . w(a) = \sum_{w \in W} a = | W | a.$$

Thus

$$\theta\left(\frac{1}{| W |} a \right) = a$$

and so a lies in $\theta(A)$.   ∎

We now prove a theorem of Weyl giving a factorization of the expression $\theta(e(S))$.

THEOREM 10.1.8.

$$\theta(e(S)) = e(-S) . \prod_{r \in \Phi^+} (e(r) - 1).$$

Before proving this theorem we give an example to illustrate it. Suppose the root system Φ is of type A_2, with fundamental roots p_1, p_2. Then $S = \frac{1}{2}(p_1 + p_2 + (p_1 + p_2)) = p_1 + p_2$. Let $e(p_1) = X$ and $e(p_2) = Y$. Then

$$\theta(e(S)) = XY - X + Y^{-1} - X^{-1}Y^{-1} + X^{-1} - Y$$

and we have the factorization

$$XY - X + Y^{-1} - X^{-1}Y^{-1} + X^{-1} - Y = X^{-1}Y^{-1}(XY - 1)(X - 1)(Y - 1).$$

PROOF.   Let

$$a = e(-S) . \prod_{r \in \Phi^+} (e(r) - 1).$$

F

We show that a is an alternating element of A. For each $p_i \in \Pi$ we have

$$w_{p_i}(a) = e(-w_{p_i}(S)) . \prod_{r \in \Phi^+} (e(w_{p_i}(r)) - 1)$$

$$= e(p_i - S) . \prod_{r \in \Phi^+} (e(r) - 1) . \frac{e(-p_i) - 1}{e(p_i) - 1}$$

$$= e(-S) . \prod_{r \in \Phi^+} (e(r) - 1) . e(p_i) . \frac{e(-p_i) - 1}{e(p_i) - 1}$$

$$= -a.$$

It follows that $w(a) = \det w . a$ for all $w \in W$ and so a is alternating. We now multiply out the product in a. We have

$$a = e(S) . \prod_{r \in \Phi^+} (1 - e(-r))$$

$$= e(S) . \sum_{\Omega \subseteq \Phi} (-1)^{|\Omega|} e\left(-\sum_{r \in \Omega} r \right)$$

$$= \sum_{\Omega \subseteq \Phi^+} (-1)^{|\Omega|} e\left(S - \sum_{r \in \Omega} r \right).$$

Now a is alternating, and so $\theta(a) = |W| a$. Thus

$$a = \frac{1}{|W|} \sum_{\Omega \subseteq \Phi} (-1)^{|\Omega|} \theta\left(e\left(S - \sum_{r \in \Omega} r \right) \right).$$

Now if

$$S - \sum_{r \in \Omega} r$$

lies in some reflecting hyperplane H_s we have

$$\theta\left(e\left(S - \sum_{r \in \Omega} r \right) \right) = 0.$$

For w and ww_s give contributions to the sum

$$\sum_w \det w \, e\left(w\left(S - \sum_{r \in \Omega} r \right) \right)$$

which are equal and opposite. Thus we need only consider subsets Ω of Φ^+ such that

$$S - \sum_{r \in \Omega} r$$

does not lie in any reflecting hyperplane, and so has the form $w'(S)$ for some $w' \in W$ by 10.1.6. In such a case we have $|\Omega| = l(w')$ by 10.1.4.

Thus

$$a = \frac{1}{|W|} \sum_{w' \in W} (-1)^{l(w')} \, \theta(e(w'(S)))$$

$$= \frac{1}{|W|} \sum_{w' \in W} \det w' \cdot \theta w'(e(S))$$

since

$$(-1)^{l(w')} = \det w'.$$

Hence

$$a = \frac{1}{|W|} \sum_{w' \in W} \theta(e(S))$$

$$= \theta(e(S)),$$

since

$$\theta w' = \det w' \cdot \theta.$$

This completes the proof of Weyl's theorem.   ∎

10.2   A Theorem of Kostant

Now the elements of $e(Q)$ are uniquely expressible in the form

$$\prod_{i=1}^{l} e(q_i)^{n_i},$$

where $n_i \in \mathbb{Z}$. We may introduce a total ordering on $e(Q)$ by means of the first difference in the exponents $n_1, \dots, n_l$. Let x, y be two non-zero elements of A. Then

$$x = \lambda e(q_1)^{m_1} \dots e(q_l)^{m_l} + \text{lower terms in } e(Q),$$

$$y = \mu e(q_1)^{n_1} \dots e(q_l)^{n_l} + \text{lower terms in } e(Q),$$

where $\lambda \neq 0$, $\mu \neq 0$. Hence

$$xy = \lambda \mu \, e(q_1)^{m_1 + n_1} \dots e(q_l)^{m_l + n_l} + \text{lower terms in } e(Q)$$

and so $xy \neq 0$. This shows that A is an integral domain.

Let F be the field of fractions of A and $F[t]$ be the polynomial ring over F in the indeterminate t. The following remarkable identity in $F[t]$ is due to B. Kostant.

THEOREM 10.2.1.

$$\sum_{w \in W} \left(\prod_{r \in \Phi^+} \frac{1 - te(-w(r))}{1 - e(-w(r))} \right) = \sum_{w \in W} t^{l(w)}.$$

We again illustrate this theorem by an example before proving it. Suppose Φ has type A_1 and that r is the single positive root. Then Kostant's identity states that

$$\frac{1-te(-r)}{1-e(-r)}+\frac{1-te(r)}{1-e(r)}=1+t.$$

PROOF. We again consider the element

$$a=e(-S).\prod_{r\in\Phi^+}(e(r)-1).$$

Since a is an alternating element we have

$$w(a)=\det w.a=e(-w(S)).\prod_{r\in\Phi^+}(e(w(r))-1)$$
$$=e(w(S)).\prod_{r\in\Phi^+}(1-e(-w(r))).$$

Thus we have

$$\prod_{r\in\Phi^+}(1-e(-w(r)))=e(-w(S)).\det w.a.$$

It follows that

$$\sum_{w\in W}\left(\prod_{r\in\Phi^+}\frac{1-t\,e(-w(r))}{1-e(-w(r))}\right)$$
$$=\frac{1}{a}\sum_{w\in W}\left(\det w\,e(w(S)).\prod_{r\in\Phi^+}(1-t\,e(-w(r)))\right)$$
$$=\frac{1}{a}\sum_{w\in W}\det w\,e(w(S)).\sum_{\Omega\subset\Phi^+}(-t)^{|\Omega|}.e\left(w\left(-\sum_{r\in\Omega}r\right)\right)$$
$$=\frac{1}{a}\sum_{\Omega\subset\Phi^+}(-t)^{|\Omega|}\sum_{w\in W}\det w.e\left(w\left(S-\sum_{r\in\Omega}r\right)\right)$$
$$=\frac{1}{a}\sum_{\Omega\subset\Phi^+}(-t)^{|\Omega|}\,\theta\left(e\left(S-\sum_{r\in\Omega}r\right)\right).$$

As in the proof of 10.1.8 we have

$$\theta\left(e\left(S-\sum_{r\in\Omega}r\right)\right)=0$$

if

$$S-\sum_{r\in\Omega}r$$

lies in any reflecting hyperplane. Moreover if

$$S - \sum_{r \in \Omega} r$$

does not lie in any reflecting hyperplane, it has the form $w'(S)$, where w' is an element of W with $l(w') = |\Omega|$ (see 10.1.6 and 10.1.4). Using this information we have

$$\sum_{w \in W} \left(\prod_{r \in \Phi^+} \frac{1 - t\,e(-w(r))}{1 - e(-w(r))} \right) = \frac{1}{a} \sum_{w' \in W} (-t)^{l(w')}\, \theta(e(w'(S)))$$

$$= \frac{1}{a} \sum_{w' \in W} t^{l(w')} . \det w' . \theta w'(e(S))$$

$$= \frac{1}{a} \sum_{w' \in W} t^{l(w')} . \theta(e(S)),$$

since $(-1)^{l(w')} = \det w'$ and $\theta w' = \det w' . \theta$. But $a = \theta(e(S))$ by 10.1.8 and so the result follows. ∎

Now the polynomial

$$\sum_{w \in W} t^{l(w)}$$

has been considered in Kostant's identity as an element of $F[t]$, where F is the field of fractions of the rational group algebra of $e(Q)$, and a factorization of this polynomial has been obtained in $F[t]$. However, one can make use of this factorization in $F[t]$ to obtain a factorization of

$$\sum_{w \in W} t^{l(w)}$$

in the very much smaller domain $\mathbb{Q}[t]$.

THEOREM 10.2.2.

$$\sum_{w \in W} t^{l(w)} = \prod_{r \in \Phi^+} \frac{t^{h(r)+1} - 1}{t^{h(r)} - 1},$$

where $h(r)$ is the height of the root r.

PROOF. Let P be the additive group generated by the fundamental roots $p_1, \ldots, p_l$ and $e(P)$ be the corresponding multiplicative group. Let h be the homomorphism from P into $\mathbb{Z}$ taking value 1 at each fundamental root. h is called the height function. Then there is a homomorphism

$$e(x) \to t^{-h(x)}$$

from $e(P)$ into the infinite cyclic group generated by an element t.

Let B be the rational group algebra of $e(P)$. Since P is a subgroup of Q (see section 7.1), B is a subalgebra of A. The above homomorphism from $e(P)$ into $\langle t \rangle$ extends to an algebra homomorphism from B into $\mathbb{Q}[t, 1/t]$, the set of rational combinations of the powers of t. ($\mathbb{Q}[t, 1/t]$ is the rational group algebra of the infinite cyclic group $\langle t \rangle$.) This algebra homomorphism may itself be extended to an algebra homomorphism from the polynomial ring $B[t]$ into $\mathbb{Q}[t, 1/t]$ under which t is mapped into t. For since t is an indeterminate over B, its image may be chosen arbitrarily in $\mathbb{Q}[t, 1/t]$ and an algebra homomorphism is then uniquely defined. We denote this homomorphism by ψ. Then $\psi \colon B[t] \to \mathbb{Q}[t, 1/t]$ satisfies

$$\psi(e(r)) = t^{-h(r)}, \qquad r \in \Phi,$$

$$\psi(t) = t.$$

Now Kostant's identity

$$\sum_{w \in W} \left(\prod_{r \in \Phi^+} \frac{1 - t\, e(-w(r))}{1 - e(-w(r))} \right) = \sum_{w \in W} t^{l(w)}$$

may be interpreted as an identity in $B[t]$. For we may remove the denominators to obtain an identity in $A[t]$, and then observe that all the elements of Q appearing in this identity are in P. We now apply the algebra homomorphism ψ to both sides, and obtain an identity

$$\sum_{w \in W} \left(\prod_{r \in \Phi^+} \frac{1 - t^{1 + h(w(r))}}{1 - t^{h(w(r))}} \right) = \sum_{w \in W} t^{l(w)}.$$

However if $w \neq 1$ there is some $r \in \Phi^+$ such that $h(w(r)) = -1$. Thus the contributions to the left-hand side from all non-identity elements of W are zero. Hence

$$\prod_{r \in \Phi^+} \frac{1 - t^{1 + h(r)}}{1 - t^{h(r)}} = \sum_{w \in W} t^{l(w)}$$

and the theorem is proved. ∎

Now it is clear that in general a large number of terms in the product

$$\prod_{r \in \Phi^+} \frac{t^{h(r)+1} - 1}{t^{h(r)} - 1}$$

will cancel, and in order to see which terms remain after cancellation we consider the number of positive roots of a given height. Let k_i be the number of roots of height i. Then inspection of the root systems (see section 3.6) shows that

$$l = k_1 \geqslant k_2 \geqslant k_3 \geqslant \ldots.$$

Now $(k_1, k_2, \ldots)$ is a partition of N. The dual partition of N has l parts and will be denoted by $(m_1, m_2, \ldots, m_l)$.

THEOREM 10.2.3.

$$\sum_{w \in W} t^{l(w)} = \prod_{i=1}^{l} \left(\frac{t^{m_i+1} - 1}{t - 1} \right).$$

PROOF. All the terms in the expression

$$\prod_{r \in \Phi^+} \frac{t^{h(r)+1} - 1}{t^{h(r)} - 1}$$

cancel with the exception of terms $t^{m_1+1} - 1$, $t^{m_2+1} - 1, \ldots, t^{m_l+1} - 1$ in the numerator and $(t-1)^l$ in the denominator. ∎

COROLLARY 10.2.4. *The degrees of the basic polynomial invariants of W are given by $d_i = m_i + 1$, $i = 1, \ldots, l$.*

PROOF. This follows from 9.4.9 and 10.2.3. ∎

It is easy to determine the integers $m_1, \ldots, m_l$ in the individual cases by inspecting the root systems.

PROPOSITION 10.2.5. *The integers $m_1, \ldots, m_l$ are as follows:*

	$m_1, m_2, \ldots, m_l$
A_l	$1, 2, \ldots, l$
B_l	$1, 3, 5, \ldots, 2l-1$
C_l	$1, 3, 5, \ldots, 2l-1$
D_l	$1, 3, 5, \ldots, 2l-3, l-1$
G_2	$1, 5$
F_4	$1, 5, 7, 11$
E_6	$1, 4, 5, 7, 8, 11$
E_7	$1, 5, 7, 9, 11, 13, 17$
E_8	$1, 7, 11, 13, 17, 19, 23, 29$

PROOF. Calculate the numbers $k_1, k_2, \ldots$ from section 3.6 and form the dual partition. ∎

It will be observed that the integers $m_1, \ldots, m_l$ determined in 10.2.5 satisfy a condition of duality. If the m_i are arranged in increasing order we have

$$m_1 + m_l = m_2 + m_{l-1} = \ldots = 1 + h(R),$$

where R is the (unique) root of maximum height. We have not yet seen any reason why this duality should exist. However, this becomes clear by giving an entirely different definition of the m_i, based on properties of the 'Coxeter elements' of the Weyl group. The approach via Coxeter elements will also explain another property of the integers m_i observable from 10.2.5, the fact that they exhibit a definite tendency to be prime.

10.3 The Class of Coxeter Elements

A Coxeter element of the Weyl group W is an element of form

$$w_{r_1} w_{r_2} \ldots w_{r_l},$$

where $r_1, r_2, \ldots, r_l$ is a fundamental system in Φ.

THEOREM 10.3.1. *The Coxeter elements of W form a conjugacy class in W.*

PROOF. It is clear that any conjugate of a Coxeter element is a Coxeter element. For

$$w w_{r_1} \ldots w_{r_l} w^{-1} = w_{w(r_1)} \ldots w_{w(r_l)}$$

and if $r_1, \ldots, r_l$ is a fundamental system so is $w(r_1), \ldots, w(r_l)$.

In showing that any two Coxeter elements are conjugate it is sufficient to consider Coxeter elements corresponding to a fixed fundamental system, since any two fundamental systems are equivalent under the Weyl group (2.2.4). We take the fundamental system $\Pi = \{p_1, \ldots, p_l\}$ and consider the Coxeter elements corresponding to this. $w_{p_1} w_{p_2} \ldots w_{p_l}$ is one such element and the others are obtained by changing the order of the factors. The fact that they are all conjugate follows from the following lemma.

LEMMA 10.3.2. *Suppose the fundamental roots $p_1, \ldots, p_l$ are written round a circle, as shown in Figure 3. Then any permutation of $p_1, \ldots, p_l$ may be obtained by performing a succession of interchanges $(p_i p_j)$, where p_i, p_j are not joined in the Dynkin diagram and are adjacent on the circle, and then reading clockwise around the circle beginning from a suitable point.*

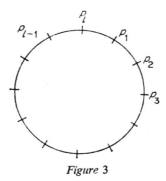

Figure 3

PROOF. We use induction on l. The result is clear if $l=1$ or $l=2$, since no interchanges have to be carried out at all. Thus we assume $l \geqslant 3$. Now the Dynkin diagram contains some node joined to at most one other node. Let p_l correspond to such a node. On omitting p_l from the circle we may by induction obtain any permutation of $p_1, \ldots, p_{l-1}$ by interchanging adjacent pairs not joined in the Dynkin diagram and then reading round from a suitable point. We show that all these interchanges can be carried out even when p_l is present. It is sufficient to show that p_i, p_j can be interchanged if p_i, p_j are not joined in the Dynkin diagram and if p_l is adjacent to both. (See Figure 4.)

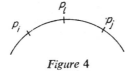

Figure 4

Now p_l is not joined in the Dynkin diagram to at least one of p_i, p_j— assume without loss of generality that p_l is not joined to p_i. Then by interchanging p_i, p_l and then p_l, p_j we have succeeded in interchanging the order of p_i, p_j when p_l is removed. Thus we may make the required arrangement of all the roots other than p_l. But then p_l, since it is joined to at most one other node in the Dynkin diagram, may be moved around the circle by a succession of steps in either a clockwise or an anticlockwise direction into its required position.   ■

Now given any arrangement of the roots $p_1, \ldots, p_l$ around the circle and any starting point on the circle, a Coxeter element $w_{r_1} \ldots w_{r_l}$ is determined. r_1 is the starting point, and $r_1, r_2, \ldots, r_l$ appear clockwise around the circle starting from r_1. If we take the same arrangement around the circle, but a different starting point, we obtain a conjugate Coxeter element. If we change the arrangement around the circle by interchanging adjacent roots not linked in the Dynkin diagram, the Coxeter element is unchanged since the corresponding reflections commute. (If we have interchanged the first and last terms, a conjugate Coxeter element is obtained.) Thus 10.3.2 shows that any Coxeter element of the form $w_{r_1} \ldots w_{r_l}$ can be obtained from $w_{p_1} \ldots w_{p_l}$ by a succession of operations which either leave the element unchanged or give a conjugate element. This completes the proof of 10.3.1.   ■

The conjugacy class containing the Coxeter elements of W is called the Coxeter class.

10.4 A Dihedral Subgroup of the Weyl Group

We shall obtain further information about the Coxeter elements by showing that each Coxeter element can be embedded in a dihedral subgroup of W which operates faithfully on a certain 2-dimensional subspace of $\mathfrak{P}$.

LEMMA 10.4.1.   *The set of fundamental roots $p_1, \ldots, p_l$ may be divided into two disjoint subsets, each of which contains roots which are all orthogonal to one another.*

PROOF.   Remove a node from the Dynkin diagram which is joined to at most one other node, and use induction on the number of nodes.   ■

We suppose that the fundamental system $\Pi = \{p_1, \ldots, p_l\}$ is decomposed as in 10.4.1 into two subsets

$$p_1, \ldots, p_k; \; p_{k+1}, \ldots, p_l.$$

Let $f_1, \ldots, f_l$ be vectors of unit length in the directions of $p_1, \ldots, p_l$ so that $p_i = |p_i| f_i$. Then $f_1, \ldots, f_l$ is a basis for $\mathfrak{P}$, although not an orthonormal basis. However, there is a uniquely determined dual basis $\hat{f}_1, \ldots, \hat{f}_l$ satisfying

$$(f_i, \hat{f}_j) = \begin{cases} 1 & \text{if } i = j, \\ 0 & \text{if } i \neq j. \end{cases}$$

Let $(f_i, f_j) = m_{ij}$ and M be the $l \times l$ matrix $M = (m_{ij})$. Then $(\hat{f}_i, \hat{f}_j) = (M^{-1})_{ij}$. Now M is a symmetric matrix whose diagonal coefficients are 1 and whose non-diagonal coefficients are non-positive (2.1.4). Thus $I - M$ is a symmetric matrix whose coefficients are all non-negative. This matrix plays a useful role in deriving the properties of the Coxeter elements.

Let λ be a non-zero eigenvalue of $I - M$ and

$$u = \begin{pmatrix} \xi_1 \\ \cdot \\ \cdot \\ \cdot \\ \xi_l \end{pmatrix}$$

be a corresponding eigenvector. Then

$$(I - M) u = \lambda u$$

and, since $I - M$ is symmetric, λ is real and $\xi_1, \ldots, \xi_l$ may be chosen real. We define two elements a, b of $\mathfrak{P}$ by

$$a = \sum_{i=1}^{k} \xi_i \hat{f}_i, \qquad b = \sum_{i=k+1}^{l} \xi_i \hat{f}_i.$$

LEMMA 10.4.2. (i) $(a, a) = (b, b)$.
(ii) *The angle θ between a, b is given by* $\cos \theta = \lambda$.

PROOF. M may be written in the form of a block matrix

$$M = \begin{pmatrix} I_k & A \\ A' & I_{l-k} \end{pmatrix},$$

where A is a certain $k \times (l-k)$ matrix and A' is the transpose of A. M^{-1}, which is also symmetric, may be written in the form

$$M^{-1} = \begin{pmatrix} B & C \\ C' & D \end{pmatrix}.$$

If we write

$$u = \begin{pmatrix} u^1 \\ u^2 \end{pmatrix}$$

the equation $(I - M)u = \lambda u$ gives

$$-Au^2 = \lambda u^1, \qquad -A'u^1 = \lambda u^2.$$

It follows that

$$-BAu^2 = \lambda Bu^1, \qquad -DA'u^1 = \lambda Du^2.$$

Now since $M^{-1}M = I_l$ we have

$$BA + C = 0, \qquad C' + DA' = 0.$$

Therefore

$$(*) \quad Cu^2 = \lambda Bu^1, \qquad C'u^1 = \lambda Du^2.$$

However

$$(a, a) = \sum_{i,j=1}^{k} \xi_i \xi_j (M^{-1})_{ij} = (u^1)' Bu^1,$$

$$(b, b) = \sum_{i,j=k+1}^{l} \xi_i \xi_j (M^{-1})_{ij} = (u^2)' Du^2.$$

The equations (*) now show

$$(u^1)' Bu^1 = \lambda^{-1} . (u^1)' Cu^2 = \lambda^{-1} . (u^2)' C'u^1 = (u^2)' Du^2.$$

Therefore

$$(a, a) = (b, b).$$

The angle between a, b is given by

$$|a| . |b| . \cos \theta = (a, b) = \sum_{i=1}^{k} \sum_{j=k+1}^{l} \xi_i \xi_j (M^{-1})_{ij}$$

$$= (u^1)' Cu^2 = \lambda (u^1)' Bu^1 = \lambda . (a, a).$$

Since $(a, a) = (b, b)$ we have $\cos \theta = \lambda$. ∎

We now define elements w_1, w_2 of W by

$$w_1 = w_{p_1} \ldots w_{p_k}, \qquad w_2 = w_{p_{k+1}} \ldots w_{p_l}.$$

Then w_1, w_2 are involutions, being products of reflections with respect to mutually orthogonal roots. Their product $w = w_1 w_2$ is a Coxeter element. We consider the group $\langle w_1, w_2 \rangle$ generated by w_1, w_2. This is a dihedral subgroup of W, since it is generated by two involutions. This dihedral group operates in a particularly simple way on the 2-dimensional subspace of $\mathcal{V}$ containing the points a and b.

PROPOSITION 10.4.3. *Let λ be a non-zero eigenvalue of $I - M$ and u a corresponding eigenvector. Let a, b be the vectors defined above and Γ be the circle with centre the origin passing through a and b (see Figure 5). Then the group $\langle w_1, w_2 \rangle$ operates on the points of Γ. In particular w_1 is the reflection in the line Ob, w_2 is the reflection in the line Oa and w is a rotation around Γ through an angle 2θ, where $\cos \theta = \lambda$.*

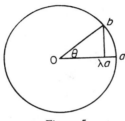

Figure 5

PROOF. It was shown in 10.4.2 that $\lambda Bu^1 - Cu^2 = 0$ and this implies that $(\hat{f}_i, \lambda a - b) = 0$ for $i = 1, \ldots, k$. Thus $\lambda a - b$ is a linear combination of $f_{k+1}, \ldots, f_l$, so also a linear combination of $p_{k+1}, \ldots, p_l$. Hence $w_2(\lambda a - b) = -(\lambda a - b)$. However, $w_2(a) = a$ and, since $\cos \theta = \lambda$, λa is the projection of b on Oa. Thus w_2 leaves Γ invariant and operates on it as the reflection in the line Oa.

Similarly it was shown that $C'u^1 - \lambda Du^2 = 0$ and this implies that $(\hat{f}_i, a - \lambda b) = 0$ for $i = k+1, \ldots, l$. Thus $a - \lambda b$ is a linear combination of $p_1, \ldots, p_k$ and so $w_1(a - \lambda b) = -(a - \lambda b)$. Also $w_1(b) = b$ and λb is the projection of a on Ob. Thus w_1 leaves Γ invariant and operates on it as the reflection in the line Ob.

Finally, since Oa and Ob are inclined at an angle θ, $w = w_1 w_2$ is a rotation through 2θ. The whole dihedral group $\langle w_1, w_2 \rangle$ therefore operates on Γ in an obvious way.   ∎

10.5 Eigenvalues of the Coxeter Elements

Before applying the results of section 10.4 to give further information about the Coxeter elements, we state and prove a classical result on real symmetric matrices which we shall need, known as the Frobenius–Perron theorem.

A real symmetric matrix $M = (m_{ij})$ is called positive semi-definite if $xMx' \geqslant 0$ for all $x \in \mathbb{R}^l$, and indecomposable if it is impossible to split up the set $1, 2, \ldots, l$ into two non-empty complementary subsets I, J such that $m_{ij} = 0$ whenever $i \in I$, $j \in J$.

PROPOSITION 10.5.1. *Let M be a real symmetric matrix such that $m_{ij} \leqslant 0$ for all $i \neq j$, and suppose that M is positive semi-definite and indecomposable. Then the eigenvalues of M are all real and non-negative. The smallest eigenvalue has multiplicity 1 and has an eigenvector whose coefficients are all positive.*

PROOF. The eigenvalues of M are real since M is symmetric, and non-negative since M is positive semi-definite. There is an orthogonal matrix T such that TMT' is a diagonal matrix whose coefficients are the eigenvalues of M, by a well-known theorem of linear algebra.

Let $\mathfrak{V}_0$ be the null-space of M, i.e. the set of $x \in \mathbb{R}^l$ such that $xM = 0$. $\mathfrak{V}_0$ is also the set of $x \in \mathbb{R}^l$ such that $xMx' = 0$, as is easily seen by considering the diagonal matrix TMT'. We shall show that dim $\mathfrak{V}_0 \leqslant 1$.

Suppose dim $\mathfrak{V}_0 > 0$ and let $x = (\alpha_1, \alpha_2, \ldots, \alpha_l)$ be a non-zero vector in $\mathfrak{V}_0$. Let $y = (\,|\,\alpha_1\,|, \,|\,\alpha_2\,|, \ldots, |\,\alpha_l\,|\,)$. Then we have

$$0 \leqslant yMy' \leqslant xMx' = 0,$$

since $m_{ij} \leqslant 0$ if $i \neq j$. It follows that $yMy' = 0$ and so $y \in \mathfrak{V}_0$. Thus $\mathfrak{V}_0$ contains a non-zero vector whose coordinates are all non-negative, and we have

$$\sum_{i=1}^{l} |\,\alpha_i\,|\, m_{ij} = 0.$$

Let I be the set of i with $\alpha_i \neq 0$ and J be the set of i with $\alpha_i = 0$. Suppose $j \in J$. Then all the terms $|\,\alpha_i\,|\, m_{ij}$ in the above sum are non-positive, and so $|\,\alpha_i\,|\, m_{ij} = 0$ for all i. If $i \in I$ we have $|\,\alpha_i\,| \neq 0$ and hence $m_{ij} = 0$. Thus $m_{ij} = 0$ for all $i \in I$, $j \in J$. However, M is indecomposable and I is non-empty, therefore J must be empty. Thus each coefficient of x is non-zero. Since this holds for each non-zero vector in $\mathfrak{V}_0$ we must have dim $\mathfrak{V}_0 \leqslant 1$. Furthermore if dim $\mathfrak{V}_0 = 1$, then $\mathfrak{V}_0$ contains a vector whose coefficients are all positive.

Let μ be the smallest eigenvalue of M. Then $M - \mu I$ satisfies the hypotheses of the proposition, as can be seen by considering again the diagonal matrix TMT'. $M - \mu I$ is singular, so its null-space has dimension 1 and contains a vector whose coefficients are all positive. Thus μ occurs as eigenvalue of M with multiplicity 1, and has an eigenvector whose coefficients are all positive.　∎

We apply the Frobenius–Perron theorem to the situation discussed in 10.4.3. Let $M = (m_{ij})$ be the matrix defined by $m_{ij} = (f_i, f_j)$. Then M is positive definite, since the Killing form is positive definite (section 3.3). Moreover $m_{ij} \leqslant 0$ when $i \neq j$ by 2.1.4.

COROLLARY 10.5.2. *If the vectors a, b are chosen as in section* 10.4 *with respect to the largest eigenvalue of $I - M$, then each point of the circle Γ lying strictly between a and b is in the fundamental chamber.*

PROOF. The eigenvector $(\xi_1, \ldots, \xi_l)$ corresponding to the smallest eigenvalue of M corresponds to the largest eigenvalue of $I - M$. Thus we may assume each $\xi_i > 0$.

Now we have

$$a = \sum_{i=1}^{k} \xi_i \hat{f}_i, \qquad b = \sum_{i=k+1}^{l} \xi_i \hat{f}_i;$$

thus every point c on Γ strictly between a and b has the form

$$c = \sum_{i=1}^{l} \lambda_i \hat{f}_i,$$

where $\lambda_i > 0$ for $i = 1, \ldots, l$. This means that $(c, f_i) > 0$ and so $(c, p_i) > 0$ for each i. Thus c lies in the fundamental chamber.　∎

We are now in a position to acquire further information about the Coxeter elements.

THEOREM 10.5.3. *The order of the Coxeter elements is $2N/l$.*

PROOF. Let w be a Coxeter element and h be the order of w. Let Γ be the circle defined in section 10.4 with respect to the largest eigenvalue of $I - M$. Then w operates as a rotation on Γ. Now the cyclic group $\langle w \rangle$ generated by w operates faithfully on Γ. For if an element of W operates trivially on Γ it fixes some point in the fundamental chamber by 10.5.2, so is the identity by 2.3.2. Thus w has order h on Γ.

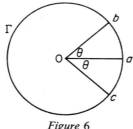

Let c be the point on Γ which is the reflection of b in Oa (see Figure 6). Then the angle between Ob and Oc is 2θ. Since w is a rotation through 2θ, there exists an integer $i > 0$ such that $w^i(a)$ lies on the arc bc. Let i be the least such positive integer. If $w^i(a)$ lies between a and b, both a and $w^i(a)$ lie in $\bar{C}$, the closure of the fundamental chamber (10.5.2). Thus $w^i(a) = a$ by 2.3.5. On the other hand, if $w^i(a)$ lies between a and c it is evident that $w^i(b)$ lies between a and b. Thus both b and $w^i(b)$ lie in $\bar{C}$, and so $w^i(b) = b$. In either case we have $w^i = 1$ since $\langle w \rangle$ operates faithfully on Γ. Hence $i = h$ and $\theta = \Pi/h$.

Now

$$a = \sum_{i=1}^{k} \xi_i f_i$$

and so a is orthogonal to the roots $p_{k+1}, \ldots, p_l$ and their negatives. We show that these are the only roots orthogonal to a. Suppose

$$\sum_{i=1}^{l} \eta_i f_i$$

is a root orthogonal to a. Then

$$\sum_{i=1}^{k} \xi_i \eta_i = 0.$$

Since each $\xi_i > 0$ and all η_i have the same sign, this implies that

$$\eta_1 = \eta_2 = \ldots = \eta_k = 0.$$

Thus the root

$$\sum_{i} \eta_i f_i$$

is a linear combination of $p_{k+1}, \ldots, p_l$. By 2.5.1 the roots which are linear combinations of $p_{k+1}, \ldots, p_l$ form a system in which $p_{k+1}, \ldots, p_l$

is a fundamental system. Since $p_{k+1}, \ldots, p_l$ are mutually orthogonal, the only such roots are $\pm p_l$. Thus a lies in exactly $l-k$ reflecting hyperplanes.

Since a lies in $l-k$ reflecting hyperplanes the same is true of all transforms $w^i(a)$, $i=1, 2, \ldots$. Similarly b and all its transforms $w^i(b)$ lie in exactly k reflecting hyperplanes. However, each reflecting hyperplane intersects the plane of Γ in a line meeting Γ at one of the transforms $w^i(a)$ or $w^i(b)$. For otherwise there would be a reflecting hyperplane intersecting Γ at a point strictly between a and b, contrary to the fact that such points lie in the fundamental chamber. Now there are h transforms of a and h transforms of b alternating round Γ. Thus the total number of reflecting hyperplanes is

$$\tfrac{1}{2}[(l-k)\,h+kh]=\tfrac{1}{2}lh$$

since each hyperplane meets Γ in two points. Thus $N=\tfrac{1}{2}lh$, and so $h=2N/l$, as required.   ∎

COROLLARY 10.5.4.   *A Coxeter element w has an eigenvalue $\mathrm{e}^{2\pi i/h}$.*

PROOF.   As w operates on Γ as a rotation through $2\Pi/h$, w has an eigenvalue $\mathrm{e}^{2\pi i/h}$ in the plane of Γ.

COROLLARY 10.5.5.   *A Coxeter element w has an eigenvector v with eigenvalue $\mathrm{e}^{2\pi i/h}$ such that v is not orthogonal to any root.*

PROOF.   Let v be an eigenvector with eigenvalue $\mathrm{e}^{2\pi i/h}$ lying in the plane of Γ. Since w operates as a rotation in this plane v is not real, so $v \neq \bar{v}$. Suppose $(v, r)=0$, where $r \in \Phi$. Then $(\bar{v}, r)=0$ since r is real, and it follows that r is orthogonal to every vector in the plane of Γ. Since this plane contains points in the fundamental chamber we have a contradiction. Thus v is not orthogonal to any root.   ∎

PROPOSITION 10.5.6.   *A Coxeter element has no eigenvalue 1.*

PROOF.   This is much easier to prove than the preceding results. Suppose a Coxeter element w fixes a vector v. Then

$$w_{p_1} w_{p_2} \ldots w_{p_l}(v) = v$$

and so

$$w_{p_2} \ldots w_{p_l}(v) = w_{p_1}(v).$$

Now $w_{p_2} \ldots w_{p_l}(v) - v$ is a linear combination of $p_2, \ldots, p_l$ and $w_{p_1}(v) - v$ is a scalar multiple of p_1. Since $p_1, p_2, \ldots, p_l$ are linearly independent we have

$$w_{p_2} \ldots w_{p_l}(v) = w_{p_1}(v) = v.$$

Thus v is orthogonal to p_1 and $w_{p_2} \ldots w_{p_l}$ fixes v. Repeating the argument we see that v is orthogonal to $p_1, p_2, \ldots, p_l$, hence $v = 0$. Thus w fixes no non-zero vector and so has no eigenvalue 1.   ∎

10.6 A Theorem of Coleman

We now prove a theorem of Coleman relating the eigenvalues of the Coxeter elements to the degrees of the basic invariants.

THEOREM 10.6.1. *Let the eigenvalues of a Coxeter element be*

$$\zeta^{m_1}, \zeta^{m_2}, \ldots, \zeta^{m_l},$$

where $\zeta = e^{2\pi i/h}$ *and* $m_1, m_2, \ldots, m_l$ *are positive integers less than* h. *Then the degrees of the basic invariants* $I_1, I_2, \ldots, I_l$ *of* W *are*

$$m_1 + 1, m_2 + 1, \ldots m_l + 1.$$

PROOF. The eigenvalues of a Coxeter element w are hth roots of unity, where h is the order of w, so are powers of ζ. By 10.5.6 ζ^0 is not an eigenvalue, so the eigenvalues are of form ζ^{m_i}, where $0 < m_i < h$. Let $f_1, \ldots, f_l$ be a basis for the complexification $\mathfrak{P}_{\mathbb{C}}$ of $\mathfrak{P}$ such that

$$w(f_i) = \zeta^{m_i} f_i.$$

Let

$$x_1 e_1 + \ldots + x_l e_l = y_1 f_1 + \ldots + y_l f_l,$$

where $e_1, \ldots, e_l$ is an orthonormal basis of $\mathfrak{P}$. Then $y_1, \ldots, y_l$ are linear functions in $x_1, \ldots, x_l$ and vice versa. Now we may assume that $m_1 = 1$ by 10.5.4 and that f_1 is not orthogonal to any root by 10.5.5.

Let

$$J = \left| \frac{\partial(I_1, \ldots, I_l)}{\partial(y_1, \ldots, y_l)} \right|$$

be the Jacobian determinant of a set of basic polynomial invariants of W expressed in terms of $y_1, \ldots, y_l$. By 9.3.5 J factorizes into a product of linear factors representing the reflecting hyperplanes. Since f_1 is not in any reflecting hyperplane, $J \neq 0$ at $(y_1, 0, \ldots, 0)$. Thus we may choose the numbering of the invariants $I_1, \ldots, I_l$ so that $\partial I_i / \partial y_i \neq 0$ at

$$(y_1, 0, \ldots, 0).$$

Thus we have

$$\frac{\partial I_i}{\partial y_i} = \lambda_i y_1{}^{d_i-1} + \text{terms involving } y_i \text{ with } i > 1.$$

Hence

$$I_i = \lambda_i y_1{}^{d_i-1} y_i + \text{terms involving different monomials},$$

where $\lambda_i \neq 0$.

We now apply w and use the fact that I_i is an invariant. Let

$$y = y_1 f_1 + \ldots + y_l f_l.$$

Then

$$w^{-1}(y) = y_1 \zeta^{-m_1} f_1 + \ldots + y_l \zeta^{-m_l} f_l.$$

Since for any polynomial $P(y)$ we have $w(P(y)) = P(w^{-1}(y))$ it follows that

$$w(y_i) = y_i \cdot \zeta^{-m_i}.$$

Hence

$$w(I_i) = \lambda_i \zeta^{1-d_i-m_i} y_1{}^{d_i-1} y_i + \text{terms involving different monomials}.$$

But $w(I_i) = I_i$ and, since $\lambda_i \neq 0$, we obtain

$$\zeta^{1-d_i-m_i} = 1.$$

Now ζ^{h-m_i} occurs as an eigenvalue of w whenever ζ^{m_i} does, since w is a real transformation. We now renumber the basis vectors $f_1, \ldots, f_l$ so that the eigenvalue ζ^{m_i} is replaced by ζ^{h-m_i}. With this new numbering we have

$$\zeta^{1-d_i+m_i-h} = 1$$

and therefore

$$\zeta^{d_i-1} = \zeta^{m_i}.$$

It follows that $d_i - 1 \equiv m_i \bmod h$.

Now the numbers $h - m_i$, $i = 1, \ldots, l$, are a permutation of the numbers m_i, $i = 1, \ldots, l$. (We have again used 10.5.6 here.) Thus

$$\sum_{i=1}^{l} (h - m_i) = \sum_{i=1}^{l} m_i,$$

which gives

$$\sum_{i=1}^{l} m_i = \tfrac{1}{2} lh.$$

But $\tfrac{1}{2} lh = N$ by 10.5.3, and so

$$\sum_{i=1}^{l} m_i = N.$$

Also we have

$$\sum_{i=1}^{l} (d_i - 1) = N$$

by 9.3.4. Since $d_i - 1 \equiv m_i$ mod h and $0 < m_i < h$, these equations imply that $d_i - 1 = m_i$ for $i = 1, \ldots, l$. ∎

Coleman's theorem shows that one can determine the degrees of the basic invariants of W merely by looking at the operation of a Coxeter element on $\mathfrak{P}$. This is one of several indications that the class of Coxeter elements is of particular significance in the Weyl group. It appears to be the conjugacy class which is 'as far removed as possible' from the unit class.

COROLLARY 10.6.2. *Let k_i be the number of positive roots of height i and let $(m_1, m_2, \ldots, m_l)$ be the partition of N dual to the partition $(k_1, k_2, \ldots)$. Then the eigenvalues of a Coxeter element are*

$$\zeta^{m_1}, \zeta^{m_2}, \ldots, \zeta^{m_l},$$

where $\zeta = e^{2\pi i / h}$.

PROOF. This follows from 10.2.4 and 10.6.1. ∎

The duality

$$m_1 + m_l = m_2 + m_{l-1} = \ldots$$

now follows immediately from the fact that w is a real transformation. Since the common value of these sums is $1 + h(R)$, where R is the highest root, we also obtain the following result.

COROLLARY 10.6.3. *The following three integers are equal:*

(i) *The order h of the Coxeter elements,*
(ii) $2N/l$,
(iii) $1 + h(R)$, *where R is the highest root.*

Note. The equivalence of (i) and (iii) can be shown directly by an argument of Steinberg [3].

The integers $m_1, \ldots, m_l$ are called the exponents of the Weyl group. We have now obtained three equivalent definitions for these exponents. They can also be defined in terms of the topology of the corresponding Lie group, i.e. in terms of the Betti numbers of the compact Lie group with Weyl group W. The Betti numbers are the coefficients of the Poincaré polynomial of this Lie group, which factorizes as

$$\prod_{i=1}^{l} (1 + t^{2m_i+1})$$

(cf. Bott [1]).

CHAPTER 11

Further Properties of the Chevalley Groups

We now return to prove some further properties of the Chevalley groups. We shall show that these groups are all simple, apart from a few exceptions when the underlying field is very small. We shall also prove that the Chevalley groups of type A_l, B_l, C_l, D_l are isomorphic to certain classical groups.

11.1 The Simplicity of the Chevalley Groups

We first prove a criterion for simplicity valid for any group with a (B, N)-pair. We recall from section 8.2 that in such a group G the quotient group $W = N/B \cap N$ is generated by a set of elements w_i, $i \in I$, such that $w_i^2 = 1$. Let G' denote the commutator subgroup of G.

THEOREM 11.1.1. *Let G be a group with a (B, N)-pair satisfying the following conditions:*

(a) $G = G'$,

(b) B is soluble,

(c) $\bigcap\limits_{g \in G} gBg^{-1} = 1$,

(d) *the set I cannot be decomposed into two non-empty complementary subsets J, K such that w_j commutes with w_k for all $j \in J$, $k \in K$.*

Then G is simple.

PROOF. Let G_1 be a normal subgroup of G. Then $G_1 B$ is a subgroup of G containing B, thus $G_1 B = P_J$ for some subset J of I, by 8.3.2. Let K be the subset of I complementary to J. Let $j \in J$ and $k \in K$, and let n_j, n_k be elements of N corresponding to w_j, $w_k \in W$ under the natural homomorphism. If we define the length function $l(w)$ as in the proof of 8.2.3 we have $l(w_j w_k) > l(w_k)$. By 8.2.4 we see that

$$Bn_j B . Bn_k B \subseteq Bn_j n_k B.$$

Now we have
$$G_1 B = P_J = B N_J B$$
and therefore
$$B n_j B \cap G_1 \neq \phi.$$

Since G_1 is a normal subgroup of G this implies
$$n_k B n_j B n_k \cap G_1 \neq \phi.$$

However
$$n_k B n_j B n_k \subseteq n_k B n_j n_k B \subseteq B n_j n_k B \cap B n_k n_j n_k B$$

by axiom $BN\,4$. Thus we have
$$B n_j n_k B \cap G_1 \neq \phi \quad \text{or} \quad B n_k n_j n_k B \cap G_1 \neq \phi.$$

The former condition implies that $n_j n_k \in N_J$, whence $n_k \in N_J$ and $w_k \in W_J$. This contradicts the fact, shown in the proof of 8.3.4, that the elements w_i form a minimal set of generators for W. Hence we have
$$B n_k n_j n_k B \cap G_1 \neq \phi.$$

This means that $n_k n_j n_k \in N_J$ and $w_k w_j w_k \in W_J$. Thus
$$w_k w_j w_k \in W_J \cap W_{\{j,\,k\}} = W_{\{j\}}$$

by 8.3.4. It follows that $w_k w_j w_k$ is either 1 or w_j. $w_k w_j w_k = 1$ implies that $w_j = 1$, which is impossible; thus we have $w_k w_j w_k = w_j$. Hence for each $j \in J$, $k \in K$ we have $w_j w_k = w_k w_j$. Since I does not decompose into two non-empty complementary subsets with this property, either J or K must be empty.

Suppose K is empty. Then $J = I$ and $G_1 B = G$. Thus we have
$$G/G_1 = G_1 B / G_1 \cong B / G_1 \cap B.$$

Now B is soluble and so G/G_1 is soluble also. However, the fact that G coincides with its commutator subgroup G' means that G has no nontrivial soluble factor group. Hence $G_1 = G$.

Suppose J is empty. Then $G_1 B = B$ and so G_1 is contained in B. Since G_1 is normal in G we have
$$G_1 \subseteq \bigcap_{g \in G} g B g^{-1} = 1.$$

Therefore $G_1 = 1$.

Thus G_1 is either G or 1, and so the group G is simple. ∎

Theorem 11.1.1 shows that groups with a (B, N)-pair have a definite tendency to be simple. Now the Chevalley groups $G = \mathfrak{L}(K)$ all have

(B, N)-pairs by 8.2.1. Thus we may use 11.1.1 to try to prove the simplicity of the Chevalley groups. The proof goes through in all but a few cases.

THEOREM 11.1.2. (i) *Let $\mathfrak{L}$ be a simple Lie algebra over $\mathbb{C}$ and K be an arbitrary field. Then the Chevalley group $G = \mathfrak{L}(K)$ is simple, except for $A_1(2)$, $A_1(3)$, $B_2(2)$, $G_2(2)$.*
(ii) *Each Chevalley group (even a non-simple one) has trivial centre.*

PROOF. By 11.1.1 it will be sufficient to prove that

$$G = G' \text{ and } \bigcap_{g \in G} gBg^{-1} = 1.$$

For the Borel subgroup B is certainly soluble, being the semi-direct product of the nilpotent group U with the abelian group H. Moreover, the set w_i of distinguished generators of W are the fundamental reflections, and these cannot be decomposed into two non-empty complementary commuting subsets since $\mathfrak{L}$ is simple.

We show first that G has no non-trivial normal subgroup contained in B. Let G_1 be such a normal subgroup. Now by 2.2.6 the Weyl group W contains an element w_0 which transforms every positive root into a negative root. Let n_0 be a corresponding element of N.
Then we have

$$n_0 U n_0^{-1} = V$$

by 7.2.1, and also

$$n_0 U H n_0^{-1} = VH.$$

Now G_1 is contained in UH, so also in VH since it is normal. Thus

$$G_1 \subseteq UH \cap VH = H$$

by 7.1.3. However, H normalizes U and so we have

$$[U, H] \subseteq U$$

and

$$[U, G_1] \subseteq U \cap G_1 \subseteq U \cap H = 1.$$

Thus every element of G_1 commutes with every element of U. Let $h(\chi) \in G_1$. Then

$$h(\chi) \, x_r(1) \, h(\chi)^{-1} = x_r(\chi(r)) = x_r(1)$$

for all $r \in \Phi^+$. It follows that $\chi(r) = 1$ for all $r \in \Phi^+$. Hence $\chi = 1$ and $h(\chi) = 1$. Thus $G_1 = 1$. We have therefore shown that

$$\bigcap_{g \in G} gBg^{-1} = 1.$$

We show next that G has trivial centre. Let Z be the centre of G. The argument used in the proof of 11.1.1 shows that either

$$G=ZB \text{ or } Z \subseteq \bigcap_{g \in G} gBg^{-1}.$$

If $G=ZB$ then B is normal in G. Since by 8.3.3 B is its own normalizer in G we have $ZB=B$. Therefore $Z \subseteq B$, and since Z is normal we have

$$Z \subseteq \bigcap_{g \in G} gBg^{-1}.$$

Hence $Z=1$.

In order to show that G is simple it remains to prove that $G=G'$. We shall require the following lemma.

LEMMA 11.1.3.   (i) *Let* $r \in \Phi$ *and* t *be a non-zero element of* K. *Let* Q *be the additive group of weights. Then there is a* K-*character* χ *of* Q *such that* $\chi(r)=t^2$.

(ii) *There is a* K-*character* χ *of* Q *such that* $\chi(r)=t$, *unless* $\mathfrak{L}=A_1$ *or* $\mathfrak{L}=C_l$ *and* r *is a long root.*

PROOF.   (i) Let $\chi_{r,\,t}$ be the K-character of Q defined, as in 7.1.1, by

$$\chi_{r,\,t}(a)=t^{2(r,\,a)/(r,\,r)}.$$

Then $\chi_{r,\,t}(r)=t^2$.

(ii) Let P be the additive group generated by the roots. We recall from section 7.1 that P, Q are free abelian groups of rank l and that P is a subgroup of Q of finite index. Let m be the greatest integer such that $(1/m)r \in Q$. (m is the highest common factor of the coefficients of r when expressed as an integral combination of $q_1, \ldots, q_l$.) Let

$$r_1 = \frac{1}{m}\,r.$$

Then

$$\frac{1}{n}\,r_1 \notin Q$$

for any $n>1$, and therefore r_1 forms part of some basis for Q. Hence a K-character χ of Q can be chosen so that $\chi(r_1)$ is an arbitrary non-zero element of K. Let

$$h_r = \frac{2r}{(r,\,r)}$$

be the co-root corresponding to r. Then $(h_r, r_1) \in \mathbb{Z}$ since $r_1 \in Q$, by 7.1. Thus

$$2 \frac{(r, r_1)}{(r, r)} = \frac{2}{m}$$

is an integer, hence $m = 1$ or 2.

If $m = 1$ we can certainly choose χ so that $\chi(r) = t$, and this is so unless $\frac{1}{2} r \in Q$. We first consider what happens when r is a fundamental root. Let $r = p_i$. Then

$$r = \sum_{j=1}^{l} A_{ji} q_j$$

by 7.1, and $\frac{1}{2} r \in Q$ only if each A_{ji} for $j = 1, \ldots, l$ is divisible by 2. A glance at the Cartan matrices of the simple Lie algebras (listed in section 3.6) shows that this can happen only if $\mathfrak{L} = A_1$ or if $\mathfrak{L} = C_l$, $(l \geq 2)$, and r is the long fundamental root. If r is not a fundamental root it can be transformed into one by an element of W and the property $\frac{1}{2} r \in Q$ is preserved. Thus $\frac{1}{2} r \in Q$ holds only for the roots of A_1 and the long roots of C_l.   ∎

We may now complete the proof of 11.1.2 and to do so we must show that $G = G'$.

Suppose first that K has at least four elements. Then there is a non-zero element $t \in K$ such that $t^2 \neq 1$. By 11.1.3 there exists for each $r \in \Phi$ a K-character χ of Q such that $\chi(r) \neq 1$. Then $h(\chi) \in H$ by 7.1.1 and we have

$$h(\chi) x_r(t) h(\chi)^{-1} = x_r(\chi(r) t).$$

It follows that $x_r((\chi(r) - 1) t) \in G'$ for all $t \in K$. Let $u \in K$ and choose

$$t = \frac{u}{\chi(r) - 1}.$$

Then $x_r(u) \in G'$. Since the elements $x_r(u)$ for all $r \in \Phi$, $u \in K$ generate G, we have $G' = G$.

Next suppose that $K = GF(3)$. If $\mathfrak{L}$ is not of type A_1 or C_l, 11.1.3 shows that there exists for each $r \in \Phi$ a K-character χ of Q with $\chi(r) \neq 1$. Then $G' = G$ as above. Thus suppose that $G = C_l(3)$, where $l \geq 2$. Now G is generated by its root subgroups X_r for all $r \in \Phi$. Since $X_r, X_{w(r)}$ are conjugate subgroups of G by 7.2.1, and so equivalent modulo G', the factor group G/G' is generated by the images of the root subgroups X_r and it is sufficient to take one root from each orbit of Φ under W, i.e. one root of each length. If r is a short root of C_l there is a K-character χ of Q such that $\chi(r) \neq 1$, by 11.1.3. Thus X_r is contained in G' in this case. Let s be the long fundamental root of C_l and r the fundamental root joined to it in the Dynkin diagram. Then we have

$$[x_s(1), \, x_r(1)] = x_{r+s}(\pm 1) \, x_{2r+s}(\pm 1)$$

by the commutator formula. Now $r+s$ is a short root and $2r+s$ is a long root. Thus $X_{r+s} \subseteq G'$ and $x_{r+s}(\pm 1) \, x_{2r+s}(\pm 1) \in G'$. It follows that $X_{2r+s} \subseteq G'$ and that $G' = G$.

Suppose finally that $K = GF(2)$. If all the roots of $\mathfrak{L}$ have the same length, G/G' is generated by the image of $x_r(1)$ for any single root r. If $\mathfrak{L} \neq A_1$ we can choose r, $s \in \Phi$ such that $r+s \in \Phi$. Then we have

$$[x_s(1), \, x_r(1)] = x_{r+s}(1)$$

by the commutator formula. Thus $x_{r+s}(1) \in G'$ and so $G' = G$.

Now suppose there are roots of two different lengths. Then G/G' is generated by $x_r(1)$, $x_s(1)$, where r, s are fundamental roots of different lengths. We may assume r is a short root, s a long root and that $r+s \in \Phi$. We assume also that the nodes corresponding to r, s are joined by a double bond in the Dynkin diagram, i.e. that $\mathfrak{L} \neq G_2$. Then the commutator formula gives

$$[x_s(1), \, x_r(1)] = x_{r+s}(1) \, x_{2r+s}(1)$$

and so $x_{r+s}(1) \, x_{2r+s}(1) \in G'$. Here $r+s$ is a short root and $2r+s$ a long one. Suppose $\mathfrak{L} \neq B_2$. Then the Dynkin diagram of $\mathfrak{L}$ has two nodes joined by a single bond. Let r_1, r_2 be the corresponding fundamental roots. Then

$$[x_{r_2}(1), \, x_{r_1}(1)] = x_{r_1+r_2}(1)$$

and so $x_{r_1+r_2}(1) \in G'$. If $r_1 + r_2$ is a short root this implies that all the root subgroups corresponding to short roots are in G'. Thus $x_{r+s}(1) \in G'$. But then $x_{2r+s}(1) \in G'$ and G' contains all the root subgroups corresponding to long roots also. Hence $G' = G$. A similar argument clearly applies if $r_1 + r_2$ is a long root.

The only Chevalley groups not covered in the above argument are $A_1(2)$, $A_1(3)$, $B_2(2)$, $G_2(2)$ and so the theorem is proved. ∎

The four groups which have not been proved to be simple are all in fact not simple. $A_1(2)$ has order 6 and is isomorphic to the symmetric group S_3. $A_1(3)$ has order 12 and is isomorphic to the alternating group

A_4. $B_2(2)$ has order 720 and is isomorphic to the symmetric group S_6. Finally $G_2(2)$ has order 12096 and has a simple subgroup of index 2 isomorphic to the unitary group $PSU_3(3^2)$.

11.2 Classical Lie Algebras in Matrix Form

We wish to show now that the Chevalley groups $A_l(K)$, $B_l(K)$, $C_l(K)$, $D_l(K)$ are isomorphic to certain classical groups. In order to do this we first describe a matrix representation of each of the simple Lie algebras A_l, B_l, C_l, D_l. We refer to Jacobson's book [1] for proofs of the statements about the simple Lie algebras which we require in this connection.

11.2.1 The algebra A_l

The Lie algebra $\mathfrak{L}$ of all $(l+1) \times (l+1)$ matrices of trace 0 over $\mathbb{C}$ is isomorphic to A_l. Let $\mathfrak{H}$ be the subalgebra of all diagonal matrices in $\mathfrak{L}$ and e_{ij} be the elementary matrix with (i, j)-coefficient 1 and other coefficients 0. Then $\mathfrak{H}$ is a Cartan subalgebra of $\mathfrak{L}$ and

$$\mathfrak{L} = \mathfrak{H} \oplus \sum_{i \neq j} \mathbb{C}e_{ij}$$

is a Cartan decomposition. Let $h \in \mathfrak{H}$ be defined by

$$h = \text{diag} (\lambda_0, \lambda_1, \ldots, \lambda_l).$$

Then we have

$$[he_{ij}] = (\lambda_i - \lambda_j) e_{ij}.$$

Thus we evidently have a root system of type A_l (cf. section 3.6) and the root space negative to $\mathbb{C}e_{ij}$ is $\mathbb{C}e_{ji}$. Let

$$h_{ij} = [e_{ij}e_{ji}] = e_{ii} - e_{jj}.$$

Then h_{ij} is the co-root corresponding to the root space $\mathbb{C}e_{ij}$. For h_{ij} is certainly a scalar multiple of this co-root, however we have

$$[h_{ij}, e_{ij}] = 2e_{ij}$$

and so h_{ij} must be the co-root itself.

Let θ be the map of $\mathfrak{L}$ into itself given by

$$\theta(x) = -x',$$

where x' is the transpose of x. Then θ is an automorphism of $\mathfrak{L}$. For

$$[-x', -y'] = x'y' - y'x' = (yx)' - (xy)' = -[xy]'.$$

Also we have

$$\theta(e_{ij}) = -e_{ji}.$$

Thus if r is the root whose root space is $\mathbb{C}e_{ij}$ and if we define the root vector e_r by $e_r = e_{ij}$, then

$$\theta(e_r) = -e_{-r}$$

for all $r \in \Phi$. Now we have

$$[e_r e_s] = N_{r,\ s} e_{r+s}$$

and applying θ this gives

$$[e_{-r} e_{-s}] = -N_{r,\ s} e_{-r-s}.$$

Hence $N_{-r,\ -s} = -N_{r,\ s}$. Since by 4.1.2 we have

$$N_{r,\ s} N_{-r,\ -s} = -(p+1)^2$$

it follows that

$$N_{r,\ s} = \pm(p+1).$$

Thus the elements h_{ij} corresponding to the fundamental roots and the elements e_{ij} corresponding to all roots form a Chevalley basis of $\mathfrak{L}$.

LEMMA 11.2.2. *Let A be an $n \times n$ matrix over $\mathbb{C}$. Then the $n \times n$ matrices T satisfying*

$$T'A + AT = 0$$

form a Lie algebra. If T is a nilpotent matrix satisfying this condition then

$$(\exp T)'\, A(\exp T) = A.$$

PROOF. The set of matrices T satisfying $T'A + AT = 0$ is closed under addition and scalar multiplication. Let T_1, T_2 satisfy this condition and consider the Lie product $[T_1 T_2] = T_1 T_2 - T_2 T_1$. We have

$$T_2' T_1' A - T_1' T_2' A = -T_2' A T_1 + T_1' A T_2,$$

$$AT_1 T_2 - AT_2 T_1 = -T_1' A T_2 + T_2' A T_1.$$

Thus

$$[T_1 T_2]'\, A + A[T_1 T_2] = 0.$$

Now suppose that T is nilpotent. Thus $T^k = 0$ for some k and

$$\exp T = \sum_{i=0}^{k-1} \frac{1}{i!} T^i.$$

Then we have

$$(\exp T)'A = \exp (T') . A = \sum_i \frac{1}{i!} (T')^i A$$

$$= \sum_i \frac{1}{i!} (-1)^i A T^i = A . \exp (-T).$$

It follows that

$$(\exp T)'A(\exp T) = A.$$ ∎

11.2.3 The algebra D_l

We now take

$$A = \begin{pmatrix} 0 & I_l \\ I_l & 0 \end{pmatrix}$$

in 11.2.2 and consider $2l \times 2l$ matrices

$$T = \begin{pmatrix} T_{11} & T_{12} \\ T_{21} & T_{22} \end{pmatrix}$$

satisfying $T'A + AT = 0$. T satisfies this condition if and only if $T_{22} = -T'_{11}$ and T_{12}, T_{21} are skew-symmetric. Let $\mathfrak{L}$ be the Lie algebra of all such matrices and $\mathfrak{H}$ be the set of diagonal matrices in $\mathfrak{L}$. The elements of $\mathfrak{H}$ have form

$$h = \begin{pmatrix} \lambda_1 & & & & & & & \\ & \lambda_2 & & & & & & \\ & & \cdot & & & & & \\ & & & \cdot & & & & \\ & & & & \lambda_l & & & \\ & & & & & -\lambda_1 & & \\ & & & & & & -\lambda_2 & \\ & & & & & & & \cdot \\ & & & & & & & & -\lambda_l \end{pmatrix}$$

Numbering the rows and columns $1, 2, \ldots, l, -1, -2, \ldots, -l$ we have

$$\mathfrak{L} = \mathfrak{H} \oplus \sum_r \mathbb{C}e_r,$$

where

$$e_r = \begin{cases} e_{ij} - e_{-j, -i}, \\ -e_{-i, -j} + e_{ji}, \\ e_{i, -j} - e_{j, -i}, \\ -e_{-i, j} + e_{-j, i}. \end{cases} \qquad 0 < i < j,$$

h transforms these root vectors according to the following formulae:

$$[h, e_{ij} - e_{-j, -i}] = (\lambda_i - \lambda_j)(e_{ij} - e_{-j, -i}),$$

$$[h, -e_{-i, -j} + e_{ji}] = (\lambda_j - \lambda_i)(-e_{-i, -j} + e_{ji}),$$

$$[h, e_{i, -j} - e_{j, -i}] = (\lambda_i + \lambda_j)(e_{i, -j} - e_{j, -i}),$$

$$[h, -e_{-i, j} + e_{-j, i}] = (-\lambda_i - \lambda_j)(-e_{-i, j} + e_{-j, i}).$$

Thus $\mathfrak{L}$ is a simple algebra of type D_l (compare section 3.6) and the above decomposition is a Cartan decomposition. The co-roots of $\mathfrak{L}$ are the elements

$$[e_{ij} - e_{-j, -i}, -e_{-i, -j} + e_j, i] = e_{ii} - e_{jj} - e_{-i, -i} + e_{-j, -j},$$

$$[e_{i, -j} - e_{j, -i}, -e_{-i, j} + e_{-j, i}] = e_{ii} + e_{jj} - e_{-i, -i} - e_{-j, -j}.$$

For in each case we have $[h_r e_r] = 2e_r$. As in 11.2.1 the map $\theta(x) = -x'$ is an automorphism of $\mathfrak{L}$, but this time it satisfies $\theta(e_r) = e_{-r}$ for each root r. To show that the fundamental co-roots and the root vectors e_r form a Chevalley basis we observe that $\theta(ie_r) = -(-ie_{-r})$ and that $[ie_r, -ie_{-r}] = h_r$. Thus the fundamental co-roots together with ie_r (r positive) and $-ie_{-r}$ (r positive) form a Chevalley basis, as in 11.2.1. It follows that the fundamental co-roots together with e_r, e_{-r} (r positive) also form a Chevalley basis, using the general ideas of section 4.2.

11.2.4 The algebra B_l

This time we take

$$A = \begin{pmatrix} 2 & 0 & \cdots & & 0 \\ 0 & 0 & & & I_l \\ \cdot & & & & \\ \cdot & & & & \\ \cdot & & & & \\ 0 & I_l & & & 0 \end{pmatrix}$$

and consider $(2l+1) \times (2l+1)$ matrices T satisfying $T'A + AT = 0$. These form a Lie algebra as in 11.2.2.

Let

$$T = \begin{pmatrix} T_{00} & T_{01} & T_{02} \\ T_{10} & T_{11} & T_{12} \\ T_{20} & T_{21} & T_{22} \end{pmatrix} \begin{matrix} 1 \\ l \\ l \end{matrix}$$

$$\begin{matrix} 1 & l & l \end{matrix}$$

be the expression of T as a block matrix. Then T satisfies $T'A + AT = 0$ if and only if $T_{22} = -T'_{11}$, T_{12} and T_{21} are skew-symmetric, $T_{10} = -2T'_{02}$, $T_{20} = -2T'_{01}$ and $T_{00} = 0$. Let $\mathfrak{L}$ be the Lie algebra of all such matrices and $\mathfrak{H}$ be the set of diagonal matrices in $\mathfrak{L}$. The elements of $\mathfrak{H}$ have form

$$h = \begin{pmatrix} 0 & & & & & & & \\ & \lambda_1 & & & & & & \\ & & \cdot & & & & & \\ & & & \cdot & & & & \\ & & & & \cdot & & & \\ & & & & & \lambda_l & & \\ & & & & & & -\lambda_1 & \\ & & & & & & & \cdot & \\ & & & & & & & & \cdot \\ & & & & & & & & & -\lambda_l \end{pmatrix}.$$

Numbering the rows and columns $0, 1, \ldots, l, -1, \ldots, -l$ we have

$$\mathfrak{L} = \mathfrak{H} \oplus \sum_r \mathbb{C}e_r,$$

where

$$e_r = \begin{cases} e_{ij} - e_{-j,\,-i}, \\ -e_{-i,\,-j} + e_{ji}, \\ e_{i,\,-j} - e_{j,\,-i}, \\ -e_{-i,\,j} + e_{-j,\,i}, \\ 2e_{i0} - e_{0,\,-i}, \\ -2e_{-i0} + e_{0i}. \end{cases} \quad 0 < i < j,$$

h transforms these root vectors according to the following formulae:

$$[h, e_{ij}-e_{-j, -i}]=(\lambda_i-\lambda_j)(e_{ij}-e_{-j, -i}),$$

$$[h, -e_{-i, -j}+e_{ji}]=(\lambda_j-\lambda_i)(-e_{-i, -j}+e_{ji}),$$

$$[h, e_{i, -j}-e_{j, -i}]=(\lambda_i+\lambda_j)(e_{i, -j}-e_{j, -i}),$$

$$[h, -e_{-i, j}+e_{-j, i}]=(-\lambda_i-\lambda_j)(-e_{-i, j}+e_{-j, i}),$$

$$[h, 2e_{i0}-e_{0, -i}]=\lambda_i(2e_{i0}-e_{0, -i}),$$

$$[h, -2e_{-i0}+e_{0i}]=-\lambda_i(-2e_{-i0}+e_{0i}).$$

$\mathfrak{L}$ is a simple Lie algebra and the above decomposition is a Cartan decomposition giving a root system of type B (cf. section 3.6). The co-roots of $\mathfrak{L}$ are the elements

$$[e_{ij}-e_{-j, -i}, -e_{-i, -j}+e_{ji}]=e_{ii}-e_{jj}-e_{-i, -i}+e_{-j, -j},$$

$$[e_{i, -j}-e_{j, -i}, -e_{-i, j}+e_{-j, i}]=e_{ii}+e_{jj}-e_{-i, -i}-e_{-j, -j},$$

$$[2e_{i0}-e_{0, -i}, -2e_{-i0}+e_{0i}]=2e_{ii}-2e_{-i, -i}.$$

For in each case we have $[h_r e_r]=2e_r$.

Let δ be the matrix diag $(2, 1, \ldots, 1)$. Then the map θ defined by

$$\theta(x)=-\delta^{-1}x'\delta$$

is an automorphism of $\mathfrak{L}$. It has the effect of transposing the matrix, changing its sign, halving the first row and doubling the first column. Thus it can be seen that

$$\theta(e_r)=e_{-r}$$

for all roots r. Hence the fundamental co-roots and root vectors e_r defined above form a Chevalley basis of $\mathfrak{L}$, as in 11.2.3.

11.2.5 The algebra C_l

This time we take

$$A=\begin{pmatrix} 0 & I_l \\ -I_l & 0 \end{pmatrix}$$

and consider the Lie algebra of all $2l \times 2l$ matrices satisfying $T'A+AT=0$. Let

$$T=\begin{pmatrix} T_{11} & T_{12} \\ T_{21} & T_{22} \end{pmatrix}.$$

G

Then T satisfies $T'A + AT = 0$ if and only if $T_{22} = -T'_{11}$ and T_{12}, T_{21} are symmetric. Let $\mathfrak{L}$ be the Lie algebra of all such matrices and $\mathfrak{H}$ be the set of diagonal matrices in $\mathfrak{L}$. The elements of $\mathfrak{H}$ have form

$$h = \begin{pmatrix} \lambda_1 & & & & & & & \\ & \cdot & & & & & & \\ & & \cdot & & & & & \\ & & & \lambda_l & & & & \\ & & & & -\lambda_1 & & & \\ & & & & & \cdot & & \\ & & & & & & \cdot & \\ & & & & & & & -\lambda_l \end{pmatrix}$$

Numbering the rows and columns $1, 2, \ldots, l, -1, -2, \ldots, -l$ we have

$$\mathfrak{L} = \mathfrak{H} \oplus \sum_r \mathbb{C}e_r,$$

where

$$e_r = \begin{cases} e_{ij} - e_{-j, \, -i}, \\[4pt] -e_{-i, \, -j} + e_{ji}, \\[4pt] e_{i, \, -j} + e_{j, \, -i}, \\[4pt] e_{-i, \, j} + e_{-j, \, i}, \\[4pt] e_{i, \, -i}, \\[4pt] e_{-i, \, i}. \end{cases} \qquad 0 < i < j,$$

h transforms these root vectors according to the following formulae:

$$[h, \, e_{ij} - e_{-j, \, -i}] = (\lambda_i - \lambda_j)(e_{ij} - e_{-j, \, -i}),$$

$$[h, \, -e_{-i, \, -j} + e_{ji}] = (\lambda_j - \lambda_i)(-e_{-i, \, -j} + e_{ji}),$$

$$[h, \, e_{i, \, -j} + e_{j, \, -i}] = (\lambda_i + \lambda_j)(e_{i, \, -j} + e_{j, \, -i}),$$

$$[h, \, e_{-i, \, j} + e_{-j, \, i}] = (-\lambda_i - \lambda_j)(e_{-i, \, j} + e_{-j, \, i}),$$

$$[h, \, e_{i, \, -i}] = 2\lambda_i e_{i, \, -i},$$

$$[h, \, e_{-i, \, i}] = -2\lambda_i e_{-i, \, i}.$$

$\mathfrak{L}$ is a simple Lie algebra and the above decomposition is a Cartan decomposition giving a root system of type C_l. The co-roots of $\mathfrak{L}$ are the elements

$$[e_{ij} - e_{-j, -i}, -e_{-i, -j} + e_{ji}] = e_{ii} - e_{jj} - e_{-i, -i} + e_{-j, -j},$$

$$[e_i, -j + e_j, -i, e_{-i, j} + e_{-j, i}] = e_{ii} + e_{jj} - e_{-i, -i} - e_{-j, -j},$$

$$[e_i, -i, e_{-i, i}] = e_{ii} - e_{-i, -i}.$$

For $[h_r e_r] = 2e_r$ in each case. The map $\theta(x) = -x'$ is an automorphism of $\mathfrak{L}$ such that $\theta(e_r) = e_{-r}$ for all roots r. Thus the fundamental co-roots and vectors e_r again form a Chevalley basis.

11.3 Identifications with some Classical Groups

We make two useful observations about the matrix representations of the simple Lie algebras A_l, B_l, C_l, D_l which have just been described. In the first place it is evident that for all the matrices e_r in these representations we have $e_r^3 = 0$. In fact $e_r^2 = 0$ in all cases except for

$$e_r = \begin{cases} 2e_{i0} - e_{0, -i}, \\ -2e_{-i, 0} + e_{0i} \end{cases}$$

in type B_l. In these cases we have

$$e_r^2 = \begin{cases} -2e_{i, -i}, \\ -2e_{-i, i}. \end{cases}$$

Thus

$$\exp(te_r) = 1 + te_r + \tfrac{1}{2}t^2 e_r^2.$$

Secondly we note that the coefficients of $\exp(te_r)$ are all of the form n, nt or nt^2, where $n \in \mathbb{Z}$. This is because the coefficients of e_r^2 (when this is not zero) are all divisible by 2. This fact enables us to transfer to an arbitrary field. For each matrix e_r in one of the above representations and each element t in an arbitrary field K, $\exp(te_r)$ is a well-defined non-singular matrix over K.

Let $\mathfrak{L}$ be a simple Lie algebra of type A_l, B_l, C_l, D_l and $G = \mathfrak{L}(K)$ be the Chevalley group of type $\mathfrak{L}$ over K. Let $\bar{G}$ be the group of matrices generated by the elements $\exp(te_r)$ for all $r \in \Phi$ and all $t \in K$. By 4.5.1 we have

$$\exp(t \operatorname{ad} e_r) . x = \exp(te_r) . x . \exp(te_r)^{-1}$$

for all $x \in \mathfrak{L}_K$. Thus there is a homomorphism σ of $\bar{G}$ onto $G = \mathfrak{L}(K)$ such that

$$\exp (te_r) \xrightarrow{\sigma} \exp (t \text{ ad } e_r).$$

We determine the kernel of σ.

LEMMA 11.3.1.   *The kernel of the homomorphism* $\sigma : \bar{G} \to G$ *is the centre* Z *of* $\bar{G}$.

PROOF.   Let $y \in \bar{G}$ be in the kernel of σ. Then $yxy^{-1} = x$ for all $x \in \mathfrak{L}_K$. In particular y commutes with te_r for all $r \in \Phi$. Thus y commutes with $\exp (te_r)$ and so y is in the centre Z of $\bar{G}$.

Conversely, suppose y is an element of Z. Then y commutes with $\exp e_r$ for all $r \in \Phi$. If $\mathfrak{L}$ is not of type B_l we have

$$\exp e_r = 1 + e_r;$$

thus y commutes with e_r. Since

$$[e_r e_{-r}] = h_r,$$

y also commutes with h_r for all $r \in \Phi$. Thus y is in the kernel of σ. Now suppose that $\mathfrak{L}$ is of type B_l. Then y commutes with e_r whenever $e_r^2 = 0$, and if $e_r^2 \neq 0$ y commutes with $(\exp (e_r) - 1)^2 = e_r^2$. Thus provided K does not have characteristic 2, y commutes with

$$\exp (e_r) - 1 - \tfrac{1}{2}(\exp (e_r) - 1)^2 = e_r$$

and it follows as above that y is in the kernel of σ. Finally, if K has characteristic 2 an easy matrix calculation shows that y is a scalar multiple of the identity, and so is in the kernel of σ.   ∎

We are now able to prove a theorem of Ree identifying Chevalley groups of type A_l, B_l, C_l, D_l with classical groups.

THEOREM 11.3.2.   (i) $A_l(K)$ *is isomorphic to the linear group* $PSL_{l+1}(K)$.
    (ii) $B_l(K)$ *is isomorphic to the orthogonal group* $P\Omega_{2l+1}(K, f_B)$, *where* f_B *is the quadratic form*

$$x_0^2 + x_1 x_{-1} + x_2 x_{-2} + \ldots + x_l x_{-l}.$$

    (iii) $C_l(K)$ *is isomorphic to the symplectic group* $P Sp_{2l}(K)$.
    (iv) $D_l(K)$ *is isomorphic to the orthogonal group* $P\Omega_{2l}(K, f_D)$, *where* f_D *is the quadratic form*

$$x_1 x_{-1} + x_2 x_{-2} + \ldots + x_l x_{-l}.$$

Note that f_B and f_D are both quadratic forms of maximal index in the spaces concerned (cf. section 1.4).

PROOF. (i) Let $G = A_l(K)$. Then $\bar{G}$ is the group of $(l+1) \times (l+1)$ matrices generated by

$$I + t e_{ij} \qquad (i \neq j)$$

for all $t \in K$. These matrices generate the group $SL_{l+1}(K)$. Thus, by 11.3.1,

$$G \cong \bar{G}/Z \cong PSL_{l+1}(K).$$

(iv) Let $G = D_l(K)$. Then, by 11.2.2 and 11.2.3, $\bar{G}$ is a group of $2l \times 2l$ matrices generated by elements T satisfying

$$T'AT = A,$$

where

$$A = \begin{pmatrix} 0 & I_l \\ I_l & 0 \end{pmatrix}.$$

If the characteristic of K is not 2 all such matrices represent isometries of the quadratic form

$$x_1 x_{-1} + x_2 x_{-2} + \ldots + x_l x_{-l}$$

and so $\bar{G}$ is a subgroup of $O_{2l}(K, f_D)$. Now $\bar{G}$ is generated by matrices

$$\begin{cases} I + t(e_{ij} - e_{-j, -i}), \\ I - t(e_{-i, -j} - e_{ji}), \\ I + t(e_{i, -j} - e_{j, -i}), \\ I - t(e_{-i, j} - e_{-j, i}). \end{cases} \qquad 0 < i < j,$$

It is easily seen that these generators leave invariant the above quadratic form also when K has characteristic 2. Thus $\bar{G}$ is a subgroup of $O_{2l}(K, f_D)$ in all cases. Moreover it is shown in Ree [1] that the above matrices generate $\Omega_{2l}(K, f_D)$, the commutator subgroup of $O_{2l}(K, f_D)$. Thus

$$\bar{G} \cong \Omega_{2l}(K, f_D)$$

and it follows by 11.3.1 that

$$G \cong \bar{G}/Z \cong P\Omega_{2l}(K, f_D).$$

(iii) Let $G = C_l(K)$. Then $\bar{G}$ is a group of $2l \times 2l$ matrices generated by elements T satisfying $T'AT=A$, where

$$A = \begin{pmatrix} 0 & I_l \\ -I_l & 0 \end{pmatrix}.$$

Such matrices are elements of the symplectic group $Sp_{2l}(K)$. $\bar{G}$ is generated by the matrices

$$\begin{cases} I+t(e_{ij}-e_{-j,-i}), \\ I-t(e_{-i,-j}-e_{ji}), \\ I+t(e_i,-j+e_j,-i), \\ I-t(-e_{-i,j}-e_{-j,i}), \\ I+te_{i,-i}, \\ I+te_{-i,i}. \end{cases} \quad 0<i<j,$$

But these matrices generate the symplectic group $Sp_{2l}(K)$ (cf. Ree [1]). Thus $\bar{G} \cong Sp_{2l}(K)$ and

$$G \cong \bar{G}/Z \cong PSp_{2l}(K).$$

(ii) Let $G=B_1(K)$ and suppose the characteristic of K is not 2. Then $\bar{G}$ is a group of $(2l+1)\times(2l+1)$ matrices generated by elements T satisfying $T'AT=A$, where

$$A = \begin{pmatrix} 2 & 0 & \cdots & 0 \\ 0 & 0 & & I_l \\ \cdot & & & \\ \cdot & & & \\ \cdot & & & \\ 0 & I_l & & 0 \end{pmatrix}.$$

Such matrices represent isometries of the quadratic form

$$x_0^2+x_1x_{-1}+x_2x_{-2}+ \ldots +x_lx_{-l}$$

and so $\bar{G}$ is a subgroup of $O_{2l+1}(K, f_B)$. Now $\bar{G}$ is generated by matrices

$$\begin{cases} I+t(e_{ij}-e_{-j,-i}), \\ I-t(e_{-i,-j}-e_{ji}), \\ I+t(e_i,-j-e_j,-i), \\ I-t(e_{-i,j}-e_{-j,i}), \\ I+t(2e_{i0}-e_{0,-i})-t^2e_{i,-i}, \\ I-t(2e_{-i,0}-e_{0,i})-t^2e_{-i,i} \end{cases}$$

and it is shown in Ree [1] that these matrices generate $\Omega_{2l+1}(K, f_B)$, the commutator subgroup of $O_{2l+1}(K, f_B)$. Thus $\bar{G} \cong \Omega_{2l+1}(K, f_B)$ and

$$G \cong \bar{G}/Z \cong P\Omega_{2l+1}(K, f_B).$$

Now suppose K has characteristic 2. In this case $\bar{G}$ can be considered as a group of linear transformations of a vector space $\mathfrak{V}$ with basis $v_0, v_1, \ldots, v_l, v_{-1}, \ldots, v_{-l}$, where

$$e_{\alpha\beta} \cdot v_\alpha = v_\beta.$$

Now in the set of generators for $\bar{G}$ described in 11.3.2 (ii), the terms involving e_{i0} and $e_{-i, 0}$ vanish when K has characteristic 2. Hence the subspace $\mathfrak{W}$ of $\mathfrak{V}$ with basis $v_1, \ldots, v_l, v_{-1}, \ldots, v_{-l}$ is invariant under $\bar{G}$. We show that $\bar{G}$ operates faithfully on $\mathfrak{W}$.

We remarked in 11.3.1 that the centre Z of $\bar{G}$ consists only of scalar multiples of the identity. However, the elements of $\bar{G}$ satisfy $T'AT = A$, where

$$A = \begin{pmatrix} 2 & 0 & . & . & . & 0 \\ 0 & 0 & & & & I_l \\ . & & & & & \\ . & & & & & \\ . & & & & & \\ 0 & I_l & & & & 0 \end{pmatrix}$$

and λI can only satisfy this if $\lambda^2 = 1$. In characteristic 2 this implies that $\lambda = 1$. Hence $Z = 1$ and $\bar{G}$ is isomorphic to G. It follows that $\bar{G}$ is simple. (We exclude the exceptional case $B_2(2)$.) Thus $\bar{G}$ acts faithfully on $\mathfrak{W}$.

Now the quadratic form f_B in characteristic 2 is non-degenerate but has defect 1 (see section 1.6). The vector space $\mathfrak{V}$ on which f_B is defined therefore has non-singular symplectic subspaces of codimension 1, and $\mathfrak{W}$ is such a subspace. We therefore compare the action of $\bar{G}$ on $\mathfrak{W}$ with the action of the symplectic group $Sp_{2l}(K)$.

Now the group $\bar{G}$ acting on $\mathfrak{W}$ is generated by the following matrices (using the fact that $-1 = 1$ in K):

$$\begin{cases} I + t(e_{ij} - e_{-j, -i}), \\ I - t(e_{-i, -j} - e_{ji}), \\ I + t(e_{i, -j} + e_{j, -i}), \\ I - t(e_{-i, j} - e_{-j, i}), \\ I + t^2 e_{i, -i}, \\ I + t^2 e_{-i, i}. \end{cases}$$

If these generators are compared with the ones in 11.3.2 (iii) it is clear that $\bar{G}$ is a subgroup of $Sp_{2l}(K)$. Moreover if K is a perfect field each element of K is a square and so $\bar{G} = Sp_{2l}(K)$. Thus

$$\bar{G} = Sp_{2l}(K) = O_{2l+1}(K, f_B)$$

as in section 1.6 and so

$$B_l(K) \cong P\Omega_{2l+1}(K, f_B),$$

since $\bar{G}$ is simple.

Now suppose that K is not perfect. The situation is now more complicated since the orthogonal group $O_{2l+1}(K, f_B)$ is a proper subgroup of $Sp_{2l}(K)$ (cf. section 1.6). However it is easily checked that the given generators of the group $\bar{G}$ lie in $O_{2l+1}(K, f_B)$. The commutator subgroup $\Omega_{2l+1}(K, f_B)$ is generated by 'orthogonal transvections' corresponding to elements $x \in \mathcal{Y}$ with $f_B(x) \in K^2$, except possibly when $l = 2$ (cf. Dieudonne [1], p. 59). The generators $I + t^2 e_{i, -i}$ and $I + t^2 e_{-i, i}$ are orthogonal transvections of this kind and it is shown by Dieudonne [4] that the group $\bar{G}$ contains all such orthogonal transvections, so must contain $\Omega_{2l+1}(K, f_B)$. Thus $\bar{G}$ lies between $\Omega_{2l+1}(K, f_B)$ and $O_{2l+1}(K, f_B)$. Since $\bar{G}$ is simple we have $\bar{G} = \Omega_{2l+1}(K, f_B)$ and so

$$B_l(K) \cong P\Omega_{2l+1}(K, f_B)$$

in this case also.   ∎

CHAPTER 12

Generators, Relations and Automorphisms in Chevalley Groups

12.1 A Theorem of Steinberg

The Chevalley group $G = \mathfrak{L}(K)$ is generated by a set of elements $x_r(t)$ for all $r \in \Phi$, $t \in K$. We consider the problem of finding relations involving these generators which are sufficient to define G as an abstract group. Such a system of relations has been discovered by Steinberg.

Let $h_r(t)$, $n_r(t)$ be the elements of G defined as in chapter 6 by

$$h_r(t) = \phi_r \begin{pmatrix} t & 0 \\ 0 & t^{-1} \end{pmatrix},$$

$$n_r(t) = \phi_r \begin{pmatrix} 0 & t \\ -t^{-1} & 0 \end{pmatrix}.$$

By 6.4.4 we have

$$n_r(t) = x_r(t) \, x_{-r}(-t^{-1}) \, x_r(t),$$

$$h_r(t) = n_r(t) \, n_r(-1)$$

and these equations show how to express $n_r(t)$ and $h_r(t)$ in terms of the generators of G.

Now the following relations have been shown to hold in G:

$$x_r(t_1) \, x_r(t_2) = x_r(t_1 + t_2),$$

$$[x_s(u), x_r(t)] = \prod_{i,j>0} x_{ir+js}(C_{ijrs}(-t)^i u^j),$$

$$h_r(t_1) \, h_r(t_2) = h_r(t_1 t_2), \qquad\qquad t_1 t_2 \neq 0,$$

where C_{ijrs} are integers defined as in 5.2.2 for linearly independent roots r, s. Steinberg's theorem shows that these relations are almost sufficient to define G abstractly, the only additional relations required being ones to ensure that the group so defined has a trivial centre.

189

THEOREM 12.1.1. *Let $\mathfrak{L}$ be a simple Lie algebra with $\mathfrak{L} \neq A_1$ and let K be a field. For each root r of $\mathfrak{L}$ and each element t of K introduce a symbol $\bar{x}_r(t)$. Let $\bar{G}$ be the abstract group generated by the elements $\bar{x}_r(t)$ subject to relations*

$$\bar{x}_r(t_1)\, \bar{x}_r(t_2) = \bar{x}_r(t_1 + t_2),$$

$$[\bar{x}_s(u), \bar{x}_r(t)] = \prod_{i,j>0} \bar{x}_{ir+js}(C_{ijrs}(-t)^i u^j),$$

$$h_r(t_1)\, h_r(t_2) = h_r(t_1 t_2), \qquad\qquad t_1 t_2 \neq 0,$$

where
$$h_r(t) = \bar{n}_r(t)\, \bar{n}_r(-1)$$

and
$$\bar{n}_r(t) = \bar{x}_r(t)\, \bar{x}_{-r}(-t^{-1})\, \bar{x}_r(t).$$

Let Z be the centre of $\bar{G}$. Then $\bar{G}/Z$ is isomorphic to the Chevalley group $G = \mathfrak{L}(K)$.

PROOF. (a) There is a homomorphism θ of $\bar{G}$ onto G such that

$$\theta(\bar{x}_r(t)) = x_r(t).$$

We must show that the kernel of θ is Z.

As usual, let Φ be the set of roots of $\mathfrak{L}$ and Φ^+ be the a positive system in Φ. Let $\bar{U}$ be the subgroup of $\bar{G}$ generated by the elements $\bar{x}_r(t)$ for $r \in \Phi^+$. Then the commutator relation shows that each element of $\bar{U}$ can be expressed in the form

$$\bar{x}_{r_1}(t_1)\, \bar{x}_{r_2}(t_2) \ldots \bar{x}_{r_N}(t_N),$$

where $r_1, \ldots, r_N$ is the set of positive roots. Thus

$$\theta(\bar{x}_{r_1}(t_1) \ldots \bar{x}_{r_N}(t_N)) = x_{r_1}(t_1) \ldots x_{r_N}(t_N).$$

However, by 5.3.3 each element of U is uniquely expressible in the form $x_{r_1}(t_1) \ldots x_{r_N}(t_N)$. Thus the map $\theta : \bar{U} \to U$ is an isomorphism.

(b) We now consider the expression

$$n_r(t)\, x_s(u)\, n_r(t)^{-1}$$

in G. Since $n_r(t) = h_r(t)\, n_r$ we have

$$
\begin{aligned}
n_r(t)\, x_s(u)\, n_r(t)^{-1} &= h_r(t)\, n_r x_s(u)\, n_r^{-1}\, h_r(t)^{-1} \\
&= h_r(t)\, x_{w_r(s)}(\eta_r,\, su)\, h_r(t)^{-1}, \qquad \text{by 7.2.1,} \\
&= x_{w_r(s)}(\eta_r,\, st^{A_r\rho}u), \quad \text{where } \rho = w_r(s), \text{ by section 7.1,} \\
&= x_{w_r(s)}(\eta_r,\, st^{-A_{rs}}u).
\end{aligned}
$$

We shall show that the corresponding relation

$$\bar{n}_r(t)\, \bar{x}_s(u)\, \bar{n}_r(t)^{-1} = \bar{x}_{w_r(s)}(\eta r,\, st^{-A_{rs}}u)$$

holds in $\bar{G}$. Suppose r, s are linearly independent. Then

$$\bar{n}_r(t)\, \bar{x}_s(u)\, \bar{n}_r(t)^{-1}$$
$$= \bar{x}_r(t)\, \bar{x}_{-r}(-t^{-1})\, \bar{x}_r(t)\, \bar{x}_s(u)\, \bar{x}_r(t)^{-1}\, \bar{x}_{-r}(-t^{-1})^{-1}\, \bar{x}_r(t)^{-1}.$$

The commutator relation shows that this can be expressed as a product of terms of the form $\bar{x}_{ir+js}(v)$, where $j > 0$. Now there is a positive system Φ^+ in Φ containing all roots of form $ir + js$, where i, j are integers and $j > 0$. With respect to such a positive system we have

$$\bar{n}_r(t)\, \bar{x}_s(u)\, \bar{n}_r(t)^{-1} \in \bar{U},$$
$$\bar{x}_{w_r(s)}(\eta r,\, st^{-A_{rs}}u) \in \bar{U}.$$

But θ is an isomorphism from $\bar{U}$ onto U and we have

$$\theta(\bar{n}_r(t)\, \bar{x}_s(u)\, \bar{n}_r(t)^{-1}) = \theta(\bar{x}_{w_r(s)}(\eta r,\, st^{-A_{rs}}u)).$$

Hence
$$\bar{n}_r(t)\, \bar{x}_s(u)\, \bar{n}_r(t)^{-1} = \bar{x}_{w_r(s)}(\eta r,\, st^{-A_{rs}}u)$$

as required.

If r, s are not linearly independent we argue differently. Suppose $r = s$. Since Φ is a simple root system of rank greater than 1 (we have assumed $\mathfrak{L} \neq A_1$), there exists $r_1 \in \Phi$, $r_1 \neq r$ such that $(r, r_1) > 0$. Then r does not begin the r_1-chain of roots through it. Let r_2 begin the r_1-chain of roots through r. Then $r = ir_1 + r_2$ for some $i > 0$. Also

$$C_{t1r_1r_2} = M_{r_1r_2i} = \pm 1$$

by 5.2.2. Now the commutator relation shows that

$$[\bar{x}_{r_2}(t_2),\, \bar{x}_{r_1}(t_1)] = \prod_{i,j > 0} \bar{x}_{ir_1+jr_2}(C_{ijr_1r_2}(-t_1)^i\, t_2^j).$$

We now tranform each term in this relation by $\bar{n}_r(t)$. Let Φ^+ be a positive system in Φ containing all roots of form

$$iw_r(r_1) + jw_r(r_2), \qquad i \geqslant 0,\ j \geqslant 0.$$

Since
$$\bar{n}_r(t)\, \bar{x}_s(u)\, \bar{n}_r(t)^{-1} = \bar{x}_{w_r(s)}(\eta r,\, st^{-A_{rs}}u)$$

whenever $s \neq \pm r$, it is evident that all the transformed terms lie in $\bar{U}$ with the possible exception of

$$\bar{n}_r(t)\, \bar{x}_r(C_{t1r_1r_2}(-t_1)^i\, t_2)\, \bar{n}_r(t)^{-1}.$$

Hence this term lies in $\bar{U}$ also. Since $C_{i1r_1r_2} \neq 0$ it follows that

$$\bar{n}_r(t)\, \bar{x}_r(u)\, \bar{n}_r(t)^{-1} \in \bar{U}$$

for all $u \in K$. Also we have

$$\bar{x}_{w_r(r)}(\eta r,\, rt^{-A_{rr}}u) \in \bar{U}.$$

But

$$\theta(\bar{n}_r(t)\, \bar{x}_r(u)\, \bar{n}_r(t)^{-1}) = \theta(\bar{x}_{w_r(r)}(\eta r,\, rt^{-A_{rr}}u))$$

and $\theta : \bar{U} \to U$ is an isomorphism. Thus

$$\bar{n}_r(t)\, \bar{x}_r(u)\, \bar{n}_r(t)^{-1} = \bar{x}_{-r}(-t^{-2}u),$$

since $\eta r,\, r = -1$, $A_{rr} = 2$.

Finally suppose $s = -r$. Then, from the above formula we have

$$\bar{n}_r(t)^{-1}\, \bar{x}_{-r}(u)\, \bar{n}_r(t) = \bar{x}_r(-t^2 u).$$

But

$$\bar{n}_r(t)^{-1} = \bar{n}_r(-t)$$

and so

$$\bar{n}_r(t)\, \bar{x}_{-r}(u)\, \bar{n}_r(t)^{-1} = \bar{x}_r(-t^2 u).$$

Thus

$$\bar{n}_r(t)\, \bar{x}_{-r}(u)\, \bar{n}_r(t)^{-1} = \bar{x}_{w_r(-r)}(\eta r,\, -rt^{-A_{r,\,-r}}u),$$

as required, since $\eta r,\, -r = -1$, $A_{r,\,-r} = -2$. The relation

$$\bar{n}_r(t)\, \bar{x}_s(u)\, \bar{n}_r(t)^{-1} = \bar{x}_{w_r(s)}(\eta r,\, st^{-A_{rs}}u)$$

is therefore valid in $\bar{G}$ in all cases.

(c) We now derive some consequences of the relation proved in (b). Let $r,\, s \in \Phi$ and $t,\, u$ be non-zero elements of K. Then

$$\bar{n}_r(t)\, \bar{n}_s(u)\, \bar{n}_r(t)^{-1}$$

$$= \bar{n}_r(t)\, \bar{x}_s(u)\, \bar{x}_{-s}(-u^{-1})\, \bar{x}_s(u)\, \bar{n}_r(t)^{-1}$$

$$= \bar{x}_{w_r(s)}(\eta r,\, st^{-A_{rs}}u)\, \bar{x}_{w_r(-s)}(-\eta r,\, -st^{-A_r,\,-s}u^{-1})\, \bar{x}_{w_r(s)}(\eta r,\, st^{-A_{rs}}u)$$

$$= \bar{x}_{w_r(s)}(\eta r,\, st^{-A_{rs}}u)\, \bar{x}_{-w_r(s)}(-\eta_r^{-1},\, st^{A_{rs}}u^{-1})\, \bar{x}_{w_r(s)}(\eta r,\, st^{-A_{rs}}u)$$

$$= \bar{n}_{w_r(s)}(\eta r,\, st^{-A_{rs}}u)$$

by 6.4.3. Thus

$$\bar{n}_r(t)\, \bar{n}_s(u)\, \bar{n}_r(t)^{-1} = \bar{n}_{w_r(s)}(\eta r,\, st^{-A_{rs}}u).$$

Now consider the expression

$$\bar{n}_r(t)\, \bar{h}_s(u)\, \bar{n}_r(t)^{-1}.$$

Using the relation just proved we have

$$\bar{n}_r(t)\, \bar{h}_s(u)\, \bar{n}_r(t)^{-1} = \bar{n}_r(t)\, \bar{n}_s(u)\, \bar{n}_s(-1)\, \bar{n}_r(t)^{-1}$$

$$= \bar{n}_{w_r(s)}(\eta_r,\, st^{-A_{rs}}u)\, \bar{n}_{w_r(s)}(-\eta_r,\, st^{-A_{rs}})$$

$$= \bar{h}_{w_r(s)}(\eta_r,\, st^{-A_{rs}}u)\, \bar{n}_{w_r(s)}(1)\, \bar{n}_{w_r(s)}(1)^{-1}\, \bar{h}_{w_r(s)}(\eta_r^{-1},\, st^{A_{rs}})$$

$$= \bar{h}_{w_r(s)}(\eta_r,\, st^{-A_{rs}}u)\, \bar{h}_{w_r(s)}(\eta_r^{-1},\, st^{A_{rs}})$$

$$= \bar{h}_{w_r(s)}(u).$$

Hence
$$\bar{n}_r(t)\, \bar{h}_s(u)\, \bar{n}_r(t)^{-1} = \bar{h}_{w_r(s)}(u).$$

(d) Let $\bar{n}_r = \bar{n}_r(1)$ for each $r \in \Phi$. Let $\bar{H}$ be the subgroup of $\bar{G}$ generated by the elements $\bar{h}_r(t)$ for all $r \in \Phi$, $t \neq 0 \in K$, and $\bar{N}$ be the subgroup generated by $\bar{H}$ and $\bar{n}_r$ for all $r \in \Phi$. Since $\bar{n}_r(t) = \bar{h}_r(t)\, \bar{n}_r$ it is clear that $\bar{n}_r(t) \in \bar{N}$ for all $t \neq 0$. Since

$$\bar{n}_r \bar{h}_s(u)\, \bar{n}_r^{-1} = \bar{h}_{w_r(s)}(u)$$

it is evident that $\bar{H}$ is a normal subgroup of $\bar{N}$.

Now $\theta(\bar{N}) = N$ and $\theta(\bar{H}) = H$. Thus θ induces a homomorphism from $\bar{N}/\bar{H}$ onto N/H, which is isomorphic to W by 7.2.2. This homomorphism maps $\bar{n}_r \bar{H}$ to w_r. However it follows from relations proved in (c) that

$$\bar{n}_r \bar{n}_s \bar{n}_r^{-1} = \bar{n}_{w_r(s)}(\eta_r,\, s) = \bar{h}_{w_r(s)}(\eta_r,\, s)\, \bar{n}_{w_r(s)}.$$

This implies that
$$\bar{n}_r \bar{H} . \bar{n}_s \bar{H} . (\bar{n}_r \bar{H})^{-1} = \bar{n}_{w_r(s)} \bar{H}.$$

Also
$$\bar{n}_r^{-1} = \bar{n}_r(-1) = \bar{h}_r(-1)\, \bar{n}_r$$

and so
$$\bar{n}_r^2 = \bar{h}_r(-1)$$

and
$$(\bar{n}_r \bar{H})^2 = \bar{H}.$$

Now the Weyl group W is defined as an abstract group by generators w_r, $r \in \Phi$, subject to relations

$$w_r^2 = 1, \qquad w_r w_s w_r^{-1} = w_{w_r(s)},$$

by 2.4.3. Also we have a homomorphism from $\bar{N}/\bar{H}$ onto W mapping $\bar{n}_r \bar{H}$ to w_r, where

$$(\bar{n}_r \bar{H})^2 = \bar{H} \qquad \bar{n}_r \bar{H} . \bar{n}_s \bar{H} . (\bar{n}_r \bar{H})^{-1} = \bar{n}_{w_r(s)} \bar{H}.$$

It follows that this homomorphism is an isomorphism, and so $\bar{N}/\bar{H}$ is isomorphic to W.

(e) Let $\bar{B}$ be the subgroup of $\bar{G}$ generated by $\bar{U}$ and $\bar{H}$. We show $\bar{H}$ is in the normalizer of $\bar{U}$, so that $\bar{B} = \bar{U}\bar{H}$. Now

$$\bar{h}_r(t)\,\bar{x}_s(u)\,\bar{h}_r(t)^{-1} = \bar{n}_r(t)\,\bar{n}_r(-1)\,\bar{x}_s(u)\,\bar{n}_r(-1)^{-1}\,\bar{n}_r(t)^{-1}$$
$$= \bar{n}_r(t)\,\bar{x}_{w_r(s)}(\eta_{r,\,s}(-1)^{A_{rs}}u)\,\bar{n}_r(t)^{-1}$$
$$= \bar{x}_s(\eta_{r,\,s}\eta_{r,\,w_r(s)}t^{-A_{r\rho}}(-1)^{A_{rs}}u),\ \text{where}\ \rho = w_r(s),$$
$$= \bar{x}_s(t^{A_{rs}}u),$$

since $\eta_{r,\,s}\eta_{r,\,w_r(s)} = (-1)^{A_{rs}}$ by 6.4.3 and $A_{r,\,w_r(s)} = -A_{rs}$. Thus

$$\bar{h}_r(t)\,\bar{x}_s(u)\,\bar{h}_r(t)^{-1} = \bar{x}_s(t^{A_{rs}}u)$$

for all $s \in \Phi$, and so $\bar{H}$ is in the normalizer of $\bar{U}$.

Let $\bar{X}_r$ be the subgroup of $\bar{G}$ generated by $\bar{x}_r(t)$ for all $t \in K$. Let Π be the fundamental system contained in Φ^+. For each root $r \in \Pi$ let $\bar{U}_r$ be the subgroup of $\bar{U}$ generated by $\bar{X}_s$ for all $s \in \Phi^+$ with $s \ne r$. The commutator relations show that

$$\bar{U}_r = \prod_{\substack{s \in \Phi^+ \\ s \ne r}} \bar{X}_s,$$

where the product can be taken in any order. Moreover, $\bar{X}_r$ and $\bar{X}_{-r}$ normalize $\bar{U}_r$ as in 8.1.1. Thus $\bar{n}_r$ normalizes $\bar{U}_r$ also.

(f) We now show that $\bar{B} \cup \bar{B}\bar{n}_r\bar{B}$ is a subgroup of $\bar{G}$ for all $r \in \Pi$. As in 8.1.4 it is sufficient to show that

$$\bar{n}_r\bar{B}\bar{n}_r \subseteq \bar{B} \cup \bar{B}\bar{n}_r\bar{B}.$$

Now we have

$$\bar{n}_r\bar{B}\bar{n}_r = \bar{n}_r\bar{B}\bar{n}_r^{-1} = \bar{n}_r\bar{U}\bar{H}\bar{n}_r^{-1}$$
$$= \bar{n}_r\bar{X}_r\bar{U}_r\bar{H}\bar{n}_r^{-1} = \bar{X}_{-r}\bar{U}_r\bar{H} \subseteq \bar{X}_{-r}\bar{B}.$$

Moreover $\bar{X}_{-r} \subseteq \bar{B} \cup \bar{B}n_r\bar{B}$; for $\bar{x}_{-r}(t) \in \bar{B}$ if $t = 0$, and if $t \ne 0$

$$\bar{x}_{-r}(t) = \bar{x}_r(t^{-1})\,\bar{n}_r(-t^{-1})\,\bar{x}_r(t^{-1})$$
$$= \bar{x}_r(t^{-1})\,\bar{h}_r(-t^{-1})\,\bar{n}_r\bar{x}_r(t^{-1}) \in \bar{B}\bar{n}_r\bar{B}.$$

Hence $\bar{B} \cup \bar{B}\bar{n}_r\bar{B}$ is a subgroup of $\bar{G}$.

(g) It now follows exactly as in 8.1.5 that

$$\bar{B}\bar{n}\bar{B}.\bar{B}\bar{n}_r\bar{B} \subseteq \bar{B}\bar{n}\bar{n}_r\bar{B} \cup \bar{B}\bar{n}\bar{B}$$

for all $\bar{n} \in \bar{N}$, $r \in \Pi$. Arguing as in 8.2.2 we then obtain $\bar{N}\bar{B}\bar{N} \subseteq \bar{B}\bar{N}\bar{B}$ and finally $\bar{G} = \bar{B}\bar{N}\bar{B}$.

For each $w \in W$ let

$$\Psi_1 = \{r \in \Phi^+; \ w(r) \in \Phi^+\},$$

$$\Psi_2 = \{r \in \Phi^+; \ w(r) \in \Phi^-\}$$

and let

$$\bar{U}_w^+ = \prod_{r \in \Psi_1} \bar{X}_r,$$

$$\bar{U}_w^- = \prod_{r \in \Psi_2} \bar{X}_r.$$

Then, since the map $\theta : \bar{U} \to U$ is an isomorphism, $\bar{U} = \bar{U}_w^+ \bar{U}_w^-$ by 8.4.1. Also, using the relations proved in (b), we have

$$\bar{n}_w \bar{U}_w^+ \bar{n}_w^{-1} \subseteq \bar{U},$$

where $\bar{n}_w$ is an element of $\bar{N}$ corresponding to $w \in W$ under the natural homomorphism. Hence

$$\bar{B} \bar{n}_w \bar{B} = \bar{B} \bar{n}_w \bar{H} \bar{U}_w^+ \bar{U}_w^- = \bar{B} \bar{H} \bar{n}_w \bar{U}_w^+ \bar{U}_w^-$$

$$= \bar{B} \bar{n}_w \bar{U}_w^-.$$

Thus each element $\bar{g} \in \bar{G}$ can be written in the form

$$\bar{g} = \bar{b} \bar{n}_w \bar{u},$$

where $\bar{b} \in \bar{B}$ and $\bar{u} \in \bar{U}_w^-$.

(h) For each $w \in W$ we choose a fixed coset representative $\bar{n}_w \in \bar{N}$ and assume that $\bar{n}_1 = 1$. We show that each element of $\bar{G}$ has a unique expression of the form

$$\bar{g} = \bar{u}' \bar{h} \bar{n}_w \bar{u}$$

with $\bar{u}' \in \bar{U}$, $\bar{h} \in \bar{H}$, $w \in W$, $\bar{u} \in \bar{U}_w^-$.

Suppose that

$$\bar{u}_1' \bar{h}_1 \bar{n}_{w_1} \bar{u}_1 = \bar{u}_2' \bar{h}_2 \bar{n}_{w_2} \bar{u}_2.$$

Then

$$\theta(\bar{u}_1') \ \theta(\bar{h}_1) \ \theta(\bar{n}_{w_1}) \ \theta(\bar{u}_1) = \theta(\bar{u}_2') \ \theta(\bar{h}_2) \ \theta(\bar{n}_{w_2}) \ \theta(\bar{u}_2).$$

By the uniqueness of expression in G proved in 8.4.4 we have

$$\theta(\bar{u}_1') = \theta(\bar{u}_2'),$$

$$\theta(\bar{h}_1) = \theta(\bar{h}_2),$$

$$w_1 = w_2,$$

$$\theta(\bar{u}_1) = \theta(\bar{u}_2).$$

Thus $\bar{n}_{w_1} = \bar{n}_{w_2}$ and, since $\theta : \bar{U} \to U$ is an isomorphism, we have $\bar{u}_1' = \bar{u}_2'$ and $\bar{u}_1 = \bar{u}_2$. It follows that $\bar{h}_1 = \bar{h}_2$ and so the expression for $\bar{g}$ is unique.

(i) We now consider the subgroup $\bar{H}$. Let $\bar{H}_r$ be the subgroup of $\bar{G}$ generated by $\bar{h}_r(t)$ for all $t \neq 0$ in K. We show that $\bar{H}$ is abelian and that

$$\bar{H} = \bar{H}_{p_1} \bar{H}_{p_2} \ldots \bar{H}_{p_l},$$

where $\Pi = \{p_1, p_2, \ldots, p_l\}$. Now

$$\bar{h}_r(t) \, \bar{h}_s(u) \, \bar{h}_r(t)^{-1} = \bar{n}_r(t) \, \bar{n}_r(-1) \, \bar{h}_s(u) \, \bar{n}_r(-1)^{-1} \, \bar{n}_r(t)^{-1}$$
$$= \bar{n}_r(t) \, \bar{h}_{w_r(s)}(u) \, \bar{n}_r(t)^{-1}$$
$$= \bar{h}_s(u)$$

by the relations proved in (c). Thus $\bar{H}$ is abelian.

In order to show that

$$\bar{H} = \prod_{i=1}^{l} \bar{H}_{p_i}$$

we need the relation

$$\bar{h}_r(t) \, \bar{n}_s(u) \, \bar{h}_r(t)^{-1} = \bar{n}_s(t^{A_{rs}}u).$$

This can be derived from the relations proved in (e) as follows:

$$\bar{h}_r(t) \, \bar{n}_s(u) \, \bar{h}_r(t)^{-1} = \bar{h}_r(t) \, \bar{x}_s(u) \, \bar{x}_{-s}(-u^{-1}) \, \bar{x}_s(u) \, \bar{h}_r(t)^{-1}$$
$$= \bar{x}_s(t^{A_{rs}}u) \, \bar{x}_{-s}(-t^{A_r, -s} u^{-1}) \, \bar{x}_s(t^{A_{rs}}u)$$
$$= \bar{n}_s(t^{A_{rs}}u).$$

Let $\bar{h}_r(t)$ be one of the generators of $\bar{H}$. If $r \in \Pi$ then certainly

$$\bar{h}_r(t) \in \prod_{i=1}^{l} \bar{H}_{p_i}.$$

Otherwise $r = w(p_i)$ for some $p_i \in \Pi$ and we use induction on $l(w)$. There exists $s \in \Pi$ such that $w = w_s w'$, where $l(w') = l(w) - 1$. Let $\rho = w'(p_i)$. Then

$$\bar{h}_r(t) = \bar{h}_{w_s(\rho)}(t) = \bar{n}_s \bar{h}_\rho(t) \, \bar{n}_s^{-1}$$
$$= \bar{n}_s \bar{n}_s(t^{A_{\rho s}})^{-1} \, \bar{h}_\rho(t)$$
$$= \bar{h}_s(t^{-A_{\rho s}}) \, \bar{h}_\rho(t).$$

Both $\bar{h}_s(t^{-A_{\rho s}})$ and $\bar{h}_\rho(t)$ may be assumed to lie in

$$\prod_{i=1}^{l} \bar{H}_{p_i}$$

by induction, thus $\bar{h}_r(t)$ is in

$$\prod_{i=1}^{l} \bar{H}_{p_i}$$

also. It follows that

$$\bar{H} = \prod_{i=1}^{l} \bar{H}_{p_i}.$$

(j) We consider finally the kernel of θ. Let $\bar{u}'\bar{h}\bar{n}_w\bar{u}$ be in this kernel. Then

$$\theta(\bar{u}')\, \theta(\bar{h})\, \theta(\bar{n}_w)\, \theta(\bar{u}) = 1.$$

By uniqueness of expression in G we have

$$\theta(\bar{u}') = 1,$$
$$\theta(\bar{h}) = 1,$$
$$\theta(\bar{n}_w) = 1,$$
$$\theta(\bar{u}) = 1.$$

Hence $w = 1$ and so $\bar{n}_w = 1$. Also $\bar{u}' = 1$ and $\bar{u} = 1$ since $\theta : \bar{U} \to U$ is an isomorphism. Thus $\bar{u}'\bar{h}\bar{n}_w\bar{u} = \bar{h}$, and so the kernel of θ is contained in $\bar{H}$. Let

$$h = \prod_{i=1}^{l} \bar{h}_{p_i}(t_i)$$

be in the kernel of θ. Then

$$h\bar{x}_r(t)\, h^{-1} = \bar{x}_r\!\left(t \prod_{i=1}^{l} t_i^{n_i} \right), \text{ where } n_i = A_{p_i r},$$

by the relations proved in (e). Applying θ we have

$$x_r(t) = x_r\!\left(t \prod_{i=1}^{l} t_i^{n_i} \right).$$

Hence

$$\prod_{i=1}^{l} t_i^{n_i} = 1$$

and so h commutes with $\bar{x}_r(t)$. Since the elements $\bar{x}_r(t)$ generate $\bar{G}$, h must be in the centre Z of $\bar{G}$. Thus the kernel of θ is contained in Z. However the Chevalley group G has trivial centre by 11.1.2. Thus Z is the kernel of θ and the proof is complete. ∎

The group $\bar{G}$ defined in 12.1.1 is called the *universal Chevalley group* of type $\mathfrak{L}$ over K. It can be shown that in the universal Chevalley group each element of $\bar{H}$ has a unique expression of the form

$$\bar{h}_{p_1}(t_1)\, \bar{h}_{p_2}(t_2) \ldots \bar{h}_{p_l}(t_l)$$

and that $\bar{H}$ is therefore isomorphic to a direct product of copies of K^*, the multiplicative group of K. A system of relations defining the Chevalley group G may be obtained by adding to the relations defining the universal group $\bar{G}$ further relations defining the subgroup H in terms of the generators $h_{p_i}(t)$.

If $\mathfrak{L} = A_1$, when the argument of 12.1.1 breaks down, it has been shown by Steinberg that the result remains valid if the commutator relations are replaced by the relations

$$\bar{n}_r(t)\, \bar{x}_r(u)\, \bar{n}_r(t)^{-1} = \bar{x}_{-r}(-t^{-2}u),$$

where t, $u \in K$ and $t \neq 0$.

Let $\bar{G}$ be a universal Chevalley group and N be a subgroup of its centre $\bar{Z}$. Then N is normal in $\bar{G}$ and the factor group $\bar{G}/N$ is a group generated by elements $\bar{x}_r(t)$ in which the relations of theorem 12.1.1 are valid. Such groups $\bar{G}/N$ are also often called Chevalley groups. Thus all the Chevalley groups for a given $\mathfrak{L}$ and K are factor groups of the universal Chevalley group $\bar{G}$, and they all contain as factor groups the 'ordinary' Chevalley group $G \cong \bar{G}/\bar{Z}$. The group G is often called the *adjoint Chevalley group* of type $\mathfrak{L}$ over K.

It is not difficult to determine which elements

$$\prod_{i=1}^{l} \bar{h}_{p_i}(t_i)$$

of $\bar{H}$ are in the centre $\bar{Z}$. We know that these are precisely the elements for which

$$\prod_{i=1}^{l} h_{p_i}(t_i) = 1$$

in G. Now

$$\prod_{i=1}^{l} h_{p_i}(t_i) = \prod_{i=1}^{l} h(\chi_{p_i,\, t_i}) = h\left(\prod_{i=1}^{l} \chi_{p_i,\, t_i} \right)$$

as in section 7.1. Thus

$$\prod_{i=1}^{l} h_{p_i}(t_i) = 1 \text{ if and only if } \prod_{i=1}^{l} \chi_{p_i,\, t_i} = 1.$$

This is so if and only if this character takes value 1 at each fundamental root. Now

$$\chi_{p_i,\, t_i}(p_j) = t_i^{2(p_i, p_j)/(p_i, p_i)} = t_i^{A_{ij}}.$$

Thus

$$\prod_{i=1}^{l} \chi_{p_i,\, t_i} = 1$$

if and only if

$$\prod_{i=1}^{l} t_i^{A_{ij}} = 1$$

for $j = 1, \ldots, l$. Now it was shown in section 7.1 that if P is the additive group generated by the fundamental roots and Q is the additive group generated by the fundamental weights then Q/P is generated by elements $\bar{q}_1, \ldots, \bar{q}_l$ subject to relations

$$\sum_{i=1}^{l} A_{ij} \bar{q}_i = 0.$$

If we compare this with the relation

$$\prod_{i=1}^{l} t_i^{A_{ij}} = 1$$

we see that

$$\prod_{i=1}^{l} h_{p_i}(t_i) \in Z$$

if and only if $t_1, \ldots, t_l$ determines a K-character of Q/P. In fact Z is isomorphic to the group of K-characters of Q/P.

12.2 Diagonal, Field and Graph Automorphisms

We turn now to a discussion of automorphisms of Chevalley groups. An automorphism is uniquely determined by its effect on the generators $x_r(t)$.

PROPOSITION 12.2.1.   *Let* $G = \mathfrak{L}(K)$ *be an adjoint Chevalley group. A bijective map of G onto itself is an isomorphism provided it preserves the relations*

$\mathcal{R}_1.$ $x_r(t_1) \, x_r(t_2) = x_r(t_1 + t_2)$,

$\mathcal{R}_2.$ $[x_s(u), x_r(t)] = \prod_{i,j \neq 0} x_{ir+js}(C_{ijrs}(-t)^i u^j)$,

$\mathcal{R}_3.$ $h_r(t_1) \, h_r(t_2) = h_r(t_1 t_2)$, $t_1, t_2 \neq 0$,

where $h_r(t) = n_r(t) \, n_r(-1)$ *and* $n_r(t) = x_r(t) \, x_{-r}(-t^{-1}) \, x_r(t)$.

PROOF.   Such a map determines an automorphism of the universal Chevalley group $\bar{G}$ by 12.1.1. Such an automorphism leaves Z invariant since this is a characteristic subgroup of $\bar{G}$. It therefore induces an automorphism of $\bar{G}/Z$, which is isomorphic to G.   ∎

We next describe some particular kinds of automorphism.

DIAGONAL AUTOMORPHISMS. It was shown in section 7.1 that G is normalized by $\hat{H}$ in the group of all automorphisms of $\mathfrak{L}_K$. Thus if $h(\chi) \in \hat{H}$ the map

$$g \rightarrow h(\chi)\, gh(\chi)^{-1}$$

is an automorphism of G. If $h(\chi)$ is in $\hat{H}$ but not in H the automorphism is called a diagonal automorphism. If the automorphisms of $\mathfrak{L}_K$ are represented by matrices with respect to the Chevalley basis the diagonal automorphisms of G are obtained by transforming by suitable diagonal matrices.

FIELD AUTOMORPHISMS. Let f be an automorphism of the field K. Then the map

$$x_r(t) \rightarrow x_r(f(t)), \qquad r \in \Phi, \ t \in K,$$

can be extended to an automorphism of G. To show this we must verify that the relations $\mathscr{R}_1$, $\mathscr{R}_2$, $\mathscr{R}_3$ are preserved. This is clear for $\mathscr{R}_1$ and $\mathscr{R}_2$ and, since $h_r(t)$ is mapped into $h_r(f(t))$, for $\mathscr{R}_3$ also. The automorphisms obtained in this way are called field automorphisms of G.

GRAPH AUTOMORPHISMS. Automorphisms of this type arise from symmetries of the Dynkin diagram. A symmetry of the Dynkin diagram of $\mathfrak{L}$ is a permutation ρ of the nodes of the diagram such that the number of bonds joining nodes i, j is the same as the number of bonds joining nodes $\rho(i)$, $\rho(j)$ for all $i \neq j$. The non-trivial symmetries of the connected Dynkin diagrams are indicated in the figure.

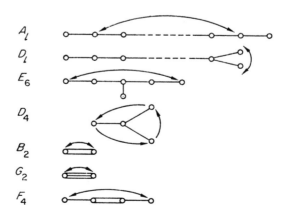

PROPOSITION 12.2.2.   *Suppose all the roots of $\mathfrak{L}$ have the same length and let ρ be a symmetry of the Dynkin diagram of $\mathfrak{L}$. Then ρ determines a permutation of the corresponding fundamental system Π. Let τ be the linear transformation of $\mathfrak{P}$ into itself which coincides with this permutation on Π. Then τ is an isometry of $\mathfrak{P}$ and $\tau(\Phi) = \Phi$.*

PROOF.   Let n_{ij} be the number of bonds joining the nodes corresponding to p_i, $p_j \in \Pi$. Then

$$n_{ij} = \frac{4(p_i, p_j)^2}{(p_i, p_i)(p_j, p_j)}$$

by section 3.4. Now $n_{ij} = n_{\rho(i)\rho(j)}$ and, since all the roots have the same length, this implies $(p_i, p_j)^2 = (p_{\rho(i)}, p_{\rho(j)})^2$. Since $(p_i, p_j) \leqslant 0$ for all i, j with $i \neq j$ we have

$$(p_i, p_j) = (p_{\rho(i)}, p_{\rho(j)}),$$

i.e.

$$(p_i, p_j) = (\tau(p_i), \tau(p_j)).$$

Thus τ extends by linearity to an isometry of $\mathfrak{P}$. Now $\tau(p_i) = p_{\rho(i)}$ and therefore $\tau w_{p_i} \tau^{-1} = w_{p_{\rho(i)}}$. It follows that $\tau w \tau^{-1} \in W$ for all $w \in W$, since W is generated by its fundamental reflections. Let $r \in \Phi$. Then $r = w(p_i)$ for some $w \in W$, $p_i \in \Pi$. Thus

$$\tau(r) = \tau w(p_i) = \tau w \tau^{-1} \cdot \tau(p_i) \in W(\Pi) = \Phi.$$

Hence $\tau(\Phi) = \Phi$.   ∎

If τ is as in 12.2.2 we write $\tau(r) = \bar{r}$ for $r \in \Phi$.

PROPOSITION 12.2.3.   *Suppose $\mathfrak{L}$ is a simple Lie algebra whose roots all have the same length, and let $r \to \bar{r}$ be a map of Φ into itself arising from a symmetry of the Dynkin diagram of $\mathfrak{L}$. Then there exist numbers $\gamma_r = \pm 1$ such that the map*

$$x_r(t) \to x_{\bar{r}}(\gamma_r t)$$

can be extended to an automorphism of G. The γ_r can be chosen so that $\gamma_r = 1$ if $r \in \Pi$ or $-r \in \Pi$.

PROOF.   We use the isomorphism theorem for simple Lie algebras given in 3.5.2. Take $\mathfrak{L}' = \mathfrak{L}$ and $p_i' = p_{\rho(i)}$. Then there exists an automorphism of $\mathfrak{L}$ such that

$$e_r \to e_{\bar{r}},$$

$$e_{-r} \to e_{-\bar{r}},$$

$$h_r \to h_{\bar{r}}$$

for $r \in \Pi$. We show that, for each $r \in \Phi$, $e_r \to \gamma_r e_{\bar{r}}$ under this automorphism, where $\gamma_r = \pm 1$. This is true for the fundamental roots. If r is positive but not fundamental we use induction on $h(r)$. r can be written as $r = r_1 + r_2$ where r_1, $r_2 \in \Phi^+$ and $h(r_1) < h(r)$, $h(r_2) < h(r)$. By induction we have

$$e_{r_1} \to \gamma_{r_1} e_{\bar{r}_1}, \qquad e_{r_2} \to \gamma_{r_2} e_{\bar{r}_2}.$$

Thus

$$N_{r_1, r_2} e_r \to \gamma_{r_1} \gamma_{r_2} N_{\bar{r}_1, \bar{r}_2} e_{\bar{r}}$$

and so

$$\gamma_r = \frac{\gamma_{r_1} \gamma_{r_2} N_{\bar{r}_1, \bar{r}_2}}{N_{r_1, r_2}} = \pm 1.$$

For negative roots we use the fact that $[e_r e_{-r}] = h_r$. This shows that $\gamma_r \gamma_{-r} = 1$, so that $\gamma_{-r} = \pm 1$ also.

We now transfer to an arbitrary field K. Since we have an automorphism of $\mathfrak{L}$ under which

$$e_r \to \gamma_r e_{\bar{r}}, \qquad h_r \to h_{\bar{r}}$$

for all $r \in \Phi$, the same map determines an automorphism of $\mathfrak{L}_K$ for any field K. Let this automorphism of $\mathfrak{L}_K$ be denoted by θ. Then

$$\theta x_r(t)\, \theta^{-1} = \theta \exp \operatorname{ad} (te_r)\, \theta^{-1}$$

$$= \exp \operatorname{ad} \theta(te_r) = \exp \operatorname{ad} (\gamma_r t e_{\bar{r}}) = x_{\bar{r}}(\gamma_r t)$$

by 5.1.1. Since the elements $x_r(t)$ generate G it follows that θ normalizes G in the automorphism group of $\mathfrak{L}_K$. Hence the map

$$g \to \theta g \theta^{-1}$$

is an automorphism of G which transforms $x_r(t)$ into $x_{\bar{r}}(\gamma_r t)$.

We obtain in this way automorphisms of order 2 of $A_l(K)$, $l \geqslant 2$; $D_l(K)$, $l \geqslant 4$; and $E_6(K)$. We also obtain automorphisms of order 3 of $D_4(K)$. ∎

12.3 Graph Automorphisms of $B_2(K)$ and $F_4(K)$

If the roots of $\mathfrak{L}$ are not all of the same length the situation is more complicated, since the argument of 12.2.2 and 12.2.3 no longer applies. The linear transformation of $\mathfrak{H}$ which extends the permutation of the fundamental roots is no longer an isometry, nor does it map the root system into itself. For example, consider the case $\mathfrak{L} = B_2$ and

$$\Phi^+ = \{a, b, a+b, 2a+b\}.$$

Then the transformation maps a into b, b into a and $2a+b$ into $2b+a$, which is not a root. Thus the symmetries of the Dynkin diagrams of type B_2, G_2 and F_4 do not in general give rise to graph automorphisms. However, it is in fact possible to find such graph automorphisms over certain fields K, although the form of these automorphisms is more complicated than that given in 12.2.3. We shall describe graph automorphisms of the groups $B_2(K)$ and $F_4(K)$ when K is a perfect field of characteristic 2, and a graph automorphism of $G_2(K)$ if K is a perfect field of characteristic 3.

LEMMA 12.3.1. *Suppose* Φ *is an indecomposable self-dual root system of type* B_2, G_2 *or* F_4. *Let* ρ *be the non-trivial symmetry of the Dynkin diagram. Then if* $\Pi = \{p_1, \ldots, p_l\}$ *is a fundamental system of* Φ,

$$\{p_{\rho(1)}, \ldots, p_{\rho(l)}\}$$

is a fundamental system of Φ *dual to* Π (cf. 3.6.1).

PROOF. The angle between $p_{\rho(i)}$, $p_{\rho(j)}$ is the same as the angle between p_i, p_j since ρ is a symmetry of the diagram. Also $p_{\rho(i)}$ is a long root if and only if p_i is a short one. ∎

LEMMA 12.3.2. *Let* $r \in \Phi$, *where* Φ *is as in* 12.3.1. *Suppose the co-root* h_r *satisfies*

$$h_r = \sum_{i=1}^{l} n_i h_{p_i}.$$

Then the element

$$\bar{r} = \sum_{i=1}^{l} n_i p_{\rho(i)}$$

is in Φ.

PROOF. Φ is self-dual and $\{p_1, \ldots, p_l\}$; $\{p_{\rho(1)}, \ldots, p_{\rho(l)}\}$ are dual fundamental systems. Thus

$$\sum_{i=1}^{l} n_i p_{\rho(i)}$$

is in Φ if and only if

$$\sum_{i=1}^{l} n_i h_{p_i}$$

has the form h_r for some $r \in \Phi$. ∎

Since $\Phi^{**}=\Phi$, the map $r\to\bar{r}$ of Φ into itself defined by 12.3.2 has order 2. We observe that if

$$r=\sum_{i=1}^{l} n_i p_i$$

then

$$\bar{r}=\sum_{i=1}^{l} n_i \frac{(p_i,\,p_i)}{(r,\,r)} p_{\rho(i)}.$$

For example, suppose $\mathfrak{L}=B_2$ and $p_1=a$, $p_2=b$ as shown in Figure 7. Then for each $r\in\Phi$, $\bar{r}$ is the root in the direction obtained by reflecting r in the line bisecting a, b.

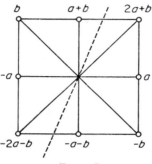

Figure 7

PROPOSITION 12.3.3. *Suppose G is a Chevalley group* $B_2(K)$ *or* $F_4(K)$, *where K is a perfect field of characteristic 2. For each root* $r\in\Phi$ *define* $\lambda(r)$ *to be 1 if r is a short root and 2 if r is a long root. Then the map*

$$x_r(t)\to x_{\bar{r}}(t^{\lambda(\bar{r})}),\qquad r\in\Phi,\ t\in K,$$

can be extended to an automorphism of G.

PROOF. We show that this map preserves the relations $\mathscr{R}_1$, $\mathscr{R}_2$, $\mathscr{R}_3$. To prove that $\mathscr{R}_1$ is preserved we must have

$$x_{\bar{r}}(t_1^{\lambda(\bar{r})})\,x_{\bar{r}}(t_2^{\lambda(\bar{r})})=x_{\bar{r}}((t_1+t_2)^{\lambda(\bar{r})}).$$

This is clear if $\lambda(\bar{r})=1$, and if $\lambda(\bar{r})=2$ it is valid over a field of characteristic 2 since $t_1^2+t_2^2=(t_1+t_2)^2$.

Consider now the relation $\mathscr{R}_3$. Under the above map we have

$$n_r(t)\to x_{\bar{r}}(t^{\lambda(\bar{r})})\,x_{-\bar{r}}((-t^{-1})^{\lambda(\bar{r})})\,x_{\bar{r}}(t^{\lambda(\bar{r})})=n_{\bar{r}}(t^{\lambda(\bar{r})}),$$

$$h_r(t)\to n_{\bar{r}}(t^{\lambda(\bar{r})})\,n_{\bar{r}}(-1)=h_{\bar{r}}(t^{\lambda(\bar{r})}),$$

again using the fact that K has characteristic 2. It is now clear that $\mathscr{R}_3$ is preserved.

Finally consider relation $\mathscr{R}_2$. If r, s are roots such that $r+s$ is a root, the roots in the subspace spanned by r, s form a system of type A_2 or B_2. (The former possibility cannot occur if $\mathfrak{L}=B_2$.) Thus it is sufficient to show that the following relations are preserved:

(a) $[x_s(u), x_r(t)]=x_{r+s}(-N_r, stu)$,

where r, s are of equal length and inclined at $2\pi/3$.

(b) $[x_s(u), x_r(t)]=x_{r+s}(-N_r, stu)$,

where r, s are short roots inclined at $\pi/2$.

(c) $[x_s(u), x_r(t)]=x_{r+s}(-N_r, stu)\, x_{2r+s}(C_{21rs}t^2u)$,

where r is short, s is long, and r, s are inclined at $3\pi/4$.

Now K has characteristic 2, so (a) becomes

$$[x_s(u), x_r(t)]=x_{r+s}(tu).$$

Applying the given map we obtain

$$[x_{\bar{s}}(u^{\lambda(\bar{s})}), x_{\bar{r}}(t^{\lambda(\bar{r})})]=x_{\overline{r+s}}((tu)^{\lambda(\overline{r+s})}).$$

However

$$[x_{\bar{s}}(u^{\lambda(\bar{s})}), x_{\bar{r}}(t^{\lambda(\bar{r})})]=x_{\bar{r}+\bar{s}}(t^{\lambda(\bar{r})}u^{\lambda(\bar{s})})$$

and so we have a valid relation, since $\overline{r+s}=\bar{r}+\bar{s}$ and $\lambda(\bar{r})=\lambda(\bar{s})=\lambda(\overline{r+s})$.

(b) becomes $[x_s(u), x_r(t)]=1$ since $N_{r, s}=\pm 2=0$. Thus (b) is preserved.

(c) becomes

$$[x_s(u), x_r(t)]=x_{r+s}(tu)\, x_{2r+s}(t^2u).$$

Applying the given map we obtain

$$[x_{\bar{s}}(u), x_{\bar{r}}(t^2)]=x_{\overline{r+s}}(t^2u^2)\, x_{\overline{2r+s}}(t^2u).$$

However

$$[x_{\bar{s}}(u), x_{\bar{r}}(t^2)]=x_{\bar{r}+\bar{s}}(t^2u)\, x_{\bar{r}+2\bar{s}}(t^2u^2).$$

Also $\overline{r+s}=\bar{r}+2\bar{s}$, $\overline{2r+s}=\bar{r}+\bar{s}$ and so (c) is preserved.

Since K is perfect, every element of K is a square. Hence the monomorphism of G into itself which extends the above map is bijective, so is an automorphism of G. ∎

12.4 A Graph Automorphism of $G_2(K)$

We now consider the group $G_2(K)$ and show that this has a graph auto-morphism when K is a perfect field of characteristic 3. Let $p_1 = a$, $p_2 = b$ be fundamental roots of G_2. Then, for each $r \in \Phi$, $\bar{r}$ is the root obtained by reflecting r in the line bisecting a, b.

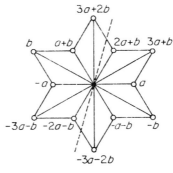

Figure 8

It is considerably more difficult to prove the existence of the graph automorphism for $G_2(K)$ than for $B_2(K)$ and $F_4(K)$. There are two reasons for this difficulty. Firstly, since K has characteristic 3 instead of 2, care must be taken over the signs of the structure constants $N_{r,\,s}$. Secondly, the graph automorphism of $G_2(K)$ has a 'nice form' only if the structure constants are chosen in a special way, whereas for $B_2(K)$ and $F_4(K)$ they could be chosen arbitrarily.

PROPOSITION 12.4.1. *Let* $G = G_2(K)$, *where* K *is a perfect field of characteristic 3. For each* $r \in \Phi$ *define* $\lambda(r)$ *to be* 1 *if* r *is short and* 3 *if* r *is long. Then the structure constants* $N_{r,\,s}$ *of* G_2 *can be chosen in such a way that the map*

$$x_r(t) \to x_{\bar{r}}(t^{\lambda(\bar{r})}), \qquad r \in \Phi, \ t \in K,$$

extends to an automorphism of G.

PROOF. It is sufficient to show that the given map transforms all relations $\mathcal{R}_1$, $\mathcal{R}_2$, $\mathcal{R}_3$ into valid relations, since the perfectness of K then implies that the map induced on G is surjective.

Now the relations $\mathscr{R}_1$ are certainly preserved. Since

$$n_r(t) \to n_{\bar{r}}(t^{\lambda(\bar{r})}),$$
$$h_r(t) \to h_{\bar{r}}(t^{\lambda(\bar{r})}),$$

the relations $\mathscr{R}_3$ are also preserved. The non-trivial relations $\mathscr{R}_2$ to be considered have the following forms:

(a) $[x_s(u), x_r(t)] = x_{r+s}(-N_{rs}tu),$

where r, s are short roots inclined at $\pi/3$.

(b) $[x_s(u), x_r(t)] = x_{r+s}(-N_{rs}tu),$

where r, s have the same length and are inclined at $2\pi/3$.

(c) $[x_s(u), x_r(t)] = x_{r+s}(-C_{11rs}tu) \, x_{2r+s}(C_{21rs}t^2u)$
$$\times x_{3r+s}(-C_{31rs}t^3u) \, x_{3r+2s}(-C_{32rs}t^3u^2),$$

where r is short, s is long, and r, s are inclined at $5\pi/6$.

Now (a) becomes
$$[x_s(u), x_r(t)] = 1,$$

since $N_{r,\,s} = \pm 3 = 0$. Hence (a) is preserved.

Applying the given map to (b) we obtain

$$[x_{\bar{s}}(u^{\lambda(\bar{s})}), x_{\bar{r}}(t^{\lambda(\bar{r})})] = x_{\overline{r+s}}(-N_{rs}t^{\lambda(\overline{r+s})}u^{\lambda(\overline{r+s})}).$$

But $$[x_{\bar{s}}(u^{\lambda(\bar{s})}), x_{\bar{r}}(t^{\lambda(\bar{r})})] = x_{\bar{r}+\bar{s}}(-N_{\bar{r},\,\bar{s}}t^{\lambda(\bar{r})}u^{\lambda(\bar{s})}).$$

Since $\overline{r+s} = \bar{r} + \bar{s}$ and $\lambda(\bar{r}) = \lambda(\bar{s}) = \lambda(\overline{r+s})$ we obtain a valid relation by transforming (b) if and only if

$$N_{\bar{r},\,\bar{s}} = N_{r,\,s}.$$

We now apply the given map to (c). We obtain

$$[x_{\bar{s}}(u), x_{\bar{r}}(t^3)] = x_{\overline{r+s}}(-C_{11rs}t^3u^3) \, x_{\overline{2r+s}}(C_{21rs}t^6u^3)$$
$$\times x_{\overline{3r+s}}(-C_{31rs}t^3u) \, x_{\overline{3r+2s}}(-C_{32rs}t^3u^2).$$

We wish to know when this is a valid relation. However

$$[x_{\bar{s}}(u), x_{\bar{r}}(t^3)] = [x_{\bar{r}}(t^3), x_{\bar{s}}(u)]^{-1}$$
$$= \{x_{\bar{s}+\bar{r}}(-C_{11\bar{s}\bar{r}}t^3u) \, x_{2\bar{s}+\bar{r}}(C_{21\bar{s}\bar{r}}t^3u^2)$$
$$\times x_{3\bar{s}+\bar{r}}(-C_{31\bar{s}\bar{r}}t^3u^3) \, x_{3\bar{s}+2\bar{r}}(-C_{32\bar{s}\bar{r}}t^6u^3)\}^{-1}$$
$$= x_{3\bar{s}+2\bar{r}}(C_{32\bar{s}\bar{r}}t^6u^3) \, x_{3\bar{s}+\bar{r}}(C_{31\bar{s}\bar{r}}t^3u^3)$$
$$\times x_{2\bar{s}+\bar{r}}(-C_{21\bar{s}\bar{r}}t^3u^2) \, x_{\bar{s}+\bar{r}}(C_{11\bar{s}\bar{r}}t^3u)$$
$$= x_{3\bar{s}+\bar{r}}(C_{31\bar{s}\bar{r}}t^3u^3) \, x_{3\bar{s}+2\bar{r}}(C_{32\bar{s}\bar{r}}t^6u^3)$$
$$\times x_{\bar{s}+\bar{r}}(C_{11\bar{s}\bar{r}}t^3u) \, x_{2\bar{s}+\bar{r}}(-C_{21\bar{s}\bar{r}}t^3u^2),$$

the terms involving $\bar{s}+\bar{r}$, $2\bar{s}+\bar{r}$ commuting since

$$N_{\bar{s}+\bar{r},\ 2\bar{s}+\bar{r}} = \pm 3 = 0.$$

Now we have

$$\overline{r+s} = 3\bar{s}+\bar{r},$$

$$\overline{2r+s} = 3\bar{s}+2\bar{r},$$

$$\overline{3r+s} = \bar{s}+\bar{r},$$

$$\overline{3r+2s} = 2\bar{s}+\bar{r}.$$

Thus the transform of (c) is a valid relation provided

$$-C_{11rs} = C_{31\bar{s}\bar{r}},$$

$$C_{31rs} = -C_{11\bar{s}\bar{r}},$$

$$C_{32rs} = C_{21\bar{s}\bar{r}},$$

$$C_{21rs} = C_{32\bar{s}\bar{r}}.$$

The first two sets of equations are clearly equivalent, since whenever r, s are a pair of roots related as in (c) $\bar{s}$, $\bar{r}$ are also. Similarly the last two sets of equations are equivalent. Thus the given map preserves all the relations provided

$$N_{r,\ s} = N_{\bar{r},\ \bar{s}} \text{ whenever } r, s \text{ are as in (b).}$$

$$\left\{ \begin{array}{l} -C_{31rs} = C_{11\bar{s}\bar{r}} \text{ whenever } r, s \text{ are as in (c).} \\ C_{32rs} = C_{21\bar{s}\bar{r}} \end{array} \right.$$

It is more convenient to write the first of these conditions in the form

$$-\tfrac{1}{2}N_{r,\ s} = N_{\bar{r},\ \bar{s}} \text{ whenever } r, s \text{ are short roots inclined at } 2\pi/3,$$

since both sides of this equation, when interpreted as rational integers, now have absolute value 1. (Note that $-\tfrac{1}{2} = 1$ in K!)

In order to complete the proof of 12.4.1 it is sufficient to prove the following lemma.

LEMMA 12.4.2. *The structure constants of the Lie algebra G_2 over the complex field can be chosen so that*

$$-\tfrac{1}{2}N_{r,\ s} = N_{\bar{r},\ \bar{s}}$$

whenever r, s are short roots inclined at $2\pi/3$,

$$-C_{31rs} = C_{11\bar{s}\bar{r}}$$

whenever r is short, s is long, and r, s are inclined at $5\pi/6$.

They then also satisfy the condition

$$C_{32rs} = C_{21\bar{s}\bar{r}}$$

whenever r is short, s is long, and r, s are inclined at $5\pi/6$.

PROOF. We show that the third condition follows from the other two. We have

$$C_{21\bar{s}\bar{r}} = \tfrac{1}{2} N_{\bar{s}, \bar{r}} N_{\bar{s}, \bar{s}+\bar{r}}$$

$$= \tfrac{1}{2} . - \tfrac{1}{6} N_{r, s} N_{r, r+s} N_{r, 2r+s} . -2 N_{s, 3r+s}$$

using relations of the first and second types. However the relation

$$N_{r, s} N_{r, 2r+s} = - N_{r+s, 2r+s} N_{s, 3r+s}$$

is easily deduced from the fact that

$$(r+s)+(2r+s)+(-s)+(-3r-s)=0,$$

as in 5.2.1. Therefore we have

$$C_{21\bar{s}\bar{r}} = -\tfrac{1}{6} N_{r, r+s} N_{r+s, 2r+s} N_{s, 3r+s}^2$$

$$= \tfrac{1}{6} N_{r+s, r} N_{r+s, 2r+s}$$

$$= \tfrac{1}{3} M_{r+s, r, 2}$$

$$= C_{32rs}.$$

We now show that it is possible to choose the signs of the structure constants $N_{r, s}$ so that the given conditions are satisfied. The result of 4.2.2 shows that the signs of $N_{a, b}$, $N_{a, a+b}$, $N_{a, 2a+b}$ and $N_{b, 3a+b}$ can

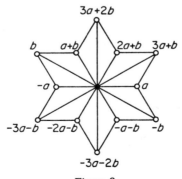

Figure 9

be chosen arbitrarily and that the remaining $N_{r,\,s}$ are then determined. Define $\epsilon_i = \pm 1$ by

$$N_{a,\,b} = \epsilon_1,$$

$$N_{a,\,a+b} = 2\epsilon_2,$$

$$N_{a,\,2a+b} = 3\epsilon_3,$$

$$N_{b,\,3a+b} = \epsilon_4.$$

Then

$$N_{a+b,\,2a+b} = -3\,\frac{\epsilon_1\epsilon_3}{\epsilon_4}$$

by 4.1.2 (iv) and the remaining $N_{r,\,s}$ may be calculated using 4.1.2 (i), (ii), (iii).

Now if r, s are short roots inclined at $2\pi/3$ so are $-r$, $-s$ and the relation $-\tfrac{1}{2}N_{r,\,s} = N_{\bar{r},\,\bar{s}}$ is equivalent to $-\tfrac{1}{2}N_{-r,\,-s} = N_{-\bar{r},\,-\bar{s}}$. Thus it is sufficient to check this relation when (r, s) is $(a, a+b)$, $(2a+b, -a)$ or $(a+b, -2a-b)$. In all these three cases the relation holds if and only if $\epsilon_2 = -\epsilon_4$.

Similarly, if r is short, s is long, and r, s are inclined at $5\pi/6$ the same is true of $-r$, $-s$; and the relation $-C_{31rs} = C_{11\bar{s}\bar{r}}$ is equivalent to $-C_{3,\,1,\,-r,\,-s} = C_{1,\,1,\,-\bar{s},\,-\bar{r}}$. Thus it is sufficient to check this relation when (r, s) is (a, b) $(2a+b, -3a-b)$ $(a, -3a-b)$ $(2a+b, -3a-2b)$ $(a+b, -3a-2b)$ or $(a+b, -b)$. In the first four cases the relation holds if and only if $\epsilon_2 = -\epsilon_3$ and in the last two cases if and only if $\epsilon_2 = -\epsilon_4$. Thus the structure constants satisfy the required conditions if and only if $-\epsilon_2 = \epsilon_3 = \epsilon_4$. This completes the proof. ∎

It may be useful to give an explicit choice of the structure constants for which the graph automorphism has the form given in 12.4.1. This is done in the table below. The structure constant $N_{r,\,s}$ appears in the row indexed by r and the column indexed by s. In this presentation we have chosen $\epsilon_1 = -1$, $\epsilon_2 = -1$, $\epsilon_3 = 1$, $\epsilon_4 = 1$.

12.5 Automorphisms of Finite Chevalley Groups

The automorphisms so far described, the diagonal, field and graph automorphisms, together with the inner automorphisms, are sufficient to generate the whole group of automorphisms of $G = \mathfrak{L}(K)$ if the field K is finite. We shall now prove this result, which is due to Steinberg.

	a	b	$a+b$	$2a+b$	$3a+b$	$3a+2b$	$-a$	$-b$	$-a-b$	$-2a-b$	$-3a-b$	$-3a-2b$
a		-1	-2	3					3	2	-1	
b	1				1				-1			-1
$a+b$	2			3			3	-1		-2		-1
$2a+b$	-3		-3				2		-2		1	1
$3a+b$		-1					-1			1		1
$3a+2b$								-1	-1	1	1	
$-a$			-3	-2	1			1	2	-3		
$-b$			1			1	-1				-1	
$-a-b$	-3	1	2			1	-2			-3		
$-2a-b$	-2		2		-1	-1	3		3			
$-3a-b$	1			-1		-1		1				
$-3a-2b$		1	1	-1	-1							

THEOREM 12.5.1. *Let $G=\mathfrak{L}(K)$ when $\mathfrak{L}$ is simple and $K=GF(q)$. Let θ be an automorphism of G. Then there exist inner, diagonal, graph and field automorphisms i, d, g, f such that $\theta=idgf$.*

PROOF. Let q be a power of the prime p. Then U, V are Sylow p-subgroups of G, by 9.4.10. By Sylow's theorem, any two Sylow p-subgroups are conjugate. Now $\theta(U)$ is also a Sylow p-subgroup of G, so is conjugate to U. Thus there exists an inner automorphism i_1 of G such that $\theta(U)=i_1(U)$. Let $\theta_1=i_1^{-1}\theta$. Then $\theta_1(U)=U$. Now $\theta_1(V)$ is a Sylow p-subgroup of G so is conjugate to U. Thus $\theta_1(V)=x^{-1}Ux$ for some $x\in G$. However, x can be written in the form $x=bn_wu$, where $b\in B$, $w\in W$, $u\in U_w^-$ as in 8.4.3. Since b normalizes U we have

$$\theta_1(V)=u^{-1}n_w^{-1}Un_wu.$$

Now $U\cap V=1$ and so $\theta_1(U)\cap\theta_1(V)=1$. Hence $U\cap u^{-1}n_w^{-1}Un_wu=1$ and it follows that $n_wUn_w^{-1}\cap U=1$. By 7.2.1 this implies that w transforms every positive root into a negative root, thus $w=w_0$ by 2.2.6. Therefore

$$n_w^{-1}Un_w=n_{w_0}^{-1}Un_{w_0}=V.$$

It follows that $\theta_1(V)=u^{-1}Vu$. Let i_2 be the inner automorphism $y\to u^{-1}yu$ of G and let $\theta_2=i_2^{-1}\theta_1$. Then $\theta_2(U)=U$ and $\theta_2(V)=V$.

Now $\mathfrak{N}_G(U)=UH$ by 8.5.2 (iii) and similarly we have $\mathfrak{N}_G(V)=VH$. Thus θ_2 leaves UH and VH invariant. Also $UH\cap VH=H$ by 7.1.3 and so $\theta_2(H)=H$. Since θ_2 leaves $UH=B$ invariant, it permutes the minimal parabolic subgroups containing B. By 8.3.2 these have the

form
$$P_{\{r\}} = B \cup Bn_rB, \qquad r \in \Pi.$$

We shall show now that $P_{\{r\}} \cap V = X_{-r}$. We have

$$P_{\{r\}} = B \cup Bn_rB$$
$$= (B \cup Bn_rB)\, n_r^{-1}$$
$$= Bn_r^{-1} \cup Bn_rBn_r^{-1}$$
$$= Bn_r^{-1} \cup BX_{-r}.$$

Consider the subset $Bn_r^{-1} \cap V$. We have

$$Bn_r^{-1} \cap V = Bn_r^{-1} \cap n_{w_0}Un_{w_0}^{-1}$$
$$\subset (Bn_r^{-1}n_{w_0} \cap n_{w_0}U)\, n_{w_0}^{-1}$$
$$\subset (Bn_r^{-1}n_{w_0}B \cap Bn_{w_0}B)\, n_{w_0}^{-1}.$$

However

$$Bn_r^{-1}n_{w_0}B \cap Bn_{w_0}B = \phi$$

by 8.2.3, thus $Bn_r^{-1} \cap V = \phi$ also. Hence

$$P_{\{r\}} \cap V = BX_{-r} \cap V = (B \cap V)\, X_{-r} = X_{-r}$$

since $B \cap V = 1$ by 7.1.2.

Now θ_2 maps V into itself and permutes the subgroups $P_{\{r\}}$ for the fundamental roots $r \in \Pi$. Thus θ_2 permutes the subgroups X_{-r} for $r \in \Pi$. One can show in a similar way that θ_2 permutes the subgroups X_r for $r \in \Pi$. However, if r, s are distinct fundamental roots, the commutator formula implies that $[X_r, X_{-s}] = 1$, whereas the result of chapter 6 implies that $[X_r, X_{-r}] \neq 1$. Thus, if r, $s \in \Pi$, $[X_r, X_{-s}] = 1$ if and only if $r \neq s$. Since this relation is preserved by θ_2 we have

$$\theta_2(X_r) = X_{\rho(r)},$$
$$\theta_2(X_{-r}) = X_{-\rho(r)}, \qquad r \in \Pi,$$

where ρ is a permutation of Π.

Let $\theta_2 . x_r(1) = x_{\rho(r)}(t_r)$. Let χ be the K-character of P defined by

$$\chi(\rho(r)) = t_r, \qquad r \in \Pi.$$

Let d be the diagonal automorphism

$$y \to h(\chi)\, yh(\chi)^{-1}$$

of G. Then we have

$$\theta_2 . x_r(1) = d . x_{\rho(r)}(1).$$

Let $\theta_3 = d^{-1}\theta_2$. Then

$$\theta_3 . x_r(1) = x_{\rho(r)}(1).$$

We shall show that

$$\theta_3 . x_{-r}(1) = x_{-\rho(r)}(1).$$

Let $\theta_3 . x_{-r}(1) = x_{-\rho(r)}(\lambda)$. We prove that $\lambda = 1$ by using the homomorphism from $SL_2(K)$ into $\langle X_r, X_{-r} \rangle$. We have

$$\begin{pmatrix} 1 & 1 \\ 0 & 1 \end{pmatrix}\begin{pmatrix} 1 & 0 \\ -1 & 1 \end{pmatrix}\begin{pmatrix} 1 & 1 \\ 0 & 1 \end{pmatrix} = \begin{pmatrix} 1 & 0 \\ -1 & 1 \end{pmatrix}\begin{pmatrix} 1 & 1 \\ 0 & 1 \end{pmatrix}\begin{pmatrix} 1 & 0 \\ -1 & 1 \end{pmatrix}$$

and it follows that

$$x_r(1)\, x_{-r}(-1)\, x_r(1) = x_{-r}(-1)\, x_r(1)\, x_{-r}(-1).$$

Applying θ_3 we obtain

$$x_{\rho(r)}(1)\, x_{-\rho(r)}(-\lambda)\, x_{\rho(r)}(1) = x_{-\rho(r)}(-\lambda)\, x_{\rho(r)}(1)\, x_{-\rho(r)}(-\lambda).$$

However

$$\begin{pmatrix} 1 & 1 \\ 0 & 1 \end{pmatrix}\begin{pmatrix} 1 & 0 \\ -\lambda & 1 \end{pmatrix}\begin{pmatrix} 1 & 1 \\ 0 & 1 \end{pmatrix} = \begin{pmatrix} 1-\lambda & 2-\lambda \\ -\lambda & 1-\lambda \end{pmatrix},$$

$$\begin{pmatrix} 1 & 0 \\ -\lambda & 1 \end{pmatrix}\begin{pmatrix} 1 & 1 \\ 0 & 1 \end{pmatrix}\begin{pmatrix} 1 & 0 \\ -\lambda & 1 \end{pmatrix} = \begin{pmatrix} 1-\lambda & 1 \\ \lambda^2-2\lambda & 1-\lambda \end{pmatrix}$$

and, since the kernel of the homomorphism $SL_2(K) \rightarrow \langle X_{\rho(r)}, X_{-\rho(r)} \rangle$ contains only I_2 and possibly $-I_2$, we have $\lambda = 1$ by comparing the above matrices. Thus

$$\theta_3 . x_r(1) = x_{\rho(r)}(1),$$

$$\theta_3 . x_{-r}(1) = x_{-\rho(r)}(1)$$

and it follows that

$$\theta_3(n_r) = \theta_3(x_r(1)\, x_{-r}(-1)\, x_r(1)) = n_{\rho(r)}.$$

We now show that the permutation ρ of Π induces a symmetry of the Dynkin diagram. Since $\theta_3(H) = H$ and $\theta_3(n_r n_s) = n_{\rho(r)}n_{\rho(s)}$, the order of the coset $n_r n_s H$ is the same as the order of the coset $n_{\rho(r)}n_{\rho(s)}H$ in the group N/H. Using the isomorphism between N/H and W, the order of $w_r w_s$ is the same as the order of $w_{\rho(r)}w_{\rho(s)}$ in W. Thus the number of bonds joining the nodes r, s in the Dynkin diagram is the same as the number of bonds joining $\rho(r)$, $\rho(s)$. Hence ρ is a symmetry of the Dynkin diagram.

H

We prove next that there is a graph automorphism g of G such that $g(X_r) = X_{\rho(r)}$ for all $r \in \Pi$. By 12.3.3 and 12.4.1 this is so provided K has characteristic 2 if $\mathfrak{L} = B_2$ or F_4 and ρ is not the identity, and K has characteristic 3 if $\mathfrak{L} = G_2$ and ρ is not the identity. Suppose ρ is the non-trivial symmetry of the Dynkin diagram of $\mathfrak{L}$, where $\mathfrak{L} = B_2$ or F_4. Then there exist roots a, $b \in \Pi$ interchanged by ρ such that the roots which are linear combinations of a, b are

$$\pm a, \ \pm b, \ \pm(a+b), \ \pm(2a+b).$$

Now $n_b X_a n_b^{-1} = X_{a+b}$. Applying θ_3 we have $n_a X_b n_a^{-1} = \theta_3(X_{a+b})$. Hence $\theta_3(X_{a+b}) = X_{2a+b}$. However $[X_{2a+b}, \ X_b] = 1$. Applying θ_3^{-1} we obtain $[X_{a+b}, \ X_a] = 1$.

But
$$[x_{a+b}(1), \ x_a(1)] = x_{2a+b}(-N_{a, \ a+b}).$$

Thus $N_{a, \ a+b} = 0$. But $N_{a, \ a+b} = \pm 2$ and so K has characteristic 2.

Now suppose $\mathfrak{L} = G_2$ and ρ is the non-trivial symmetry of the Dynkin diagram of $\mathfrak{L}$. Then ρ interchanges the fundamental roots a, b of $\mathfrak{L}$ and the other roots are

$$\pm a, \ \pm b, \ \pm(a+b), \ \pm(2a+b), \ \pm(3a+b), \ \pm(3a+2b)$$

Now $n_b X_a n_b^{-1} = X_{a+b}$ and so, applying θ_3, we have $n_a X_b n_a^{-1} = \theta_3(X_{a+b})$. Thus $\theta_3(X_{a+b}) = X_{3a+b}$. Also $n_a X_{a+b} n_a^{-1} = X_{2a+b}$ and, applying θ_3, $n_b X_{3a+b} n_b^{-1} = \theta_3(X_{2a+b})$. Thus $\theta_3(X_{2a+b}) = X_{3a+2b}$. Now $[X_{3a+2b}, \ X_b] = 1$. Applying θ_3^{-1} gives $[X_{2a+b}, \ X_a] = 1$. However

$$[x_{2a+b}(1), \ x_a(1)] = x_{3a+b}(-N_{a, \ 2a+b}).$$

Thus $N_{a, \ 2a+b} = 0$. But $N_{a, \ 2a+b} = \pm 3$ and so K has characteristic 3.

Thus in each case there is a graph automorphism g of G such that $g(X_r) = X_{\rho(r)}$ for $r \in \Pi$. Let $\theta_4 = g^{-1}\theta_3$. Then $\theta_4(X_r) = X_r$ and $\theta_4(X_{-r}) = X_{-r}$ for all $r \in \Pi$. Also θ_4 fixes $x_r(1)$ and $x_{-r}(1)$, thus θ_4 fixes n_r for all $r \in \Pi$. It follows that θ_4 fixes each X_r, $r \in \Phi$. For $r = w(s)$ for some $s \in \Pi$ and $w = w_{r_1} \ldots w_{r_k}$ with $r_i \in \Pi$. Hence

$$X_r = n_{r_1} \ldots n_{r_k} X_s n_{r_k}^{-1} \ldots n_{r_1}^{-1}$$

and so $\theta_4(X_r) = X_r$.

Let r, $s \in \Pi$ be fundamental roots which are joined in the Dynkin diagram. Then $r + s \in \Phi$. Let

$$\theta_4 . x_r(t) = x_r(f(t)),$$
$$\theta_4 . x_s(t) = x_s(g(t)),$$
$$\theta_4 . x_{r+s}(t) = x_{r+s}(h(t)).$$

Since

$$[x_s(u), x_r(t)] = x_{r+s}(-N_r, {}_s tu) \ldots,$$

we have, on applying θ_4,

$$[x_s(g(u)), x_r(f(t))] = x_{r+s}(h(-N_r, {}_s tu)) \ldots.$$

But

$$[x_s(g(u)), x_r(f(t))] = x_{r+s}(-N_r, {}_s f(t) g(u)) \ldots$$

and so

$$h(-N_r, {}_s tu) = -N_r, {}_s f(t) g(u).$$

Now $N_r, {}_s = \pm 1$, hence $h(tu) = f(t) g(u)$ for all t, $u \in K$. Putting $u = 1$ we have $f(t) = h(t)$ and putting $t = 1$ we have $g(u) = h(u)$. Hence

$$f(t) = g(t) = h(t)$$

for all t. We also have $f(tu) = f(t) f(u)$ for all t, u. Since the map $f : K \to K$ is bijective and additive it must be an automorphism of K.

Now $\mathfrak{L}$ is simple and so its Dynkin diagram is connected. Thus there exists an automorphism f of K such that

$$\theta_4 . x_r(t) = x_r(f(t))$$

for all $r \in \Pi$. f can be extended to a field automorphism of G. Let $\theta_5 = f^{-1}\theta_4$. Then

$$\theta_5 . x_r(t) = x_r(t)$$

for all $r \in \Pi$, $t \in K$. Since these elements $x_r(t)$ generate G, θ_5 must be the identity.

Now

$$\theta_5 = f^{-1}\theta_4 = f^{-1}g^{-1}\theta_3 = f^{-1}g^{-1}d^{-1}\theta_2 = f^{-1}g^{-1}d^{-1}i_2^{-1}\theta_1$$

$$= f^{-1}g^{-1}d^{-1}i_2^{-1}i_1^{-1}\theta.$$

Therefore $\theta = i_1 i_2 dgf$ and the theorem is proved.   ∎

(We note that the proof that the function f on K satisfies

$$f(t_1) f(t_2) = f(t_1 t_2)$$

breaks down if $\mathfrak{L} = A_1$ since we cannot choose two distinct fundamental roots. It is not difficult to find an alternative argument to cover this case.)

The result of 12.5.1 is also valid for perfect fields which are not finite. However the proof is more difficult there as Sylow's theorem cannot be used to show that the automorphism can be modified to fix U and V.

CHAPTER 13

The Twisted Simple Groups

We have shown that the Chevalley groups of type A_l, B_l, C_l, D_l can be identified with certain classical groups. However only some of the classical groups can be interpreted as Chevalley groups. Even over a finite field there are classical groups which are not Chevalley groups, for example the unitary groups (cf. section 1.5) and the second class of orthogonal groups in even dimension (cf. section 1.4). We shall describe in this chapter how to construct certain additional simple groups, the so-called 'twisted types'; some of which can be identified with classical groups. Every classical group over a finite field can be interpreted as a Chevalley group or as a twisted group. In general, the classical groups which are Chevalley groups or twisted groups are the linear and symplectic groups, and the orthogonal and unitary groups corresponding to forms whose Witt index is sufficiently large. The remaining classical groups can be interpreted as 'non-split' groups of Lie type, but the discussion of such groups is beyond the scope of the present volume. In addition to the extra classical groups, we obtain as twisted groups several new families of exceptional groups.

The twisted simple groups were discovered independently by Steinberg, Tits and Hertzig. The development we shall give follows Steinberg's approach.

13.1 The Reflection Subgroup W^1

The twisted groups will be obtained as certain subgroups of the Chevalley groups $G = \mathfrak{L}(K)$. The twisted groups only exist in the cases when the Dynkin diagram of $\mathfrak{L}$ has a non-trivial symmetry. It will be shown that the twisted groups are also groups with a (B, N)-pair and that the Weyl group W^1 of this (B, N)-pair is a reflection group which is a subgroup of the Weyl group W of G.

Before discussing the twisted groups themselves we shall therefore first consider the Weyl group W in order to describe how the reflection subgroup W^1 arises. W operates as usual as a Euclidean reflection group on an l-dimensional space $\mathfrak{P}$.

Let ρ be a non-trivial symmetry of the Dynkin diagram of $\mathfrak{L}$. Then there is a unique isometry τ of $\mathfrak{P}$ such that $\tau(r)$ is a positive multiple of $\rho(r) = \bar{r}$ for all fundamental roots $r \in \Pi$. τ satisfies the conditions:

$\tau(r) = \bar{r}$ if all the roots of $\mathfrak{L}$ have the same length.

$$\tau(r) = \begin{cases} \dfrac{1}{\sqrt{2}}\, \bar{r} \text{ if } r \text{ is short, for } \mathfrak{L} = B_2 \text{ or } F_4, \\ \sqrt{2}\, \bar{r} \text{ if } r \text{ is long.} \end{cases}$$

$$\tau(r) = \begin{cases} \dfrac{1}{\sqrt{3}}\, \bar{r} \text{ if } r \text{ is short, for } \mathfrak{L} = G_2, \\ \sqrt{3}\, \bar{r} \text{ if } r \text{ is long.} \end{cases}$$

It is clear that the order of τ as an isometry of $\mathfrak{P}$ is equal to the order of ρ as a permutation of Π.

Definition. We denote by $\mathfrak{P}^1$ the set of $v \in \mathfrak{P}$ such that $\tau(v) = v$.

For each $v \in \mathfrak{P}$ we denote by v^1 the projection of v on to the subspace $\mathfrak{P}^1$. Then v^1 is the average of the vectors in the orbit of v under τ. For this average is certainly in $\mathfrak{P}^1$, but since

$$(v, x) = (\tau(v), x) = (\tau^2(v), x) = \ldots$$

for $x \in \mathfrak{P}^1$, the average has the same scalar product with x as v does.

We now consider the relation between τ and W. Since $\tau(r)$ is a positive multiple of $\bar{r}$ for all $r \in \Pi$ we have

$$\tau w_r \tau^{-1} = w_{\bar{r}}, \qquad r \in \Pi.$$

Since the fundamental reflections generate W, this shows that τ normalizes W in the group of all isometries of $\mathfrak{P}$.

Definition. We denote by W^1 the set of $w \in W$ such that $\tau w \tau^{-1} = w$.

LEMMA 13.1.1. W^1 *operates faithfully on* $\mathfrak{P}^1$.

PROOF. Let $w \in W^1$ and $v \in \mathfrak{P}^1$. Then

$$\tau . w(v) = w\tau(v) = w(v).$$

Thus $w(v) \in \mathfrak{P}^1$, and so W^1 transforms $\mathfrak{P}^1$ into itself.

Now suppose $w \in W^1$ and $w \neq 1$. Then there exists a root $r \in \Phi^+$ such that $w(r) \in \Phi^-$. However, the transforms of r by the powers of τ are also positive, thus $r^1 \succ 0$. Similarly $w(r^1) = w(r)^1 \prec 0$. Thus w transforms some positive element of $\mathfrak{P}^1$ into a negative element, so cannot be the identity. ∎

We now show that W^1 is generated by elements of order 2.

PROPOSITION 13.1.2. *Let J be an orbit of Π under ρ. Let W_J be the subgroup of W generated by the elements w_r for $r \in J$. Let w_0^J be the element of W_J which transforms every positive root in Φ_J into a negative root. Then $w_0^J \in W^1$, and W^1 is generated by the elements w_0^J for the different ρ-orbits of Π.*

PROOF. Since $\tau w_r \tau^{-1} = w_{\bar{r}} \in W_J$ for all $r \in J$, we have $\tau w \tau^{-1} \in W_J$ for all $w \in W_J$. Thus $\tau w_0^J \tau^{-1} \in W_J$. However, τ preserves the sign of each root; thus

$$\tau w_0^J \tau^{-1}(\Phi_J^+) = \Phi_J^-.$$

Hence $\tau w_0^J \tau^{-1} = w_0^J$, and it follows that $w_0^J \in W^1$.

Now let $w \in W^1$ with $w \neq 1$. Then there exists a root $r \in \Pi$ such that $w(r) \in \Phi^-$. Let J be the ρ-orbit of Π containing r. Then $w(s) \in \Phi^-$ for all $s \in J$ since τ preserves the sign of each root. However w_0^J changes the signs of all roots in Φ_J but of none in $\Phi - \Phi_J$. Hence

$$l(ww_0^J) = l(w) - l(w_0^J).$$

We show that W^1 is generated by the elements w_0^J by induction on $l(w)$. If $w \in W^1$, $w \neq 1$, choose J as above. Then by induction we have

$$ww_0^J = w_0^{J_1} w_0^{J_2} \ldots w_0^{J_k}$$

for certain orbits J_i of Π. Thus

$$w = w_0^{J_1} w_0^{J_2} \ldots w_0^{J_k} w_0^J$$

as required. ∎

We show next that the elements w_0^J are reflections when restricted to $\mathfrak{P}^1$.

LEMMA 13.1.3. *w_0^J coincides with w_{r^1} on $\mathfrak{P}^1$ for each root $r \in J$.*

PROOF. Let r, $s \in J$. Since r, s are in the same ρ-orbit of Π, r^1 is a positive multiple of s^1. It follows that the projections r^1 for $r \in \Phi_J^+$ are all positive multiples of one another. Now

$$w_0^J(r^1) = w_0^J(r)^1$$

and $w_0^J(r) \in \Phi_J^-$ for all $r \in \Phi_J^+$. Thus $w_0^J(r^1)$ is a negative multiple of r^1 and, being an isometry, w_0^J must transform r^1 into $-r^1$.

Now let $v \in \mathfrak{P}^1$ satisfy $(r^1, v) = 0$. Then $(s^1, v) = 0$ for all $s \in J$ and so $(s, v) = 0$ also. Hence $w_0^J(v) = v$. Thus w_0^J coincides with the reflection w_{r^1} on $\mathfrak{P}^1$. ∎

COROLLARY 13.1.4. *The reflections w_{r^1} of $\mathfrak{P}^1$, for all $r \in \Pi$, generate the group W^1 of isometries of $\mathfrak{P}^1$.*

13.2 The System Φ^1 in $\mathfrak{P}^1$

We have seen that W^1 may be regarded as a reflection group operating on $\mathfrak{P}^1$. We denote by Φ^1 the set of vectors $r^1 \in \mathfrak{P}^1$ for all $r \in \Phi$; and by Π^1 the set of vectors $r^1 \in \mathfrak{P}^1$ for all $r \in \Pi$. We shall show that Φ^1 behaves rather like a root system for W^1 and that Π^1 behaves like a fundamental system of Φ^1. However, both in Φ^1 and Π^1 there can be positive multiples of a vector distinct from the vector itself. In order to control this situation we introduce an equivalence relation on Φ.

LEMMA 13.2.1. *The sets $w(\Phi_J^+)$ form a partition of Φ as w runs through the elements of W^1 and J runs through the ρ-orbits of Π. r and s are in the same set if and only if r^1 is a positive multiple of s^1.*

PROOF. We show first that each root is in some set $w(\Phi_J^+)$. Let w_0 be the element of W which transforms every positive root into a negative root. Since τ preserves this sign of each root, $\tau w_0 \tau^{-1}$ transforms every positive root into a negative root. Thus $\tau w_0 \tau^{-1} = w_0$ and $w_0 \in W^1$.

Let $r \in \Phi^+$. Then $w_0(r) \in \Phi^-$ and by 13.1.2 we have $w_0 = w_0^{J_1} \ldots w_0^{J_k}$ for certain ρ-orbits J_i of Π. Thus there exists an integer i such that

$$w_0^{J_{i+1}} \ldots w_0^{J_k}(r) \in \Phi^+,$$

$$w_0^{J_i} w_0^{J_{i+1}} \ldots w_0^{J_k}(r) \in \Phi^-.$$

However the only positive roots transformed by $w_0^{J_i}$ into negative roots are those in $\Phi_{J_i}^+$, and therefore $w_0^{J_{i+1}} \ldots w_0^{J_k}(r) \in \Phi_{J_i}^+$. Thus

$$r \in w_0^{J_k} \ldots w_0^{J_{i+1}}(\Phi_{J_i}^+),$$
$$-r \in w_0^{J_k} \ldots w_0^{J_{i+1}} w_0^{J_i}(\Phi_{J_i}^+).$$

Hence every root is in one of the given sets.

Now if r, s are two roots in Φ_J^+, we have seen that r^1 is a positive multiple of s^1. Transforming by $w \in W^1$ we see that $w(r^1) = w(r)^1$ is a positive multiple of $w(s^1) = w(s)^1$.

Suppose conversely that $r, s \in \Phi$ and that r^1 is a positive multiple of s^1. Now we have shown that $r \in w(\Phi_J^+)$ for a suitable element $w \in W^1$ and a suitable subset J of Π. Then $w^{-1}(r) \in \Phi_J^+$ and so $w^{-1}(r)^1$ is a positive element which is a linear combination of roots in J. However

$$w^{-1}(r)^1 = w^{-1}(r^1) = w^{-1}(\lambda s^1) = \lambda w^{-1}(s)^1$$

for some $\lambda > 0$, and so $w^{-1}(s)^1$ is a positive vector which is a linear combination of roots in J. The same must be true of $w^{-1}(s)$, as J is a ρ-orbit. Hence $w^{-1}(s) \in \Phi_J^+$ and $s \in w(\Phi_J^+)$. Thus s is contained in each of the given sets in which r is contained. Therefore the given sets cover Φ and any two of them either coincide or have no elements in common. ∎

We can now describe the extent to which Φ^1 and Π^1 act as a root system and fundamental system for W^1 acting on $\mathfrak{P}^1$.

PROPOSITION 13.2.2. (i) Φ^1 *spans* $\mathfrak{P}^1$.

(ii) *Every element of* Φ^1 *is a linear combination of elements of* Π^1 *with coefficients all non-negative or all non-positive.*

(iii) *A basis of* $\mathfrak{P}^1$ *may be obtained by picking one element of* Π^1 *out of each set of positive multiples.*

(iv) *If* $r^1 \in \Phi^1$ *then there is an element of* W^1 *which coincides with* w_{r^1} *on* $\mathfrak{P}^1$.

(v) *If* $r^1, s^1 \in \Phi^1$ *then* $w_{r^1}(s^1) \in \Phi^1$.

PROOF. Since Φ spans $\mathfrak{P}$ it follows immediately that Φ^1 spans $\mathfrak{P}^1$. Since each element of Φ is a linear combination of elements of Π with coefficients all non-negative or all non-positive, the same is true for Φ^1 and Π^1. Let $J_1, \ldots, J_k$ be the ρ-orbits of Π. If we pick out one element of Π^1 from each set of positive multiples we obtain a set $r_1^1, r_2^1, \ldots, r_k^1$, where $r_i^1 \in \Phi_{J_i}^+$ by 13.2.1. These elements are linearly independent, so form a basis for $\mathfrak{P}^1$.

Let $r^1 \in \Phi^1$. We show there is an element of W^1 which coincides with w_{r^1} on $\mathfrak{P}^1$. Now r is contained in one of the sets $w(\Phi_J^+)$ of 13.2.1. We may assume that $r = w(s)$, where $s \in J$, $w \in W^1$, since r^1 will be changed only by a positive multiple by taking $s \in J$, and w_{r^1} will be unchanged. Then

$$r^1 = w(s)^1 = w(s^1).$$

Now w_{s^1} coincides with an element of W^1 on $\mathfrak{P}^1$ by 13.1.3. Hence

$$w_{r^1} = w w_{s^1} w^{-1}$$

coincides with an element of W^1 on $\mathfrak{P}^1$ also.

Finally, let $r^1, s^1 \in \Phi^1$. There is an element $w \in W^1$ such that w_{r^1} coincides with w on $\mathfrak{P}^1$. Hence we have

$$w_{r^1}(s^1) = w(s^1) = w(s)^1 \in \Phi^1$$

and the proof is complete.   ∎

13.3 The Structure of W^1

We now describe the structure of W^1 in the individual cases which arise. Let $r_1, \ldots, r_k$ be a set of roots, one from each ρ-orbit of Π. Then $r_1^1, \ldots, r_k^1$ are linearly independent vectors in $\mathfrak{P}^1$, and W^1, considered as a group of isometries of $\mathfrak{P}^1$, is generated by $w_{r_1}^1, \ldots w_{r_k}^1$. By considering the angles between the vectors $r_1^1, \ldots, r_k^1$ it is usually possible to identify W^1 with one of the Weyl groups of rank k.

13.3.1 Type A_l

Let the fundamental roots be $p_1, \ldots, p_l$, numbered as in the diagram.

Suppose l is odd and write $l = 2k - 1$. Then the vectors $r_1^1, \ldots, r_k^1$ are

$$\tfrac{1}{2}(p_1 + p_{2k-1}), \ \tfrac{1}{2}(p_2 + p_{2k-2}), \ \ldots, \ \tfrac{1}{2}(p_{k-1} + p_{k+1}), \ p_k.$$

These form a fundamental system of type C_k. Thus W^1 is isomorphic to $W(C_k)$.

Suppose l is even and write $l = 2k$. Then the vectors $r_1^1, \ldots, r_k^1$ are

$$\tfrac{1}{2}(p_1 + p_{2k}), \ \tfrac{1}{2}(p_2 + p_{2k-1}), \ \ldots, \ \tfrac{1}{2}(p_k + p_{k+1}),$$

which form a fundamental system of type B_k. Thus W^1 is isomorphic to $W(B_k)$.

13.3.2 Type D_l

The symmetry in this case is as shown in the diagram.

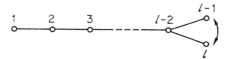

The vectors $r_1^1, \ldots, r_k^1$ are:

$$p_1, p_2, p_3, \ldots, p_{l-2}, \tfrac{1}{2}(p_{l-1} + p_l).$$

These form a fundamental system of type B_{l-1}. Thus W^1 is isomorphic to $W(B_{l-1})$.

13.3.3 Type E_6

The symmetry is as shown in the diagram.

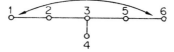

The vectors $r_1^1, \ldots, r_k^1$ are:

$$\tfrac{1}{2}(p_1 + p_6), \ \tfrac{1}{2}(p_2 + p_5), \ p_3, \ p_4.$$

These form a fundamental system of type F_4, thus W^1 is isomorphic to $W(F_4)$.

13.3.4 Type D_4

This time the symmetry has order 3.

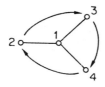

The vectors $r_1^1, \ldots r_k^1$ are:

$$p_1, \tfrac{1}{3}(p_2 + p_3 + p_4).$$

These form a fundamental system of type G_2. Thus W^1 is isomorphic to $W(G_2)$.

13.3.5 Type B_2

In this case $k = 1$ and so W is isomorphic to $W(A_1)$, a cyclic group of order 2.

13.3.6 Type G_2

$k = 1$ in this case also, and so W is isomorphic to $W(A_1)$.

13.3.7 Type F_4

The situation here is a little more complicated. Suppose p_1, p_2 are the long fundamental roots and p_3, p_4 the short ones, as shown in the diagram.

Then the vectors $r_1^1, \ldots, r_k^1$ may be taken as:

$$\tfrac{1}{2}(p_1 + \sqrt{2}p_4), \ \tfrac{1}{2}(p_2 + \sqrt{2}p_3).$$

The angle between these two vectors is given by

$$\cos \theta = \frac{(p_1 + \sqrt{2}p_4, \, p_2 + \sqrt{2}p_3)}{|\, p_1 + \sqrt{2}p_4 \,| . |\, p_2 + \sqrt{2}p_3 \,|}.$$

Now it is readily verified that

$$(p_1 + \sqrt{2}p_4, \, p_2 + \sqrt{2}p_3) = -|\, p_1 \,|^2,$$
$$|\, p_1 + \sqrt{2}p_4 \,| = \sqrt{2}\,|\, p_1 \,|,$$
$$|\, p_2 + \sqrt{2}p_3 \,| = 2\,|\, p_1 \,| \sin \pi/8.$$

Hence

$$\cos \theta = \frac{-1}{2\sqrt{2} \sin \pi/8}.$$

But $2 \sin \pi/8 \cos \pi/8 = \sin \pi/4 = 1/\sqrt{2}$, and so

$$\cos \theta = -\cos \pi/8 = \cos 7\pi/8.$$

Thus $\theta = 7\pi/8$.

It follows that in this case W^1 is not one of the Weyl groups of the simple root systems. It is generated by two reflections in a plane in axes inclined at $7\pi/8$, and is therefore a dihedral group of order 16.

13.3.8

The type of W^1 can conveniently be memorized by identifying nodes in the Dynkin diagram corresponding to fundamental roots in the same ρ-orbit. The Dynkin diagram of W^1 is given in the following table.

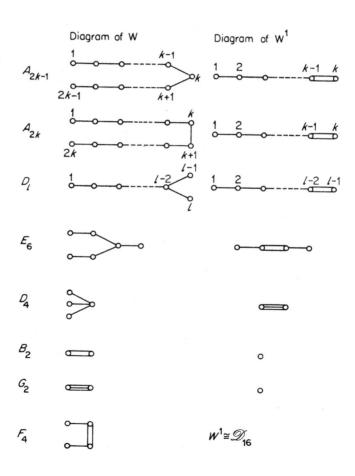

13.4 Definition of the Twisted Groups

Let G be the Chevalley group $\mathfrak{L}(K)$ and ρ be a non-trivial symmetry of the Dynkin diagram of $\mathfrak{L}$. We suppose that K is a perfect field of characteristic 2 if $\mathfrak{L} = B_2$ or F_4 and that K is perfect of characteristic 3 if $\mathfrak{L} = G_2$. Thus there is a graph automorphism g of G such that

$$g(X_r) = X_{\bar{r}}$$

for all $r \in \Pi$, where $\bar{r} = \rho(r)$ (see 12.2.3, 12.3.3 and 12.4.1). Now g commutes with each field automorphism f of G. For if all the roots have the same length we have

$$gf(x_r(t)) = g \cdot x_r(f(t)) = x_{\bar{r}}(f(t)) = f \cdot x_{\bar{r}}(t) = fg(x_r(t))$$

for $r \in \Pi$; whereas if $\mathfrak{L} = B_2$, G_2 or F_4 we have

$$gf(x_r(t)) = g \cdot x_r(f(t)) = x_{\bar{r}}(f(t)^{\lambda(\bar{r})}) = x_{\bar{r}}(f(t^{\lambda(\bar{r})}))$$
$$= f \cdot x_{\bar{r}}(t^{\lambda(\bar{r})}) = fg \cdot (x_r(t))$$

for $r \in \Pi$.

Let n be the order of the symmetry ρ. Then n is either 2 or 3. If all the roots of $\mathfrak{L}$ have the same length then

$$g^n \cdot x_r(t) = x_r(t), \qquad r \in \Pi,$$

and so $g^n = 1$. Otherwise $\mathfrak{L} = B_2$, G_2 or F_4 and $n = 2$. In this case

$$g^2 \cdot x_r(t) = g \cdot x_{\bar{r}}(t^{\lambda(\bar{r})}) = x_r(t^{\lambda(r)\lambda(\bar{r})}) = x_r(t^p),$$

where p is the characteristic of K. Thus g^2 is the field automorphism which raises every coefficient to the pth power.

Let σ be the automorphism of G defined by $\sigma = gf$, and suppose that a non-trivial field automorphism f is chosen so that $\sigma^n = 1$—then σ will have the same order as ρ. Since $\sigma^n = g^n f^n$ the condition to be satisfied by f may be stated as follows:

$f^n = 1$ if all the roots of $\mathfrak{L}$ have the same length,

$pf^2 = 1$ if $\mathfrak{L} = B_2$, G_2 or F_4, where p is interpreted as the pth-power map on K.

PROPOSITION 13.4.1. *Let $G = \mathfrak{L}(K)$ be a Chevalley group whose Dynkin diagram has a non-trivial symmetry ρ. Assume that K is perfect of characteristic 2 if $\mathfrak{L} = B_2$ or F_4, and that K is perfect of characteristic 3 if $\mathfrak{L} = G_2$. Let g be the graph automorphism corresponding to ρ and f be a non-trivial* *field automorphism chosen so that* $\sigma = gf$ *satisfies* $\sigma^n = 1$, *where* n *is the order of* ρ. *Then we have*

$$\sigma(U) = U, \qquad \sigma(V) = V, \qquad \sigma(H) = H, \qquad \sigma(N) = N$$

and σ *operates on* $N/H \cong W$ *according to the formula* $\sigma(w_r) = w_{\bar{r}}$ *for all* $r \in \Pi$.

PROOF. It is clear from the definition of σ that $\sigma(U) = U$ and $\sigma(V) = V$. Thus $\sigma(UH) = UH$ and $\sigma(VH) = VH$ since UH is the normalizer of U and VH is the normalizer of V. Also $\sigma(H) = H$ since $H = UH \cap VH$. Now N is the subgroup of G generated by H and the elements n_r for $r \in \Pi$. Also

$$\sigma(n_r) = \sigma(x_r(1)\, x_{-r}(-1)\, x_r(1)) = x_{\bar{r}}(1)\, x_{-\bar{r}}(-1)\, x_{\bar{r}}(1) = n_{\bar{r}}$$

for $r \in \Pi$. Hence $\sigma(N) = N$. It follows that σ induces an automorphism of $N/H \cong W$ such that $\sigma(w_r) = w_{\bar{r}}$ for all $r \in \Pi$. ∎

The twisted groups are defined as certain subgroups of the Chevalley groups which are fixed elementwise by automorphisms σ of the type considered in 13.4.1. Let G be a Chevalley group admitting an automorphism σ as in 13.4.1. Then σ fixes U, V, H and N. We define subgroups U^1, V^1, G^1, H^1, N^1 as follows:

Definition 13.4.2.   (i) U^1 is the set of elements $x \in U$ such that $\sigma(x) = x$.

   (ii) V^1 is the set of elements $x \in V$ such that $\sigma(x) = x$.

   (iii) G^1 is the subgroup of G generated by U^1 and V^1.

   (iv) H^1 is the intersection of G^1 and H.

   (v) N^1 is the intersection of G^1 and N.

We shall show that G^1 is (apart from a few exceptional cases) a simple group and that the subgroups U^1, V^1, H^1, N^1 play an analogous rôle in G^1 to the subgroups U, V, H, N of G. Although every element of G^1 is fixed by σ, G^1 is not necessarily the subgroup of all σ-invariant elements of G.

Now the operation of σ on W is the same as the operation of transformation by the isometry τ determined by the symmetry ρ. Thus the subgroup W^1 consisting of all elements $w \in W$ such that $\sigma(w) = w$ is just the subgroup considered in section 13.1. We shall show that W^1 plays the rôle of the Weyl group of G^1.

13.5 Existence of a (B, N)-pair in the Twisted Groups

It was shown in 13.2.1 that the subsets $w(\Phi_J^+)$ form a partition of Φ as w runs through W^1 and J runs through the ρ-orbits of Π. We shall now look at the individual equivalence classes a little more closely. The roots in the equivalence class $w(\Phi_J^+)$ are the positive roots in the system $w(\Phi_J)$ with respect to the fundamental system $w(J)$. The type of this root system is the same as the type of the fundamental system J. The various possibilities for the type of J may easily be obtained by inspection of the Dynkin diagrams, and are as follows.

If $\mathfrak{L} = A_{2k-1}$, D_l or E_6, each ρ-orbit J of Π has type A_1 or $A_1 \times A_1$. If $\mathfrak{L} = A_{2k}$, J has type $A_1 \times A_1$ or A_2. If $\mathfrak{L} = D_4$, J has type A_1 or

$$A_1 \times A_1 \times A_1.$$

If $\mathfrak{L} = B_2$, J has type B_2. If $\mathfrak{L} = G_2$, J has type G_2. Finally, if $\mathfrak{L} = F_4$, J has type $A_1 \times A_1$ or B_2.

Thus every equivalence class S in Φ may be regarded as a positive system of type A_1, $A_1 \times A_1$, $A_1 \times A_1 \times A_1$, A_2, B_2 or G_2.

LEMMA 13.5.1. *Let S be one of the equivalence classes in Φ defined in 13.2.1. Let X_S be the subgroup of G generated by the subgroups X_r for all $r \in S$. Then*

$$X_S = \prod_{r \in S} X_r,$$

where the product is taken over the roots in S in any order. Also $\sigma(X_S) = X_S$, and the subgroup X_S^1 of elements of X_S fixed by σ satisfies $X_S^1 \neq 1$.

PROOF. We have seen that S is the positive system, with respect to some ordering, of some root system contained in Φ. The subgroup X_S therefore plays the rôle of U in the Chevalley group corresponding to this subsystem of Φ. The factorization

$$X_S = \prod_{r \in S} X_r$$

therefore follows from the results of chapter 5. For each $r \in S$ we have

$$\sigma(X_r) = X_{\bar{r}}$$

for some $\bar{r} \in S$, thus it follows that $\sigma(X_S) = X_S$.

Now consider the subgroup X_S^1 of σ-invariant elements of X_S. If S is of type A_1, A_2, B_2 or G_2, there is a root $r \in S$ such that $r = \bar{r}$. If S is of type B_2 or G_2 we have for such a root

$$\sigma.x_r(1)=x_r(1)$$

and so $x_r(1)$ is a non-unit element of X_S^1. If S is of type A_1 or A_2 we have instead

$$\sigma.x_r(1)=x_{\bar{r}}(\gamma_r), \qquad \gamma_r=\pm 1,$$

by 12.2.3. If $\gamma_r=1$, $x_r(1)$ is the required non-unit element of X_S^1. If $\gamma_r=-1$, then σ has order 2 (since σ cannot have order 3) and therefore f has order 2 also. Now there is an element $t\neq 0\in K$ such that $f(t)=-t$. Thus

$$\sigma.x_r(t)=x_r(-f(t))=x_r(t)$$

and $x_r(t)$ is the required element.

Now suppose S is of type $A_1\times A_1$ or $A_1\times A_1\times A_1$. Then we have

$$\sigma.x_r(1)=x_{\bar{r}}(\gamma_r), \qquad \gamma_r=\pm 1,$$

for any root $r\in S$, by 12.2.3 and 12.3.3. We consider the elements $x_r(1)\,x_{\bar{r}}(\gamma_r)$, $x_r(1)\,x_{\bar{r}}(\gamma_r)\,x_{\bar{\bar{r}}}(\gamma_{\bar{r}})$ when S has type $A_1\times A_1$, $A_1\times A_1\times A_1$ respectively. Now

$$x_r(1)\xrightarrow{\sigma}x_{\bar{r}}(\gamma_r)\xrightarrow{\sigma}x_r(\gamma_r\gamma_{\bar{r}}),$$
$$x_r(1)\xrightarrow{\sigma}x_{\bar{r}}(\gamma_r)\xrightarrow{\sigma}x_{\bar{\bar{r}}}(\gamma_r\gamma_{\bar{r}})\xrightarrow{\sigma}x_r(\gamma_r\gamma_{\bar{r}}\gamma_{\bar{\bar{r}}})$$

in the two above cases, and so we have $\gamma_r\gamma_{\bar{r}}=1$ and $\gamma_r\gamma_{\bar{r}}\gamma_{\bar{\bar{r}}}=1$ respectively. Thus

$$\sigma.x_r(1)\,x_{\bar{r}}(\gamma_r)=x_{\bar{r}}(\gamma_r)\,x_r(\gamma_r\gamma_{\bar{r}})=x_r(1)\,x_{\bar{r}}(\gamma_r),$$
$$\sigma.x_r(1)\,x_{\bar{r}}(\gamma_r)\,x_{\bar{\bar{r}}}(\gamma_r\gamma_{\bar{r}})=x_{\bar{r}}(\gamma_r)\,x_{\bar{\bar{r}}}(\gamma_r\gamma_{\bar{r}})\,x_r(\gamma_r\gamma_{\bar{r}}\gamma_{\bar{\bar{r}}})$$
$$=x_r(1)\,x_{\bar{r}}(\gamma_r)\,x_{\bar{\bar{r}}}(\gamma_r\gamma_{\bar{r}})$$

and so we have a non-unit element of X_S fixed by σ in each case. ∎

The elements of X_S^1 will be described in detail in 13.6.3.

PROPOSITION 13.5.2. (i) *For each $w\in W^1$ there exists $n_w\in N^1$ such that n_w corresponds to w under the natural homomorphism from N into W.* (ii) *N^1/H^1 is isomorphic to W^1.*

PROOF. (i) Since by 13.1.2 W^1 is generated by the elements w_0^J for all ρ-orbits J of Π, it is sufficient to prove the existence of elements $n_w\in N^1$ with $w=w_0^J$. Now one of the equivalence classes of Φ defined in 13.2.1 is $w_0^J(\Phi_J^+)=\Phi_J^-$. Let $S=\Phi_J^-$. Then by 13.5.1 there is a non-unit element $x\in X_S^1$. We express this element in canonical form 8.4.4, and obtain

$$x=u'n_wu, \qquad u'\in U, \qquad u\in U_w^-.$$

(We have chosen the representative n_w here, as we may do, so that no element of H appears in the expression for x.) Now $w \neq 1$, for $w = 1$ would imply that $x \in UH \cap V = 1$. Since $\sigma(x) = x$ we have

$$u' n_w u = \sigma(u') \, \sigma(n_w) \, \sigma(u).$$

Comparing the double cosets BnB containing these elements we obtain $\sigma(w) = w$, then $w \in W^1$. Now x is an element of the Chevalley group generated by the subgroups X_r for all $r \in \Phi_J$. Thus the element w occurring in the above decomposition of x must be in W_J. Since $w \neq 1$, w transforms some root in J into a negative root. But J is a ρ-orbit of Π and $w \in W^1$. Thus w transforms all roots in J into negative roots. Hence $w = w_0^J$, the only element of W_J transforming all roots in J into negative roots.

We can now show that n_w is the representative we require. We have

$$\sigma(U_w^-) = \sigma(U \cap n_w^{-1} V n_w) = U \cap n_w^{-1} V n_w = U_w^-,$$

since $w \in W^1$. Thus in the decompositions

$$x = u' n_w u = \sigma(u') \, \sigma(n_w) \, \sigma(u)$$

we have u', $\sigma(u') \in U$ and u, $\sigma(u) \in U_w^-$. By the uniqueness of such a decomposition we have $\sigma(u') = u'$ and $\sigma(u) = u$. Hence

$$n_w = u'^{-1} x u^{-1} \in G^1 \cap N = N^1$$

by 13.4.2, and so n_w is the required element.

(ii) Consider the natural homomorphism from N to W. Each element of N^1 is fixed by σ, so is transformed into an element of W fixed by σ, i.e. an element of W^1. However for each $w \in W^1$ there is an element $n_w \in N^1$ which maps onto w, as shown in (i). Thus the image of N^1 under the above homomorphism is W^1. There is therefore a homomorphism from N^1 onto W^1 with kernel $N^1 \cap H = H^1$. Hence H^1 is normal in N^1 and N^1/H^1 is isomorphic to W^1. ∎

PROPOSITION 13.5.3. *Each element of G^1 has a unique expression $g = u' h n_w u$, where $u' \in U^1$, $h \in H^1$, $w \in W^1$, $n_w \in N^1$ and $u \in (U_w^-)^1$ the set of σ-invariant elements of U_w^-.*

PROOF. By 8.4.4 we may write

$$g = u' h n_w u,$$

where $u' \in U$, $h \in H$, $w \in W$, $u \in U_w^-$. Since $\sigma(g) = g$ we have

$$u' h n_w u = \sigma(u') \, \sigma(h) \, \sigma(n_w) \, \sigma(u).$$

Comparing the double cosets containing both elements we have $\sigma(w) = w$, thus $w \in W^1$. Hence n_w may be chosen to lie in N^1. Also $\sigma(u') \in U$, $\sigma(h) \in H$, $\sigma(u) \in U_w^-$. By uniqueness of the canonical form we have

$$\sigma(u') = u', \qquad \sigma(h) = h, \qquad \sigma(u) = u.$$

Thus $u' \in U^1$ and $u \in (U_w^-)^1$. Finally

$$h = (u')^{-1} g u^{-1} n_w^{-1} \in G^1 \cap H = H^1$$

by 13.4.2. Thus g has an expression of the required form, and the uniqueness is clear.   ■

We shall now establish the existence of a (B, N)-pair in the group G^1. We define $B^1 = G^1 \cap B$. Then it follows from 13.5.3 that $B^1 = U^1 H^1$ and that H^1 is in the normalizer of U^1.

THEOREM 13.5.4.   *The subgroups B^1, N^1 form a (B, N)-pair in G^1.*

PROOF. We verify the axioms $BN\,1 - BN\,5$ given in section 8.2. By 13.5.3 we have $G^1 = B^1 N^1 B^1$, thus B^1 and N^1 generate G^1. Also

$$B^1 \cap N^1 = G^1 \cap B \cap N = G^1 \cap H = H^1.$$

Thus $B^1 \cap N^1$ is normal in N^1, and the factor group $N^1/B^1 \cap N^1$ is isomorphic to W^1, so is generated by a set of elements w_0^J of order 2. The elements $w_0^J \in W^1$ play the rôle of the fundamental reflections, as J runs over the ρ-orbits of Π. We show next that $n(w_0^J) B^1 n(w_0^J)^{-1} \neq B^1$. Let S be the equivalence class Φ_J^+ of Φ. Then by 13.5.1 there exists $x \in X_S^1$ with $x \neq 1$. Now

$$n(w_0^J)\, x n(w_0^J)^{-1} \in n(w_0^J)\, X_S^1 n(w_0^J)^{-1} = X_{-S}^1.$$

Thus $n(w_0^J)\, x n(w_0^J)^{-1}$ is a non-unit element of V^1, so cannot be in B^1 since $B^1 \cap V^1 = 1$.

Finally we must show that

$$B^1 n(w_0^J)\, B^1 . B^1 n B^1 \subseteq B^1 n(w_0^J)\, n B^1 \cup B^1 n B^1$$

for all $n \in N^1$ and all ρ-orbits J of Π. The element w_0^J can be expressed as a product of reflections corresponding to roots in J. Let

$$w_0^J = w_{r_1} w_{r_2} \ldots w_{r_k}, \qquad r_i \in J.$$

Then by 8.2.1 we have

$$Bn(w_{r_i})\, B . BnB \subseteq Bn(w_{r_i})\, nB \cup BnB$$

and by applying this repeatedly we see that $B^1 n(w_0^J) B^1 . B^1 n B^1$ is in a union of double cosets $Bn(w) \, nB$, where w is an element of the form $w_{s_1} w_{s_2} \ldots w_{s_h}$ and $(s_1, s_2, \ldots, s_h)$ is a subsequence of $(r_1, r_2, \ldots, r_k)$. Also, by 13.5.3, G^1 intersects $Bn_w B$ in $B^1 n_w B^1$ if $w \in W^1$, and in the empty set if $w \notin W^1$. Thus $B^1 n(w_0^J) B^1 . B^1 n B^1$ is in a union of double cosets of form $B^1 n(w) \, nB^1$, where $w \in W^1 \cap W_J$. However, the only elements in $W^1 \cap W_J$ are 1 and w_0^J. For every non-unit element of W^1 in W_J transforms some root in J into a negative root, so transforms each root in J into a negative root since J is a ρ-orbit of Π. It must therefore be w_0^J. Thus

$$B^1 n(w_0^J) B^1 . B^1 n B^1 \subseteq B^1 n(w_0^J) \, nB^1 \cup B^1 n B^1$$

and all the axioms for a (B, N)-pair are satisfied. ∎

The results on parabolic subgroups proved in section 8.3 are now valid for the twisted group G^1. The parabolic subgroups of G^1 are the subgroups containing B^1 together with the conjugates of these subgroups. It follows from 8.3.4 that there are exactly 2^k subgroups of G^1 containing B^1, where k is the number of ρ-orbits of Π.

13.6 The Subgroup U^1

We shall now describe in detail the elements of the subgroup U^1 of G^1, and of the related subgroups V^1, $(U_w^-)^1$ and X_S^1.

PROPOSITION 13.6.1. $U^1 = \prod X_S^1$, where the product is taken over all equivalence classes S which are in Φ^+. (The terms in the product may be taken in any order.) Each element of U^1 has a unique expression as a product of elements of X_S^1 taken in this order. Similar results hold for V^1 and for $(U_w^-)^1$ for each $w \in W^1$. We have
$V^1 = \prod X_S^1$ with uniqueness, taken over all equivalence classes S which are in Φ^-, and
$(U_w^-)^1 = \prod X_S^1$ with uniqueness, taken over all equivalence classes S such that S is in Φ^+ and $w(S)$ is in Φ^-.

PROOF. The equivalence classes S in Φ^+ form a partition of Φ^+. Thus

$$\prod_{S \subset \Phi^+} X_S = \prod_{r \in \Phi^+} X_r = U.$$

Also each element of U has a unique expression of the form

$$u = \prod_{S \subset \Phi^+} x_S$$

with $x_S \in X_S$. Now $\sigma(X_S) = X_S$ for each equivalence class S. Thus $\sigma(u) = u$ if and only if $\sigma(x_S) = x_S$ for each S. If $u \in U^1$ we therefore have $x_S \in X_S^1$, and so

$$U^1 = \prod_{S \subset \Phi^+} X_S^1,$$

as required.

Similar proofs can be given for V^1 and for $(U_w^-)^1$. ∎

In order to give descriptions of the elements of X_S^1 which are as simple as possible, it is useful to make a choice of the Chevalley basis of $\mathfrak{L}$ with respect to which the constants $\gamma_r = \pm 1$ introduced in 12.2.3 can be given explicitly.

LEMMA 13.6.2. *Let $\mathfrak{L}$ be a simple Lie algebra all of whose roots have the same length, and let $r \to \bar{r}$ be the map of Φ into itself arising from a non-trivial symmetry of the Dynkin diagram of $\mathfrak{L}$. Then it is possible to choose a Chevalley basis of $\mathfrak{L}$ in such a way that the automorphism of $\mathfrak{L}$ determined by*

$$h_r \to h_{\bar{r}}, \; e_r \to e_{\bar{r}}, \; e_{-r} \to e_{-\bar{r}} \qquad (r \in \Pi)$$

satisfies $e_r \to \gamma_r e_{\bar{r}}$, where $\gamma_r = 1$ unless the equivalence class S containing r has type A_2 and $r = \bar{r}$, in which case $\gamma_r = -1$.

PROOF. We observe first that, given a Chevalley basis, the changing of sign of any set of root vectors e_r, $r \in \Phi^+$, does not affect the property of being a Chevalley basis, provided the signs of the corresponding negative root vectors e_r, $r \in \Phi^-$, are changed also.

Suppose the given symmetry has order 2. Then the square of the automorphism transforms h_r, e_r into themselves for all $r \in \Pi$, so is the identity. Thus, for each $r \in \Phi$ we have

$$e_r \to \gamma_r e_{\bar{r}} \to \gamma_r \gamma_{\bar{r}} e_r = e_r$$

and so $\gamma_r \gamma_{\bar{r}} = 1$. Thus $\gamma_r = \gamma_{\bar{r}}$. Suppose $r \neq \bar{r}$. If $e_r \to -e_{\bar{r}}$ and $e_{\bar{r}} \to -e_r$ we may change the sign of one of e_r, $e_{\bar{r}}$ but not the other. Then $\gamma_r = \gamma_{\bar{r}} = 1$. Now suppose instead that $r = \bar{r}$ and $r = s + \bar{s}$ for some $s \in \Phi$. Then

$$[e_s e_{\bar{s}}] = N_{s,\bar{s}} e_r.$$

Applying the automorphism we have

$$\gamma_s \gamma_{\bar{s}} [e_{\bar{s}} e_s] = N_{s,\bar{s}} \gamma_r e_r$$

and, since $\gamma_s\gamma_{\bar{s}}=1$, this implies that $\gamma_r=-1$. Note here that the equivalence class S containing r is $S=\{s, \bar{s}, r\}$, which has type A_2.

Now suppose that $r\in\Phi^+$, $r=\bar{r}$, but r is not expressible in the form $r=s+\bar{s}$. Then by inspection of the root systems (described in section 3.6) it can be seen that there is either a fundamental root $r_i=\bar{r}_i\in\Pi$ such that $r-r_i\in\Phi$, or a pair of fundamental roots r_i, $\bar{r}_i\in\Pi$ such that $r-r_i$, $r-\bar{r}_i$, $r-r_i-\bar{r}_i\in\Phi$, or r is itself in Π. If $r\in\Pi$ then $\gamma_r=1$. In the first case we have

$$[e_{r-r_i}, e_{r_i}]=N_{r-r_i, r_i}e_r.$$

Using induction on the height we may assume $e_{r-r_i}\to e_{r-r_i}$. Since $e_{r_i}\to e_{r_i}$ we have $e_r\to e_r$, and so $\gamma_r=1$. In the second case we have

$$[[e_{r-r_i-\bar{r}_i}, e_{r_i}]\,e_{\bar{r}_i}]=N_{r-r_i-\bar{r}_i, r_i}N_{r-\bar{r}_i, \bar{r}_i}e_r.$$

By induction on the height we may assume $e_{r-r_i-\bar{r}_i}\to e_{r-r_i-\bar{r}_i}$. Since $e_{r_i}\to e_{r_i}$ and $e_{\bar{r}_i}\to e_{\bar{r}_i}$ we have $e_r\to e_r$. Thus $\gamma_r=1$ in this case also. We have shown that $\gamma_r=1$ for all positive roots r with $r=\bar{r}$, $r\ne s+\bar{s}$, and it follows immediately that $\gamma_r=1$ for all negative roots r with this property also.

Finally suppose the given symmetry has order 3. If $r=\bar{r}$ then $e_r\to\gamma_r e_r$, where $\gamma_r^3=1$. Hence $\gamma_r=1$. If $r\ne\bar{r}$ then

$$e_r\to\gamma_r e_{\bar{r}}\to\gamma_r\gamma_{\bar{r}}e_{\bar{\bar{r}}}\to\gamma_r\gamma_{\bar{r}}\gamma_{\bar{\bar{r}}}e_r=e_r.$$

Thus $\gamma_r\gamma_{\bar{r}}\gamma_{\bar{\bar{r}}}=1$. If two of γ_r, $\gamma_{\bar{r}}$, $\gamma_{\bar{\bar{r}}}$ are -1 we can, by changing the sign of one of the root vectors e_r, $e_{\bar{r}}$, $e_{\bar{\bar{r}}}$, arrange matters so that

$$\gamma_r=\gamma_{\bar{r}}=\gamma_{\bar{\bar{r}}}=1. \qquad\blacksquare$$

If all the roots of $\mathfrak{L}$ have the same length, we shall assume that a Chevalley basis for $\mathfrak{L}$ is chosen as in 13.6.2, where this is relevant. We can now describe the elements in the subgroups X_S^1.

PROPOSITION 13.6.3. *Let S be an equivalence class in Φ. Then the elements in the subgroup X_S^1 are as shown below. (For the field automorphism f we write $f(t)=\bar{t}$ if all the roots of $\mathfrak{L}$ have the same length, and $f(t)=t^\theta$ otherwise.)*

(i) *If $S=\{r\}$ has type A_1 then X_S^1 consists of the elements $x_r(t)$ with $t=\bar{t}$.*

(ii) *If $S=\{r, \bar{r}\}$ has type $A_1\times A_1$, where r, $\bar{r}$ have the same length, then X_S^1 consists of the elements $x_r(t)\,x_{\bar{r}}(\bar{t})$ for all $t\in K$.*

(iii) *If $S=\{r, \bar{r}, \bar{\bar{r}}\}$ has type $A_1\times A_1\times A_1$ then X_S^1 consists of the elements $x_r(t)\,x_{\bar{r}}(\bar{t})\,x_{\bar{\bar{r}}}(\bar{\bar{t}})$ for all $t\in K$.*

(iv) *If* $S = \{r, \bar{r}, r + \bar{r}\}$ *has type* A_2 *then* X_S^1 *consists of the elements* $x_r(t)\, x_{\bar{r}}(\bar{t})\, x_{r+\bar{r}}(u)$, *where* $u + \bar{u} = -N_{r,\bar{r}}t\bar{t}$.

(v) *If* $S = \{r, \bar{r}\}$ *has type* $A_1 \times A_1$, *where* r *is short and* $\bar{r}$ *is long, then* X_S^1 *consists of the elements* $x_r(t^\theta)\, x_{\bar{r}}(t)$ *for all* $t \in K$.

(vi) *If* $S = \{a, b, a+b, 2a+b\}$ *has type* B_2 *then* X_S^1 *consists of the elements*

$$x_a(t^\theta)\, x_b(t)\, x_{a+b}(t^{\theta+1}+u)\, x_{2a+b}(u^{2\theta})$$

for all $t, u \in K$.

(vii) *If* $S = \{a, b, a+b, 2a+b, 3a+b, 3a+2b\}$ *has type* G_2 *then* X_S^1 *consists of the elements*

$$x_a(t^\theta)\, x_b(t)\, x_{a+b}(t^{\theta+1}+u^\theta)\, x_{2a+b}(t^{2\theta+1}+v^\theta)\, x_{3a+b}(u)\, x_{3a+2b}(v)$$

for all $t, u, v \in K$. (*The structure constants here have been chosen as in section* 12.4.)

PROOF. (i) Since $\sigma . x_r(t) = x_r(\bar{t})$, X_S^1 is as stated.

(ii) Since

$$\sigma . x_r(t)\, x_{\bar{r}}(u) = x_{\bar{r}}(\bar{t})\, x_r(\bar{u}) = x_r(\bar{u})\, x_{\bar{r}}(\bar{t}),$$

the σ-invariant elements are those for which $u = \bar{t}$.

(iii) Since

$$\sigma . x_r(t)\, x_{\bar{r}}(u)\, x_{\bar{r}}(v) = x_{\bar{r}}(\bar{t})\, x_{\bar{r}}(\bar{u})\, x_r(\bar{v}) = x_r(\bar{v})\, x_{\bar{r}}(\bar{t})\, x_{\bar{r}}(\bar{u}),$$

the σ-invariant elements are those for which $u = \bar{t}$, $v = \bar{t}$.

(iv) Since

$$\sigma . x_r(t)\, x_{\bar{r}}(v)\, x_{r+\bar{r}}(u) = x_{\bar{r}}(\bar{t})\, x_r(\bar{v})\, x_{r+\bar{r}}(-\bar{u})$$
$$= x_r(\bar{v})\, x_{\bar{r}}(\bar{t})\, x_{r+\bar{r}}(-N_{r,\bar{r}}\bar{t}\bar{v} - \bar{u}),$$

the σ-invariant elements are those for which $v = \bar{t}$ and $u + \bar{u} = -N_{r,\bar{r}}t\bar{t}$.

(v) Here K has characteristic 2 and $2\theta^2 = 1$. Now

$$\sigma . x_r(u)\, x_{\bar{r}}(t) = x_{\bar{r}}(u^{2\theta})\, x_r(t^\theta) = x_r(t^\theta)\, x_{\bar{r}}(u^{2\theta}).$$

Thus the σ-invariant elements are those for which $u = t^\theta$.

(vi) In this case also K has characteristic 2 and $2\theta^2 = 1$. We have

$$\sigma . x_a(t_a)\, x_b(t_b)\, x_{a+b}(t_{a+b})\, x_{2a+b}(t_{2a+b})$$
$$= x_b(t_a^{2\theta})\, x_a(t_b^\theta)\, x_{2a+b}(t_{a+b}^{2\theta})\, x_{a+b}(t_{2a+b}^\theta)$$
$$= x_a(t_b^\theta)\, x_b(t_a^{2\theta})\, x_{a+b}(t_a^{2\theta}t_b^\theta + t_{2a+b}^\theta)\, x_{2a+b}(t_a^{2\theta}t_b^{2\theta} + t_{a+b}^{2\theta}),$$

using the commutator formula. Thus the σ-invariant elements are those for which

$$t_a = t_b^\theta,\ t_b = t_a^{2\theta},\ t_{a+b} = t_a^{2\theta}t_b^\theta + t_{2a+b}^\theta,$$
$$t_{2a+b} = t_a^{2\theta}t_b^{2\theta} + t_{a+b}^{2\theta}.$$

Putting $t = t_b$, $u = t_{2a+b}^\theta$, we obtain the required form.

(vii) In this case K has characteristic 3 and $3\theta^2 = 1$. We choose the structure constants as in section 12.4. Then

$$\sigma \cdot x_a(t_a)\, x_b(t_b)\, x_{a+b}(t_{a+b})\, x_{2a+b}(t_{2a+b})\, x_{3a+b}(t_{3a+b})\, x_{3a+2b}(t_{3a+2b})$$
$$= x_b(t_a^{3\theta})\, x_a(t_b^\theta)\, x_{3a+b}(t_{a+b}^{3\theta})\, x_{3a+2b}(t_{2a+b}^{3\theta})\, x_{a+b}(t_{3a+b}^\theta)\, x_{2a+b}(t_{3a+2b}^\theta)$$
$$= x_a(t_b^\theta)\, x_b(t_a^{3\theta})\, x_{a+b}(t_a^{3\theta}t_b^\theta)\, x_{2a+b}(t_a^{3\theta}t_b^{2\theta})\, x_{3a+b}(-t_a^{3\theta}t_b^{3\theta})$$
$$\times x_{3a+2b}(-t_a^{6\theta}t_b^{3\theta})\, x_{3a+b}(t_{a+b}^{3\theta})\, x_{3a+2b}(t_{2a+b}^{3\theta})\, x_{a+b}(t_{3a+b}^\theta)$$
$$\times x_{2a+b}(t_{3a+2b}^\theta)$$
$$= x_a(t_b^\theta)\, x_b(t_a^{3\theta})\, x_{a+b}(t_a^{3\theta}t_b^\theta + t_{3a+b}^\theta)\, x_{2a+b}(t_a^{3\theta}t_b^{2\theta} + t_{3a+2b}^\theta)$$
$$\times x_{3a+b}(-t_a^{3\theta}t_b^{3\theta} + t_{a+b}^{3\theta})\, x_{3a+2b}(-t_a^{6\theta}t_b^{3\theta} + t_{2a+b}^{3\theta}).$$

Thus the σ-invariant elements are those for which

$$t_a = t_b^\theta,\ t_b = t_a^{3\theta},\ t_{a+b} = t_a^{3\theta}t_b^\theta + t_{3a+b}^\theta,\ t_{2a+b} = t_a^{3\theta}t_b^{2\theta} + t_{3a+2b}^\theta,$$
$$t_{3a+b} = -t_a^{3\theta}t_b^{3\theta} + t_{a+b}^{3\theta},\ t_{3a+2b} = -t_a^{6\theta}t_b^{3\theta} + t_{2a+b}^{3\theta}.$$

Putting $t = t_b$, $u = t_{3a+b}$, $v = t_{3a+2b}$ we obtain the required elements. ■

The subgroups X_S^1 of the twisted groups are of importance because they are the analogue of the root subgroups of the Chevalley groups. We describe next the way in which the elements in a subgroup X_S^1 combine together.

PROPOSITION 13.6.4. (i) If $S = \{r\}$ has type A_1 and we write $x_S(t) = x_r(t)$ then

$$x_S(t_1)\, x_S(t_2) = x_S(t_1 + t_2).$$

(ii) If $S = \{r, \bar{r}\}$ has type $A_1 \times A_1$, where $r, \bar{r}$ have the same length, and we write $x_S(t) = x_r(t)\, x_{\bar{r}}(\bar{t})$ then

$$x_S(t_1)\, x_S(t_2) = x_S(t_1 + t_2).$$

(iii) If $S = \{r, \bar{r}, \bar{\bar{r}}\}$ has type $A_1 \times A_1 \times A_1$ and we write

$$x_S(t) = x_r(t)\, x_{\bar{r}}(\bar{t})\, x_{\bar{\bar{r}}}(\bar{\bar{t}})$$

then

$$x_S(t_1)\, x_S(t_2) = x_S(t_1 + t_2).$$

(iv) If $S = \{r, \bar{r}, r + \bar{r}\}$ has type A_2 and we write

$$x_S(t, u) = x_r(t) \, x_{\bar{r}}(\bar{t}) \, x_{r+\bar{r}}(u),$$

where $u + \bar{u} = - N_{r, \bar{r}} t \bar{t}$, then

$$x_S(t_1, u_1) \, x_S(t_2, u_2) = x_S(t_1 + t_2, u_1 + u_2 - N_{r, \bar{r}} \bar{t}_1 t_2).$$

(v) If $S = \{r, \bar{r}\}$ has type $A_1 \times A_1$, where r is short and $\bar{r}$ long and if we write $x_S(t) = x_r(t^\theta) \, x_{\bar{r}}(t)$ then

$$x_S(t_1) \, x_S(t_2) = x_S(t_1 + t_2).$$

(vi) If $S = \{a, b, a + b, 2a + b\}$ has type B_2 and we write

$$\alpha(t) = x_a(t^\theta) \, x_b(t) \, x_{a+b}(t^{\theta+1}),$$
$$\beta(u) = x_{a+b}(u) \, x_{2a+b}(u^{2\theta}),$$
$$x_S(t, u) = \alpha(t) \, \beta(u)$$

then

$$x_S(t_1, u_1) \, x_S(t_2, u_2) = x_S(t_1 + t_2, u_1 + u_2 + t_1^\theta t_2).$$

(vii) If $S = \{a, b, a + b, 2a + b, 3a + b, 3a + 2b\}$ has type G_2 and we write

$$\alpha(t) = x_a(t^\theta) \, x_b(t) \, x_{a+b}(t^{\theta+1}) \, x_{2a+b}(t^{2\theta+1})$$
$$\beta(u) = x_{a+b}(u^\theta) \, x_{3a+b}(u),$$
$$\gamma(v) = x_{2a+b}(v^\theta) \, x_{3a+2b}(v),$$
$$x_S(t, u, v) = \alpha(t) \, \beta(u) \, \gamma(v)$$

then

$$x_S(t_1, u_1, v_1) \, x_S(t_2, u_2, v_2)$$
$$= x_S(t_1 + t_2, \, u_1 + u_2 - t_1 t_2^{3\theta}, \, v_1 + v_2 - t_2 u_1 + t_1 t_2^{3\theta+1} - t_1^2 t_2^{3\theta}).$$

PROOF. These are all straightforward consequences of the commutator formula 5.2.2. ∎

We have seen earlier that the Chevalley group G is generated by the root subgroups X_r, X_{-r} for all $r \in \Pi$. We now prove an analogue of this result for the twisted groups.

PROPOSITION 13.6.5. G^1 is generated by the subgroups $X^1_{\Phi_J^+}$, $X^1_{\Phi_J^-}$ for all ρ-orbits J of Π.

PROOF. G^1 is generated by U^1 and V^1 by 13.4.2. U^1 is generated by the subgroups X_S^1 for all equivalence classes S in Φ^+, by 13.6.1, and V^1 is generated by the X_S^1 for all S in Φ^-. Thus G^1 is generated by the subgroups X_S^1 for all equivalence classes S of Φ.

Now each equivalence class S has the form $w(\Phi_J^+)$ for some $w \in W^1$ and some ρ-orbit J of Π (13.2.1). For each $w \in W^1$, let n_w be an element of N^1 chosen as in 13.5.2. Let G_σ be the set of σ-invariant elements of G. Then we have

$$n_w X_S^1 n_w^{-1} = n_w (X_S \cap G_\sigma) n_w^{-1} = X_{w(S)} \cap G_\sigma = X_{w(S)}^1$$

since $n_w \in G_\sigma$. Thus G^1 is generated by the subgroups $X_{\Phi_J^+}^1$ and the elements n_w for all $w \in W^1$. However, W^1 is generated by the elements w_0^J for all ρ-orbits J of Π. Thus G^1 is generated by the $X_{\Phi_J^+}^1$ and elements $n(w_0^J)$ for all J. Now consider the subgroup $\langle X_{\Phi_J^+}^1, X_{\Phi_J^-}^1 \rangle$. This is by 13.4.2 a twisted group whose Weyl group is, by 13.1.2, $\langle w_0^J \rangle$. By 13.5.2 this twisted group contains an element $n(w_0^J) \in N$ mapping to w_0^J. Thus G^1 is generated by the subgroups $X_{\Phi_J^+}^1$, $X_{\Phi_J^-}^1$ for all J.

13.7 The Subgroup H^1

We now turn to a discussion of the subgroup $H^1 = H \cap G^1$. It turns out to be quite troublesome to decide which elements of H are contained in this subgroup. We begin by describing the operation of σ on H.

LEMMA 13.7.1. *If all the roots of $\mathfrak{L}$ have the same length then*

$$\sigma . h_r(t) = h_{\bar{r}}(\bar{t}).$$

If there are roots of different lengths then

$$\sigma . h_r(t) = h_{\bar{r}}(t^{\lambda(\bar{r})\theta}).$$

PROOF. Suppose all the roots have the same length. Then

$$\sigma . x_r(t) = x_{\bar{r}}(\gamma_r \bar{t}),$$

where $\gamma_r = \pm 1$. Thus

$$\begin{aligned}
\sigma . n_r(t) &= \sigma(x_r(t) \, x_{-r}(-t^{-1}) \, x_r(t)) \\
&= x_{\bar{r}}(\gamma_r \bar{t}) \, x_{-\bar{r}}(-\gamma_{-r} \bar{t}^{-1}) \, x_{\bar{r}}(\gamma_r \bar{t}) \\
&= n_{\bar{r}}(\gamma_r \bar{t})
\end{aligned}$$

since $\gamma_r\gamma_{-r}=1$. Therefore

$$\sigma \cdot h_r(t) = \sigma(n_r(t) \cdot n_r(-1)) = n_{\bar{r}}(\gamma_r \bar{t}) \, n_{\bar{r}}(\gamma_r)^{-1}$$
$$= h_{\bar{r}}(\gamma_r \bar{t}) \, n_{\bar{r}}(1) \, n_{\bar{r}}(1)^{-1} h_{\bar{r}}(\gamma_r)^{-1} = h_{\bar{r}}(\bar{t}).$$

Now suppose there are roots of different lengths. Then

$$\sigma \cdot x_r(t) = x_{\bar{r}}(t^{\lambda(\bar{r})\theta}).$$

Thus

$$\sigma \cdot n_r(t) = \sigma(x_r(t) \, x_{-r}(-t^{-1}) \, x_r(t))$$
$$= x_{\bar{r}}(t^{\lambda(\bar{r})\theta}) \, x_{-\bar{r}}((-t^{-1})^{\lambda(\bar{r})\theta}) \, x_{\bar{r}}(t^{\lambda(\bar{r})\theta})$$
$$= n_{\bar{r}}(t^{\lambda(\bar{r})\theta})$$

since $(-1)^{\lambda(\bar{r})} = -1$. (The latter statement is trivial if K has characteristic 2, whereas if K has characteristic 3 $\lambda(\bar{r})$ is either 1 or 3.) Therefore

$$\sigma \cdot h_r(t) = \sigma \cdot (n_r(t) \, n_r(-1)) = n_{\bar{r}}(t^{\lambda(\bar{r})\theta}) \, n_{\bar{r}}(-1) = h_{\bar{r}}(t^{\lambda(\bar{r})\theta}). \qquad \blacksquare$$

Now the subgroup H of G consists of elements $h(\chi)$, where χ is a K-character of P, the additive group generated by the fundamental roots $p_1, \ldots, p_l$. The K-characters χ which give rise to elements of H are those which can be extended to K-characters of Q, the additive group generated by the fundamental weights $q_1, \ldots, q_l$, by 7.1.1. We shall prove an analogue of this result for the subgroup H^1 of G^1 in the case when all the roots have the same length. In such a case the symmetry ρ of the Dynkin diagram determines an isometry τ of $\mathfrak{Y}$, the real vector space generated by Φ, such that $\tau(p_i) = p_{\rho(i)}$. τ transforms every root into a root and we have $\tau(r) = \bar{r}$. Now τ transforms the fundamental co-roots by $\tau(h_{p_i}) = h_{p_{\rho(i)}}$. Since $q_1, \ldots, q_l$ is the basis of $\mathfrak{Y}$ dual to the basis $h_{p_1}, \ldots, h_{p_l}$ (see section 7.1), we have $\tau(q_i) = q_{\rho(i)}$. Thus τ permutes the fundamental weights in the same way that it permutes the fundamental roots. In particular we have $\tau(Q) = Q$. We write

$$\tau(a) = \bar{a}$$

for each $a \in Q$.

Definition. Suppose all the roots of $\mathfrak{L}$ have the same length. A K-character χ of P (or Q) is said to be self-conjugate if $\chi(\bar{a}) = \overline{\chi(a)}$ for all $a \in P$ (or Q).

THEOREM 13.7.2. Suppose all the roots of $\mathfrak{L}$ have the same length. Then $h(\chi) \in H^1$ if and only if χ is a self-conjugate K-character of P which can be extended to a self-conjugate K-character of Q.

PROOF. Let $h(\chi) \in H$. By 7.1.1 χ may be regarded as a K-character of Q. Let $\chi(q_i) = \lambda_i$. Then we have

$$\chi = \chi_{p_1, \lambda_1} \chi_{p_2, \lambda_2} \cdots \chi_{p_l, \lambda_l}$$

as in 7.1.1, where $\chi_{r, \lambda}$ is the K-character of Q defined by

$$\chi_{r, \lambda}(a) = \lambda^{2(r, a)/(r, r)}, \qquad a \in Q.$$

We consider which K-characters χ of Q are self-conjugate. The necessary and sufficient condition for this is that $\chi(\bar{q}_i) = \overline{\chi(q_i)}$ for $i = 1, \ldots, l$. Now

$$\chi(\bar{q}_i) = \chi(q_{\rho(i)}) = \lambda_{\rho(i)},$$

thus we require $\lambda_{\rho(i)} = \bar{\lambda}_i$ for all i. Thus every self-conjugate K-character of Q is a product of characters corresponding to the ρ-orbits J of Π, and these characters have the form:

$$\chi_{r, \lambda}, \quad \lambda = \bar{\lambda}, \qquad \text{if } J = \{r\},$$
$$\chi_{r, \lambda} \chi_{\bar{r}, \bar{\lambda}}, \qquad \text{if } J = \{r, \bar{r}\},$$
$$\chi_{r, \lambda} \chi_{\bar{r}, \bar{\lambda}} \chi_{\bar{\bar{r}}, \lambda}, \qquad \text{if } J = \{r, \bar{r}, \bar{\bar{r}}\}.$$

Let H^2 be the group of automorphisms of $\mathfrak{L}_K$ of the form $h(\chi)$, where χ is a self-conjugate character of Q. We recall from section 7.1 that $h_r(t) = h(\chi_{r, t})$. Thus H^2 is generated by elements corresponding to the ρ-orbits J of Π as follows:

$$h_r(t), \ t = \bar{t}, \qquad \text{if } J = \{r\},$$
$$h_r(t) h_{\bar{r}}(\bar{t}), \qquad \text{if } J = \{r, \bar{r}\},$$
$$h_r(t) h_{\bar{r}}(\bar{t}) h_{\bar{\bar{r}}}(\bar{t}), \qquad \text{if } J = \{r, \bar{r}, \bar{\bar{r}}\}.$$

We shall show that H^2 is contained in H^1 by showing that each of these generators of H^2 lies in H^1.

Let $S = \Phi_J^+$, where J is a ρ-orbit of Π. If S has type A_1 and $t = \bar{t}$, then $x_r(t)$ and $x_{-r}(-t^{-1})$ are in G^1. Hence $n_r(t) = x_r(t) x_{-r}(-t^{-1}) x_r(t) \in G^1$ and $h_r(t) = n_r(t) n_r(-1) \in G^1$. If S has type $A_1 \times A_1$ or $A_1 \times A_1 \times A_1$, a similar argument shows that $h_r(t) h_{\bar{r}}(\bar{t})$ or $h_r(t) h_{\bar{r}}(\bar{t}) h_{\bar{\bar{r}}}(\bar{t})$ are in G^1, since each of X_r, X_{-r} commutes with each of $X_{\bar{r}}, X_{-\bar{r}}$ and each of $X_{\bar{\bar{r}}}, X_{-\bar{\bar{r}}}$.

Suppose S has type A_2. Here things are not so simple, for the subgroup $\langle X_r, X_{\bar{r}}, X_{-r}, X_{-\bar{r}} \rangle$ of G is isomorphic to the Chevalley group $A_2(K)$. By 11.3.2 there is a homomorphism from $SL_3(K)$ into $\langle X_r, X_{\bar{r}}, X_{-r}, X_{-\bar{r}} \rangle$

such that

$$I + te_{12} \rightarrow x_r(t),$$
$$I + te_{23} \rightarrow x_{\bar{r}}(t),$$
$$I + te_{13} \rightarrow x_{r+\bar{r}}(t),$$
$$I + te_{21} \rightarrow x_{-r}(t),$$
$$I + te_{32} \rightarrow x_{-\bar{r}}(t),$$
$$I + te_{31} \rightarrow x_{-r-\bar{r}}(t).$$

Matrix multiplication shows that

$$\begin{pmatrix} 0 & t & 0 \\ -t^{-1} & 0 & 0 \\ 0 & 0 & 1 \end{pmatrix} \rightarrow x_r(t)\, x_{-r}(-t^{-1})\, x_r(t) = n_r(t),$$

$$\begin{pmatrix} 1 & 0 & 0 \\ 0 & 0 & t \\ 0 & -t^{-1} & 0 \end{pmatrix} \rightarrow x_{\bar{r}}(t)\, x_{-\bar{r}}(-t^{-1})\, x_{\bar{r}}(t) = n_{\bar{r}}(t),$$

$$\begin{pmatrix} t & 0 & 0 \\ 0 & t^{-1} & 0 \\ 0 & 0 & 1 \end{pmatrix} \rightarrow n_r(t)\, n_r(-1) = h_r(t),$$

$$\begin{pmatrix} 1 & 0 & 0 \\ 0 & t & 0 \\ 0 & 0 & t^{-1} \end{pmatrix} \rightarrow n_{\bar{r}}(t)\, n_{\bar{r}}(-1) = h_{\bar{r}}(t),$$

$$\begin{pmatrix} t & 0 & 0 \\ 0 & t^{-1}\bar{t} & 0 \\ 0 & 0 & \bar{t}^{-1} \end{pmatrix} \rightarrow h_r(t)\, h_{\bar{r}}(\bar{t}).$$

Now $x_r(t)\, x_{\bar{r}}(\bar{t})\, x_{r+\bar{r}}(u)$ is in G^1 whenever $u + \bar{u} = -t\bar{t}$, by 13.6.3 (iv). (We have chosen the structure constants so that $N_{r,\bar{r}} = 1$ in order to obtain the matrix representation in the above form.) Also

$$x_{-r}(t)\, x_{-\bar{r}}(\bar{t})\, x_{-r-\bar{r}}(u)$$

is in G^1 whenever $u + \bar{u} = t\bar{t}$, since $N_{-r,-\bar{r}} = -1$.

Now we have

$$\begin{pmatrix} 1 & t & -\bar{u} \\ 0 & 1 & \bar{t} \\ 0 & 0 & 1 \end{pmatrix} \to x_r(t)\, x_{\bar{r}}(\bar{t})\, x_{r+\bar{r}}(u),$$

$$\begin{pmatrix} 1 & 0 & 0 \\ t & 1 & 0 \\ u & \bar{t} & 1 \end{pmatrix} \to x_{-r}(t)\, x_{-\bar{r}}(\bar{t})\, x_{-r-\bar{r}}(u).$$

Suppose λ, t are elements of K satisfying $\lambda^{-1}+\bar{\lambda}^{-1}=t\bar{t}$. Then the matrices

$$\begin{pmatrix} 1 & \lambda t & \lambda \\ 0 & 1 & \bar{\lambda}\bar{t} \\ 0 & 0 & 1 \end{pmatrix}, \quad \begin{pmatrix} 1 & 0 & 0 \\ -\bar{t} & 1 & 0 \\ \bar{\lambda}^{-1} & -t & 1 \end{pmatrix}, \quad \begin{pmatrix} 1 & \bar{\lambda}t & \lambda \\ 0 & 1 & \lambda\bar{t} \\ 0 & 0 & 1 \end{pmatrix}$$

are mapped into elements of G^1. Thus their product (from left to right)

$$\begin{pmatrix} 0 & 0 & \lambda \\ 0 & -\lambda^{-1}\bar{\lambda} & 0 \\ \bar{\lambda}^{-1} & 0 & 0 \end{pmatrix}$$

is mapped into an element of G^1. This holds for a given λ whenever there exists $t \in K$ such that $t\bar{t}=\lambda^{-1}+\bar{\lambda}^{-1}$. We show in the following lemma that each element $\lambda \neq 0$ of K is expressible in the form $\lambda = \lambda_1 \lambda_2^{-1}$, where λ_1, λ_2 are such that the above equation can be solved for t. Thus

$$\begin{pmatrix} \lambda & 0 & 0 \\ 0 & \lambda^{-1}\bar{\lambda} & 0 \\ 0 & 0 & \bar{\lambda}^{-1} \end{pmatrix} = \begin{pmatrix} 0 & 0 & \lambda_1 \\ 0 & -\lambda_1^{-1}\bar{\lambda}_1 & 0 \\ \bar{\lambda}_1^{-1} & 0 & 0 \end{pmatrix} \begin{pmatrix} 0 & 0 & \lambda_2 \\ 0 & -\lambda_2^{-1}\bar{\lambda}_2 & 0 \\ \bar{\lambda}_2^{-1} & 0 & 0 \end{pmatrix}$$

is mapped into an element of G^1. Hence $h_r(\lambda)\, h_{\bar{r}}(\bar{\lambda}) \in G^1$, as required.

LEMMA 13.7.3. *Let K be a field admitting an automorphism of order 2. Let L be the subset of K consisting of elements λ such that there exists $t \in K$ with $t\bar{t}=\lambda^{-1}+\bar{\lambda}^{-1}$. Then each element $\lambda \in K$ is expressible in the form $\lambda = \lambda_1 \lambda_2^{-1}$, where λ_1, $\lambda_2 \in L$.*

PROOF. Suppose $\lambda \neq \bar{\lambda}$. Define $\lambda_1 = \lambda - \bar{\lambda}$, $\lambda_2 = (\bar{\lambda} - \lambda)/\bar{\lambda}$. Then

$$\lambda_1^{-1} + \bar{\lambda}_1^{-1} = 0 \quad \text{and} \quad \lambda_2^{-1} + \bar{\lambda}_2^{-1} = 1.$$

Thus $\lambda_1, \lambda_2 \in L$ and $\lambda = \lambda_1 \bar{\lambda}_2^{-1}$.

Now suppose $\lambda = \bar{\lambda}$. There exists $\mu \in K$ such that $\mu \neq 0$ and $\bar{\mu} = -\mu$. Define $\lambda_1 = \lambda\mu$, $\lambda_2 = \bar{\mu}$. Then $\lambda_1^{-1} + \bar{\lambda}_1^{-1} = 0$ and $\lambda_2^{-1} + \bar{\lambda}_2^{-1} = 0$. Thus $\lambda_1, \lambda_2 \in L$ and $\lambda = \lambda_1 \bar{\lambda}_2^{-1}$. ∎

We now return to the proof of 13.7.2. We have now shown that H^2 is a subgroup of H^1 in all cases, and shall now prove that H^1 is contained in H^2.

Let N^2 be the subgroup of N^1 generated by H^2 and the elements n_w, chosen as in 13.5.2, for all $w \in W^1$. We have

$$n_w h(\chi) n_w^{-1} = h(\chi'),$$

where $\chi'(r) = \chi(w^{-1}(r))$, by 7.2.2. Suppose $h(\chi) \in H^2$. Then χ is a self-conjugate character of Q. Thus, for all $a \in Q$, we have

$$\chi'(\bar{a}) = \chi(w^{-1}(\bar{a})) = \chi(\overline{w^{-1}(a)}) = \overline{\chi(w^{-1}(a))} = \overline{\chi'(a)}$$

and so χ' is also a self-conjugate character of Q. Thus n_w normalizes H^2, and so H^2 is normal in N^2. Also $N^1 = H^1 N^2$ and $H^1 \cap N^2 = H^2$. Thus

$$N^2/H^2 = N^2/H^1 \cap N^2 \cong H^1 N^2/H^1 = N^1/H^1 \cong W^1.$$

We next define $B^2 = U^1 H^2$. Since H^2 normalizes U^1, B^2 is a subgroup of G^1. We shall show that for each ρ-orbit J of Π

$$X^1_{\Phi_J^-} \subseteq B^2 \cup B^2 n(w_0^J) B^2.$$

Since all the roots have the same length, J has type A_1, $A_1 \times A_1$, $A_1 \times A_1 \times A_1$ or A_2. If J has type A_1 the elements of $X^1_{\Phi_J^-}$ have form $x_{-r}(t)$, where $t = \bar{t}$. Now

$$x_{-r}(t) = x_r(t^{-1}) h_r(-t^{-1}) n_r x_r(t^{-1}), \qquad t \neq 0,$$

and we may choose $n(w_0^J) = n_r$. Since $h_r(-t^{-1}) \in H^2$ we have

$$x_{-r}(t) \in B^2 \cup B^2 n(w_0^J) B^2$$

whenever $t = \bar{t}$.

If J has type $A_1 \times A_1$ we have

$$x_{-r}(t) \, x_{-\bar{r}}(\bar{t}) = x_r(t^{-1}) \, x_{\bar{r}}(\bar{t}^{-1}) \, h_r(-t^{-1}) \, h_{\bar{r}}(-\bar{t}^{-1}) \, n_r n_{\bar{r}} x_r(t^{-1}) \, x_{\bar{r}}(\bar{t}^{-1})$$

and we may choose $n(w_0^J) = n_r n_{\bar{r}}$. Also $h_r(-t^{-1}) \, h_{\bar{r}}(-\bar{t}^{-1}) \in H^2$ and so

$$x_{-r}(t) \, x_{-\bar{r}}(\bar{t}) \in B^2 \cup B^2 n(w_0^J) B^2.$$

A similar argument works if J has type $A_1 \times A_1 \times A_1$.

If J has type A_2 we again make use of the homomorphism from $SL_3(K)$ to $\langle X_r, X_{\bar{r}}, X_{-r}, X_{-\bar{r}} \rangle$. Under this homomorphism

$$\begin{pmatrix} 1 & 0 & 0 \\ t & 1 & 0 \\ u & \bar{t} & 1 \end{pmatrix} \to x_{-r}(t) \, x_{-\bar{r}}(\bar{t}) \, x_{-r-\bar{r}}(u).$$

Now we have

$$\begin{pmatrix} 1 & 0 & 0 \\ t & 1 & 0 \\ u & \bar{t} & 1 \end{pmatrix} = \begin{pmatrix} 1 & \bar{t}\bar{u}^{-1} & u^{-1} \\ 0 & 1 & tu^{-1} \\ 0 & 0 & 1 \end{pmatrix} \begin{pmatrix} \bar{u}^{-1} & 0 & 0 \\ 0 & \bar{u}u^{-1} & 0 \\ 0 & 0 & u \end{pmatrix} \begin{pmatrix} 0 & 0 & 1 \\ 0 & -1 & 0 \\ 1 & 0 & 0 \end{pmatrix} \begin{pmatrix} 1 & \bar{t}u^{-1} & u^{-1} \\ 0 & 1 & t\bar{u}^{-1} \\ 0 & 0 & 1 \end{pmatrix}$$

Suppose $u + \bar{u} = t\bar{t}$. Then we have seen that the images of all the matrices in this equation are in G^1, except for

$$\begin{pmatrix} 0 & 0 & 1 \\ 0 & -1 & 0 \\ 1 & 0 & 0 \end{pmatrix}.$$

Thus the image of this matrix is in G^1 also. However, this image is in N and corresponds to the element w_0^J of W. Thus we may choose this image as $n(w_0^J)$. Since the image of

$$\begin{pmatrix} \bar{u}^{-1} & 0 & 0 \\ 0 & \bar{u}u^{-1} & 0 \\ 0 & 0 & u \end{pmatrix}$$

is $h_r(\bar{u}^{-1}) \, h_{\bar{r}}(u^{-1}) \in H^2$ we have

$$x_{-r}(t) \, x_{-\bar{r}}(\bar{t}) \, x_{-r-\bar{r}}(u) \in B^2 \cup B^2 n(w_0^J) \, B^2$$

whenever $u + \bar{u} = t\bar{t}$. Thus

$$X_{\Phi_J^-}^1 \subseteq B^2 \cup B^2 n(w_0^J) \, B^2$$

in all cases.

It now follows exactly as in 8.1.4 that $B^2 \cup B^2 n(w_0^J) \, B^2$ is a subgroup of G^1 for each p-orbit J of Π. Repeating the argument of 8.1.5 we then have

$$B^2 n B^2 . B^2 n(w_0^J) \, B^2 \subseteq B^2 n n(w_0^J) \, B^2 \cup B^2 n B^2$$

for each $n \in N^2$. It then follows as in 8.2.2 that $B^2 N^2 B^2$ is a subgroup of G^1. However $X^1_{\Phi_J^+}$ and $X^1_{\Phi_J^-}$ are contained in $B^2 N^2 B^2$ for all ρ-orbits J of Π and these subgroups generate G^1 by 13.6.5. Thus $G^1 = B^2 N^2 B^2$.

It is now easy to show that H^1 is contained in H^2. Let $h \in H^1$. Then

$$h \in B^2 N^2 B^2 = U^1 H^2 N^2 H^2 U^1 = U^1 N^2 U^1.$$

Thus we may write $h = u'nu$, where $u, u' \in U^1$ and $n \in N^2$. Hence

$$n = (u')^{-1} h u^{-1} \in B^1 \cap N^2 = B^1 \cap N^1 \cap N^2 = H^1 \cap N^2 = H^2.$$

Thus $(u')^{-1} . h u^{-1} h^{-1} . h \in H^2$. It follows that $(u')^{-1} . h u^{-1} h^{-1} = 1$ and $h \in H^2$. This completes the proof. ∎

We now discuss the subgroup H^1 in the cases where there are roots of different lengths. Then $\mathfrak{L}$ has type B_2, G_2 or F_4.

THEOREM 13.7.4. *Suppose $\mathfrak{L}$ contains roots of two different lengths. Then the elements $h(\chi) \in H$ which are fixed by σ are the ones for which*

$$\chi(\bar{r}) = \chi(r)^{\lambda(\bar{r})\theta}, \qquad r \in \Pi.$$

If $\mathfrak{L}$ has type B_2 or F_4, then an element $h(\chi)$ lies in H^1 if and only if it is fixed by σ. The same holds if $\mathfrak{L}$ has type G_2 provided the field K is finite.

(It is not known whether H^1 coincides with the set of σ-invariant elements of H if $\mathfrak{L}$ has type G_2 and K is infinite.)

PROOF. We first describe the elements $h(\chi)$ which are fixed by σ. As before, χ is a K-character of Q and we have

$$\chi = \chi_{p_1, t_1} \chi_{p_2, t_2} \cdots \chi_{p_l, t_l},$$

where $\chi(q_i) = t_i$. Thus

$$h(\chi) = \prod_{i=1}^{l} h_{p_i}(t_i).$$

Hence

$$\sigma . h(\chi) = \prod_i h_{\bar{p}_i}(t^{\lambda(\bar{p}_i)\theta}) = h\left(\prod_i \chi_{\bar{p}_i, t_i^{\lambda(\bar{p}_i)\theta}} \right).$$

Thus $h(\chi)$ is σ-invariant if and only if

$$\chi = \prod_i \chi_{\bar{p}_i, t_i^{\lambda(\bar{p}_i)\theta}}.$$

Comparing the values of these two characters at $q_{\rho(i)}$ we obtain

$$t_{\rho(i)} = t_i^{\lambda(\bar{p}_i)\theta}.$$

Thus

$$\chi(q_{\rho(i)}) = \chi(q_i)^{\lambda(\bar{p}_i)\theta}.$$

We show that an analogous result holds if the fundamental weights q_i are replaced by the fundamental roots p_i. We recall from section 7.1 that

$$p_i = \sum_{j=1}^{l} A_{ji}q_j.$$

Thus

$$\chi(p_{\rho(i)}) = \chi\left(\sum_j A_{\rho(j)\rho(i)} q_{\rho(j)}\right) = \prod_j \chi(q_{\rho(j)})^{A_{\rho(j),\,\rho(i)}}$$

$$= \prod_j \chi(q_j)^{A_{\rho(j),\,\rho(i)}\,\lambda(p_{\rho(j)})\theta} = \prod_j \chi(q_j)^{A_{ji}\,\lambda(p_{\rho(i)})\theta}$$

$$= \chi(p_i)^{\lambda(p_{\rho(i)})\theta}$$

since

$$A_{ji} = A_{\rho(i),\,\rho(j)} = A_{\rho(j),\,\rho(i)} \frac{\lambda(p_{\rho(j)})}{\lambda(p_{\rho(i)})}.$$

(We recall that ρ interchanges long and short roots.) Thus

$$\chi(p_{\rho(i)}) = \chi(p_i)^{\lambda(p_{\rho(i)})\theta}.$$

Hence $h(\chi)$ is σ-invariant if and only if

$$\chi(\bar{r}) = \chi(r)^{\lambda(\bar{r})\theta}, \qquad r \in \Pi.$$

Now every element $h(\chi)$ in H^1 is fixed by σ, and we consider whether the converse is true. It is evident that the group of elements $h(\chi)$ fixed by σ is generated by the elements of form $h(\chi_r, t\chi_{\bar{r}}, t^{\lambda(\bar{r})\theta})$ for each ρ-orbit $\{r, \bar{r}\}$ of Π. If $\mathfrak{L}$ has type B_2 or G_2 there is just one ρ-orbit of Π and if $\mathfrak{L}$ has type F_4 there are two ρ-orbits. If $\mathfrak{L}$ has type B_2 or F_4 the ρ-orbits have type $A_1 \times A_1$ or B_2.

Suppose $J = \{r, \bar{r}\}$ is a ρ-orbit of type $A_1 \times A_1$. Then we have

$$x_r(t)\, x_{\bar{r}}(t^{\lambda(\bar{r})\theta})\, x_{-r}(-t^{-1})\, x_{-\bar{r}}(-(t^{\lambda(\bar{r})\theta})^{-1})\, x_r(t)\, x_{\bar{r}}(t^{\lambda(\bar{r})\theta}) = n_r(t)\, n_{\bar{r}}(t^{\lambda(\bar{r})\theta})$$

and this is an element of G^1. Putting $t = -1$ we have $n_r(-1)\, n_{\bar{r}}(-1)$ is in G^1. (Recall that K has characteristic 2.) Thus

$$n_r(t)\, n_{\bar{r}}(t^{\lambda(\bar{r})\theta})\, n_r(-1)\, n_{\bar{r}}(-1) = h_r(t)\, h_{\bar{r}}(t^{\lambda(\bar{r})\theta}) = h(\chi_r, t\chi_{\bar{r}}, t^{\lambda(\bar{r})\theta})$$

and this is an element of $G^1 \cap H = H^1$.

I

Now suppose that $J = \{a, b\}$ is a ρ-orbit of type B_2, where a is a short root and b is long. We show that $h(\chi_a, {}_t\chi_b, t^{2\theta}) \in H^1$ by means of a matrix representation. The subgroup of G generated by X_a, X_b, X_{-a}, X_{-b} is isomorphic to the Chevalley group $B_2(K) = C_2(K)$. Then by 11.3.2 (iii) there is a homomorphism from $Sp_4(K)$ into $\langle X_a, X_b, X_{-a}, X_{-b} \rangle$ under which

$$I + t(e_{12} - e_{-2, -1}) \to \chi_a(t),$$
$$I + te_{2, -2} \to x_b(t),$$
$$I + t(e_{1, -2} + e_{2, -1}) \to x_{a+b}(t),$$
$$I + te_{1, -1} \to x_{2a+b}(t),$$
$$I - t(e_{-1, -2} - e_{21}) \to x_{-a}(t),$$
$$I + te_{-2, 2} \to x_{-b}(t),$$
$$I + t(e_{-1, 2} + e_{-2, 1}) \to x_{-(a+b)}(t),$$
$$I + te_{-1, 1} \to x_{-(2a+b)}(t).$$

Since K has characteristic 2 the signs are irrelevant and the map is in fact an isomorphism. It may be checked by a straightforward matrix multiplication that, under this isomorphism, we have

$$\begin{pmatrix} t & 0 & 0 & 0 \\ 0 & t^{2\theta-1} & 0 & 0 \\ 0 & 0 & t^{-1} & 0 \\ 0 & 0 & 0 & t^{1-2\theta} \end{pmatrix} \to h_a(t)\, h_b(t^{2\theta}).$$

Define

$$u_1(t) = x_a(t^\theta)\, x_b(t)\, x_{a+b}(t^{\theta+1})$$

and

$$v_2(t) = x_{-(a+b)}(t^\theta) x_{-(2a+b)}(t).$$

Then $u_1(t)$, $v_2(t)$ lie in G^1 for all $t \in K$, by 13.6.4 (vi). We may now verify that

$$A = \begin{pmatrix} 1 & t^{1-\theta} & t & 0 \\ 0 & 1 & t^\theta & t^{2\theta-1} \\ 0 & 0 & 1 & 0 \\ 0 & 0 & t^{1-\theta} & 1 \end{pmatrix} \to u_1(t^{2\theta-1}),$$

$$B = \begin{pmatrix} 1 & 0 & 0 & 0 \\ 0 & 1 & 0 & 0 \\ t^{-1} & t^{-\theta} & 1 & 0 \\ t^{-\theta} & 0 & 0 & 1 \end{pmatrix} \to v_2(t^{-1}).$$

Now ABA^{-1} is the matrix

$$\begin{pmatrix} 0 & 0 & t & 0 \\ 0 & 0 & 0 & t^{2\theta-1} \\ t^{-1} & 0 & 0 & 0 \\ 0 & t^{1-2\theta} & 0 & 0 \end{pmatrix}$$

and so the image of this matrix is in G^1. Putting $t=1$, the image of

$$\begin{pmatrix} 0 & 0 & 1 & 0 \\ 0 & 0 & 0 & 1 \\ 1 & 0 & 0 & 0 \\ 0 & 1 & 0 & 0 \end{pmatrix}$$

is in G^1. Multiplying these two matrices together we see that

$$h_a(t)\, h_b(t^{2\theta}) \in G^1,$$

and so is in H^1, as required. Thus H^1 coincides with the set of σ-invariant elements of H when $\mathfrak{L}$ has type B_2 or F_4.

Finally suppose $\mathfrak{L}$ has type G_2. As we lack a matrix representation of conveniently small degree (the smallest has degree 7) we argue in a different manner. We assume in this case that the field K is finite. We define two elements $u(t)$, $u_0 \in U^1$ as follows:

$$u(t) = x_{2a+b}(t^{\theta})\, x_{3a+2b}(t),$$
$$u_0 = x_a(1)\, x_b(1)\, x_{a+b}(1)\, x_{2a+b}(1).$$

The fact that these elements are in U^1 follows from 13.6.4 (vii). Consider the effect of these elements on the element e_{-3a-2b} of a Chevalley basis. We have

$$u(t)\,.\,e_{-3a+2b} = -t^2 e_{3a+2b} + a_1,$$
$$u_0\,.\,e_{-3a-2b} = N_{b,\,3a+b} M_{2a+b,\,-3a-2b,\,3} e_{3a+2b} + a_2,$$

where a_1, a_2 are linear combinations of elements of the Chevalley basis other than e_{3a+2b}. Using the structure constants given at the end of section 12.4 we have $N_{b,\,3a+b} M_{2a+b,\,-3a-2b,\,3} = 1$, hence

$$u_0\,.\,e_{-3a-2b} = e_{3a+2b} + a_2.$$

Now $G^1 = B^1 N^1 B^1 = U^1 N^1 U^1$ and we have similarly $G^1 = V^1 N^1 V^1$. Let $t \neq 0$ and $u(t) = v'nv$, with v, $v' \in V^1$ and $n \in N^1$. Then

$$v'nv\,.\,e_{-3a-2b} = -t^2 e_{3a+2b} + a_1$$

and it follows that

$$n . e_{-3a-2b} = -t^2 e_{3a+2b} + \text{terms involving different } e_r\text{'s.}$$

If we argue in a similar manner for u_0 we have $u_0 = v_0' n_0 v_0$, where

$$v_0, \; v_0' \in V^1, \; n_0 \in N^1$$

and

$$n_0 . e_{-3a-2b} = e_{3a+2b} + \text{terms involving different } e_r\text{'s.}$$

Now n and n_0 are elements of N^1 corresponding to the element w_0 of W^1. For w_0 is the only non-unit elements of W^1. Thus n, n_0 operate monomially on the root vectors and we have

$$n . e_{-3a-2b} = -t^2 e_{3a+2b},$$
$$n_0 . e_{-3a-2b} = e_{3a+2b}.$$

Also $n_0^{-1} . n$ must be an element of H^1 satisfying

$$n_0^{-1} n . e_{-3a-2b} = -t^2 e_{-3a-2b}.$$

However, the σ-invariant elements of H are those of the form $h_a(u^\theta) \, h_b(u)$ for all $u \in K$. Such an element operates on e_{-3a-2b} as follows:

$$h_a(u^\theta) \, h_b(u) . e_{-3a-2b} = (u^\theta)^{A_{a, \; -3a-2b}} u^{A_{b, \; -3a-2b}} e_{-3a-2b}$$
$$= u^{-1} e_{-3a-2b}.$$

We now compare coefficients and see that

$$n_0^{-1} n = h_a((-t^{-2})^\theta) \, h_b(-t^{-2}).$$

Thus

$$h_a((-t^{-2})^\theta) \, h_b(-t^{-2}) \in H^1.$$

Putting $t = 1$ we have

$$h_a((-1)^\theta) \, h_b(-1) \in H^1$$

and by multiplying these two elements we obtain

$$h_a((t^{-2})^\theta) \, h_b(t^{-2}) \in H^1.$$

Thus the group of σ-invariant elements of H will be generated by elements of H^1 provided the multiplicative group of K is generated by the squares in K and the element -1.

Now K is assumed finite and so the non-zero squares in K form a subgroup of the multiplicative group of K of index 2. The multiplicative group is therefore generated by the squares together with any single non-square. We show that -1 is a non-square. Suppose $t^2 = -1$. Then $t^{2\theta} = (-1)^\theta = -1$. Thus $t^{2\theta} = t^2$ and so $t^\theta = \pm t$. Hence $t^{\theta^2} = t$ and

$$t^{3\theta^2} = t^3 = -t.$$

But the automorphism θ of K satisfies $3\theta^2 = 1$. Hence $t^{3\theta^2} = t$ and we have a contradiction. This completes the proof. ∎

It is not known whether the result of $13.7.4$ holds if $\mathfrak{L} = G_2$ and K is infinite. For further information on the situation in this case see Ree [3] and Steinberg [15].

CHAPTER 14

Further Properties of the Twisted Groups

In the last chapter we defined the twisted groups and gave a detailed description of the subgroups U^1, H^1, N^1, W^1. In the present chapter we shall derive the orders of the finite groups in the families, prove the simplicity of the twisted groups, and show that certain of these groups can be identified with classical groups.

14.1 The Finite Twisted Groups

Suppose G^1 is a twisted group constructed as a subgroup of the Chevalley group $G = \mathfrak{L}(K)$, and suppose that the field K is finite. If $\mathfrak{L}$ has type A_l, where $l \geqslant 2$, then K admits an automorphism of order 2, so must be the field $GF(q^2)$ for some prime-power q. The same applies if $\mathfrak{L}$ has type D_l or E_6. If $\mathfrak{L}$ has type D_4 and the graph automorphism of G used in the construction has order 3, then K must be a field admitting an automorphism of order 3. Thus $K = GF(q^3)$ for some prime-power q. If $\mathfrak{L}$ has type B_2 or F_4, then K is a field of characteristic 2 admitting an automorphism θ such that $2\theta^2 = 1$. If $\mathfrak{L}$ has type G_2, K is a field of characteristic 3 admitting an automorphism θ such that $3\theta^2 = 1$. We determine which finite fields have these properties.

LEMMA 14.1.1. *Let* $K = GF(p^n)$ *be a finite field of characteristic* p *admitting an automorphism* θ *satisfying* $p\theta^2 = 1$. *Then* n *is odd and* θ *has the form*

$$\lambda^\theta = \lambda^{p^m},$$

where $n = 2m + 1$.

PROOF. We have $\lambda^\theta = \lambda^{p^r}$ for some r. Thus $\lambda^{p\theta^2} = \lambda^{p^{2r+1}}$, and so $\lambda^{p^{2r+1}} = \lambda$ for all $\lambda \in K$. Thus K is contained in $GF(p^{2r+1})$, hence n divides $2r + 1$. It follows that n is odd, and we write $n = 2m + 1$. Let

$$2r + 1 = (2m + 1).(2s + 1).$$

250

Then $r = s(2m+1) + m$ and we have

$$\lambda^{p^r} = \lambda^{p^{(2m+1)s+m}} = \lambda^{p^{(2m+1)s}} {}^{p^m} = (\lambda^{p^{(2m+1)s}})^{p^m}.$$

But $\lambda^{p^{2m+1}} = \lambda$, thus $\lambda^{p^{(2m+1)s}} = \lambda$ and $\lambda^{p^r} = \lambda^{p^m}$. Hence $\lambda^{\theta} = \lambda^{p^m}$ as required. ∎

Lemma 14.1.1 shows that if $\mathfrak{L}$ has type B_2 or F_4 then $K = GF(2^{2m+1})$ for some m, and if $\mathfrak{L}$ has type G_2 then $K = GF(3^{2m+1})$ for some m.

The notation $G^1 = {}^i\mathfrak{L}(K)$ will be used for the twisted groups, where the superscript i denotes the order of the symmetry of the Dynkin diagram used in the construction. The finite twisted groups will be denoted by

$$^2A_l(q^2), \qquad l \geqslant 2,$$
$$^2D_l(q^2), \qquad l \geqslant 4,$$
$$^2E_6(q^2),$$
$$^3D_4(q^3),$$
$$^2B_2(2^{2m+1}),$$
$$^2G_2(3^{2m+1}),$$
$$^2F_4(2^{2m+1}).$$

It is convenient to write the three latter groups also as $^2B_2(q^2)$, $^2G_2(q^2)$, $^2F_4(q^2)$ and so we define q in these cases to be $p^{m+1/2}$, where $p = 2$ or 3. Note however that q is an irrational number in these cases rather than a rational prime-power.

The groups $^2B_2(K)$ were discovered by Suzuki [1, 2] and are called Suzuki groups, and the groups $^2G_2(K)$, $^2F_4(K)$ were discovered by Ree [2, 3] and are called Ree groups.

In order to determine the orders of the finite twisted groups we first calculate the orders of their subgroups U^1, $(U_w^-)^1$ and H^1.

LEMMA 14.1.2. (i) $|U^1| = q^N$.

(ii) $|(U_w^-)^1| = q^{l(w)}, \qquad w \in W^1$.

(iii) $|H^1| = \dfrac{1}{d}(q - \eta_1)(q - \eta_2) \ldots (q - \eta_l)$,

where $\eta_1, \ldots, \eta_l$ are the eigenvalues of the isometry τ of $\mathfrak{P}$, and d is the order of the group of self-conjugate K-characters of Q/P in the case when all roots have the same length, and $d = 1$ otherwise.

PROOF. Let S be one of the equivalence classes in Φ defined in 13.2.1. Then 13.6.3 shows that $|X_S^1| = q^{|S|}$. By 13.6.1 it follows that $|U^1| = q^N$ and $|(U_w^-)^1| = q^{l(w)}$.

We now consider $|H^1|$. Suppose all the roots have the same length. Then $\sigma.h_r(t) = h_{\bar{r}}(\bar{t})$ by 13.7.1. Let $h(\chi)$ be an element of H, where χ is a K-character of Q. Then

$$\sigma.h(\chi) = \sigma.h(\chi_{p_1, \lambda_1} \cdots \chi_{p_l, \lambda_l})$$

as in 7.1.1 where $\lambda_i = \chi(q_i)$. It follows that

$$\sigma.h(\chi) = h(\chi_{\bar{p}_1, \bar{\lambda}_1} \cdots \chi_{\bar{p}_l, \bar{\lambda}_l}) = h(\bar{\chi}),$$

where $\bar{\chi}(\bar{q}_i) = \overline{\chi(q_i)}$. Thus $h(\chi)$ is invariant under σ if and only if χ is a self-conjugate K-character of Q.

Let $\hat{H}^1$ be the subgroup of H of elements which are invariant under σ. $|\hat{H}^1|$ is the number of self-conjugate K-characters of Q. In order to determine the number of self-conjugate K-characters we consider the ρ-orbits of Π and the corresponding ρ-orbits of the set $q_1, \ldots, q_l$ of fundamental weights. Let J be such a ρ-orbit. If $|J| = 1$ the value of a self-conjugate K-character χ at the corresponding q_i can be any non-zero element of $GF(q)$, so can be chosen in $q-1$ ways. If $|J| = 2$ the value of χ at q_i can be any non-zero element of $GF(q^2)$ and the value at $\bar{q}_i$ is then determined. Thus the values of χ on this orbit can be chosen in $q^2 - 1$ ways. Similarly if $|J| = 3$ the values of χ on the corresponding orbit can be chosen in $q^3 - 1$ ways. Thus the number of self-conjugate K-characters of Q is

$$\prod_J q^{(|J|-1)}$$

over all ρ-orbits J of Π. However, the roots of $x^{|J|} - 1$ are the eigenvalues of the isometry τ of $\mathfrak{P}$ on the subspace spanned by the fundamental roots in J. Thus

$$\prod_J (q^{|J|} - 1) = (q - \eta_1)(q - \eta_2) \ldots (q - \eta_l),$$

where $\eta_1, \ldots, \eta_l$ are the eigenvalues of τ on $\mathfrak{P}$.

Now H^1 is the set of elements $h(\chi)$ where χ is a self-conjugate K-character of P which can be extended to a self-conjugate K-character of Q (13.7.2). Thus there is a homomorphism of $\hat{H}^1$ onto H^1 obtained by restricting χ from Q to P. The kernel of this homomorphism is the set of $h(\chi)$, where χ is a self-conjugate K-character of Q which is the identity on P. Thus $|\hat{H}^1 : H^1| = d$, where d is the order of the group of self-conjugate K-characters of Q/P. Therefore

$$| H^1 | = \frac{1}{d} \prod_{i=1}^{l} (q - \eta_i),$$

as required.

Now suppose that there are roots of different lengths, i.e. that G^1 is a Suzuki or Ree group. Then H^1 consists of all σ-invariant elements of H by 13.7.4, since we have a finite field. Also $h(\chi) \in H^1$ if and only if

$$\chi(\bar{r}) = \chi(r)^{\lambda(\bar{r})\theta}, \qquad r \in \Pi.$$

As before the values of χ on a ρ-orbit J of Π can be chosen in $q^{|J|} - 1$ ways. (Note that all ρ-orbits now have two elements.) Thus

$$| H^1 | = \prod_{J} (q^{|J|} - 1) = \prod_{i=1}^{l} (q - \eta_i)$$

for the Suzuki and Ree groups.   ∎

Note. The number d in 14.1.2 takes the value $(l+1, q+1)$ for $^2A_l(q^2)$; $(4, q^l + 1)$ for $^2D_l(q^2)$; $(3, q+1)$ for $^2E_6(q^2)$; and 1 for $^3D_4(q^3)$.

For if $\mathfrak{L} = A_l$, D_{2k+1} or E_6, the group Q/P is cyclic with a generator a satisfying $\bar{a} = -a$. Thus χ is a self-conjugate K-character of Q/P if and only if $\chi(-a) = \chi(a)^q$, i.e. $\chi(a)^{q+1} = 1$.

Thus $d = (\Delta, q+1)$, where $\Delta = | Q/P |$. For the groups $^2D_{2k}(q^2)$, Q/P is elementary abelian of order 4 and the map $a \rightarrow \bar{a}$ interchanges two non-zero elements and fixes the third. Thus $d = (2, q+1)$ in this case, and the case $^2D_l(q^2)$ can be summarized by writing $d = (4, q^l + 1)$. Finally for $^3D_4(q^3)$, Q/P is elementary abelian of order 4 and the map $a \rightarrow \bar{a}$ permutes cyclically the three non-zero elements. Thus $d = 1$ in this case. These facts may easily be verified using the information about the group Q/P given in section 8.6.

PROPOSITION 14.1.3.

$$| G^1 | = \frac{1}{d} q^N \prod_{i=1}^{l} (q - \eta_i) \sum_{w \in W^1} q^{l(w)}.$$

PROOF. This follows from the double coset decomposition of G^1 with respect to B^1. The number of elements in the double coset $B^1 n_w B^1$, $w \in W^1$, is $| B^1 | . | (U_w^-)^1 |$ by 13.5.3, and this is equal to

$$\frac{1}{d} q^N \prod_{i=1}^{l} (q - \eta_i) . q^{l(w)}$$

by 14.1.2. The result follows.   ∎

14.2 Factorization of the Polynomial $P_{W^1}(t)$

As with the Chevalley groups, it is possible to simplify the expression obtained for the order of the twisted groups by finding a factorization of the polynomial

$$P_{W^1}(t) = \sum_{w \in W^1} t^{l(w)}.$$

To do this we use a modification of the argument for the Chevalley groups used in section 9.4.

Let $\mathfrak{J} = \mathbb{R}[I_1, \ldots, I_l]$ be the ring of polynomial invariants of W. The isometry τ of $\mathfrak{P}$ operates naturally on $\mathfrak{J}$ by

$$\tau(P)(x) = P(\tau^{-1}(x))$$

for $P \in \mathfrak{J}$, $x \in \mathfrak{P}$. Since P is an invariant polynomial and τ normalizes W, $\tau(P)$ must also be invariant. Now τ transforms the subspace $\mathfrak{J}_n$ of homogeneous polynomials of degree n into itself for each n. Thus by extending the base field from $\mathbb{R}$ to $\mathbb{C}$ we can choose a basis of $\mathfrak{J}_n$ consisting of eigenvectors of τ. Hence the generators $I_1, \ldots, I_l$ of $\mathfrak{J}$ as a polynomial ring can be chosen to be eigenvectors of τ. We suppose the basic invariants $I_1, \ldots, I_l$ are chosen in this way, where deg $I_i = d_i$ and

$$\tau(I_i) = \epsilon_i I_i, \qquad \epsilon_i \in \mathbb{C}.$$

PROPOSITION 14.2.1.

$$\sum_{w \in W^1} t^{l(w)} = \prod_{i=1}^{l} \left(\frac{1 - \epsilon_i t^{d_i}}{1 - \eta_i t} \right).$$

PROOF. (a) Let

$$P_{W^1}(t) = \sum_{w \in W^1} t^{l(w)} \quad \text{and} \quad \bar{P}_{W^1}(t) = \prod_{i=1}^{l} \left(\frac{1 - \epsilon_i t^{d_i}}{1 - \eta_i t} \right).$$

Let $P_{W_J^1}(t)$ and $\bar{P}_{W_J^1}(t)$ denote the corresponding polynomials for the groups W_J^1, where J is any ρ-invariant subset of Π. For each ρ-invariant subset J let O_J be the number of ρ-orbits in J. We shall show that $P_{W^1}(t)$ and $\bar{P}_{W^1}(t)$ satisfy the following identities:

$$\sum_J (-1)^{O_J} \frac{P_{W^1}(t)}{P_{W_J^1}(t)} = t^N,$$

$$\sum_J (-1)^{O_J} \frac{\bar{P}_{W^1}(t)}{\bar{P}_{W_J^1}(t)} = t^N,$$

where the sums are taken over all ρ-invariant subsets of Π. (Compare these identities with the ones in section 9.4.)

(b) The identity involving $P_{W^1}(t)$ is proved as before. We have

$$\sum_J (-1)^{O_J} \frac{P_{W^1}(t)}{P_{W_J^1}(t)} = \sum_J (-1)^{O_J} P_{D_J^1}(t)$$

$$= \sum_J (-1)^{O_J} \left(\sum_{\substack{w \in W \\ w(J) \subseteq \Phi^+}} t^{l(w)} \right)$$

$$= \sum_{w \in W^1} \left(\sum_{\substack{J \\ w(J) \subseteq \Phi^+}} (-1)^{O_J} \right) t^{l(w)}.$$

Let J_w be the set of roots $r \in \Pi$ such that $w(r) \in \Phi^+$. Then the coefficient of $t^{l(w)}$ in the above sum is

$$\sum_{J \subseteq J_w} (-1)^{O_J} = (1-1)^n, \quad \text{where } n = O_{J_w}.$$

This is 0 unless J_w is the empty set. If $J_w = \phi$ then $w = w_0$. Thus the above sum is $t^{l(w_0)} = t^N$.

(c) We now consider the polynomial $\bar{P}_{W^1}(t)$. In order to show that this polynomial satisfies the above identity we must consider the operation of the Weyl group on the Coxeter complex. Note that the isometry τ of $\mathfrak{P}$ also operates on the Coxeter complex. Now each orbit of the Coxeter complex under the operation of W contains just one element C_J by 2.6.3, where

$$C_J = \{v; \ (v, r) = 0 \text{ for } r \in J, \ (v, r) > 0 \text{ for } r \in \Pi - J\}.$$

For each ρ-invariant subset J of Π, let $n_J(w\tau)$ be the number of elements in the orbit containing C_J which are fixed by $w\tau$. We show that

$$\sum_J (-1)^{O_J} n_J(w\tau) = \det w, \quad w \in W^1.$$

Let $\mathfrak{U}$ be the subspace of $\mathfrak{P}$ of elements fixed by $w\tau$. Then the elements of the Coxeter complex which intersect $\mathfrak{U}$ are precisely those which are fixed by $w\tau$. For if an element $\mathfrak{K}$ of the complex intersects $\mathfrak{U}$ it contains a point in common with $w\tau(\mathfrak{K})$, and so $w\tau(\mathfrak{K}) = \mathfrak{K}$. Conversely, if $w\tau(\mathfrak{K}) = \mathfrak{K}$, $w\tau$ must fix some point of $\mathfrak{K}$ since $w\tau$ is an isometry. Thus $\mathfrak{K}$ intersects $\mathfrak{U}$.

We now consider the complex in $\mathfrak{U}$ whose elements have form $\mathfrak{K} \cap \mathfrak{U}$, where $\mathfrak{K}$ is an element of the Coxeter complex with $w\tau(\mathfrak{K}) = \mathfrak{K}$. We wish to find the dimension of $\mathfrak{K} \cap \mathfrak{U}$, and consider first the case in which $\mathfrak{K} = C_J$ for some J. Since $w\tau(C_J) = C_J$ we have $w(C_{\rho(J)}) = C_J$ and so $\rho(J) = J$ by 2.6.3. Thus $\tau(C_J) = C_J$ and $w(C_J) = C_J$. Therefore w fixes every element of C_J by 2.6.2 and so $w\tau$ acts on C_J in the same way as τ. It follows that the elements of C_J fixed by $w\tau$ are the elements of C_J fixed by τ. However, dim $C_J = |\Pi - J|$ and each ρ-orbit of $\Pi - J$ contributes 1 to the dimension of the τ-invariant elements of C_J. Thus we have

$$\dim (C_J \cap \mathfrak{U}) = O_{\Pi - J}.$$

We now consider an arbitrary element $\mathfrak{K}$ of the Coxeter complex with $w\tau(\mathfrak{K}) = \mathfrak{K}$. We have $\mathfrak{K} = w'(C_J)$ for some $w' \in W$ and some J, by 2.6.3. Then

$$w\tau w'(C_J) = w'(C_J)$$

and so

$$w'^{-1}w \cdot \tau w' \tau^{-1} \cdot \tau(C_J) = C_J.$$

However, $\tau w' \tau^{-1} \in W$, and so we have

$$\dim (C_J \cap \mathfrak{U}') = O_{\Pi - J},$$

where $\mathfrak{U}'$ is the subspace of $\mathfrak{P}$ of elements fixed by $w'^{-1}w\tau w'$. Now $\mathfrak{U} = w'(\mathfrak{U}')$ and so

$$O_{\Pi - J} = \dim (C_J \cap w'^{-1}(\mathfrak{U})) = \dim (w'(C_J) \cap \mathfrak{U}).$$

Thus dim $(\mathfrak{K} \cap \mathfrak{U}) = O_{\Pi - J}$.

We now apply 9.4.4 to the complex induced in $\mathfrak{U}$. We have seen that the number of elements of this complex of dimension i is given by

$$n_i = \sum_{O_{\Pi - J} = i} n_J(w\tau).$$

Thus we have

$$\sum_J (-1)^{O_{\Pi - J}} n_J(w\tau) = (-1)^{\dim \mathfrak{U}}.$$

Hence

$$\sum_J (-1)^{O_J} n_J(w\tau) = (-1)^{O_\Pi} \cdot (-1)^{\dim \mathfrak{U}}.$$

Now w, τ and $w\tau$, being orthogonal transformations, have eigenvalues 1, -1 or pairs of complex conjugates. Thus

$$\det (w\tau) = (-1)^{l - \dim \mathfrak{U}},$$

$$\det \tau = (-1)^{l - \dim \mathfrak{P}^1} = (-1)^{l - O_\Pi}.$$

Hence

$$\det w = (-1)^{O_\Pi} \cdot (-1)^{\dim \mathfrak{A}}$$

and we have

$$\sum_J (-1)^{O_J} n_J(w\tau) = \det w.$$

(d) We prove next an analogue of 9.4.7 by showing that

$$\sum_J (-1)^{O_J} \operatorname{tr}_{(\mathfrak{I}_J)_n} \tau = \operatorname{tr} \hat{\mathfrak{I}}_n \tau,$$

where $(\mathfrak{I}_J)_n$ is the space of homogeneous polynomials on $\mathfrak{P}$ of degree n invariant under W_J, $\hat{\mathfrak{I}}_n$ is the space of homogeneous alternating polynomials of degree n, and the sum extends over all ρ-invariant subsets J of Π. Note that the terms $\dim (\mathfrak{I}_J)_n$ and $\dim \hat{\mathfrak{I}}_n$ in 9.4.7 have been replaced here by the trace of τ acting on these two subspaces.

Consider the group $\langle W, \tau \rangle$ generated by W and τ. The remarks in (c) show that the stabilizer of C_J in $\langle W, \tau \rangle$ is $\langle W_J, \tau \rangle$. Therefore the unit character of $\langle W_J, \tau \rangle$ induced up to $\langle W, \tau \rangle$ satisfies

$$1\langle^{W, \tau}_{W_J, \tau}\rangle(w\tau) = n_J(w\tau),$$

just as in 9.4.2. Now let χ be the trace function of $\langle W, \tau \rangle$ acting on $\mathfrak{H}_n$, the space of homogeneous polynomials of degree n. Then we have

$$\chi\langle^{W, \tau}_{W_J, \tau}\rangle(w\tau) = 1\langle^{W, \tau}_{W_J, \tau}\rangle(w\tau) \cdot \chi(w\tau)$$

$$= n_J(w\tau) \chi(w\tau),$$

as in the proof of 9.4.7. By (c) it follows that

$$\sum_J (-1)^{O_J} \chi\langle^{W, \tau}_{W_J, \tau}\rangle(w\tau) = \det w \cdot \chi(w\tau).$$

We now average over W and obtain

$$\sum_J (-1)^{O_J} \frac{1}{|W|} \sum_{w \in W} \chi\langle^{W, \tau}_{W_J, \tau}\rangle(w\tau) = \frac{1}{|W|} \sum_{w \in W} \det w \cdot \chi(w\tau).$$

But

$$\frac{1}{|W|} \sum_{w \in W} \chi\langle^{W, \tau}_{W_J, \tau}\rangle(w\tau) = \frac{1}{|W_J|} \sum_{w \in W_J} \chi(w\tau),$$

as in the proof of 9.4.7. Thus

$$\sum_J (-1)^{O_J} \chi\left(\frac{1}{|W_J|} \sum_{w \in W_J} w\tau\right) = \frac{1}{|W|} \sum_{w \in W} \det w \cdot \chi(w\tau).$$

Let M be any $\langle W, \tau \rangle$-module and let

$$T = \frac{1}{|W|} \sum_{w \in W} w.$$

Then M has a direct decomposition $M = M_0 \oplus M_1$, as in 9.3.2, where M_0 is the set of elements annihilated by T and M_1 is the set of elements fixed by T. Since T commutes with τ, $T\tau$ leaves M_0 and M_1 invariant. In fact $T\tau = 0$ on M_0 and $T\tau = \tau$ on M_1. Thus the trace of $T\tau$ on M is equal to the trace of τ on M_1.

Applying this with $M = \mathfrak{S}_n$ we have

$$\chi\left(\frac{1}{|W_J|} \sum_{w \in W_J} w\tau\right) = \mathrm{tr}_{(\mathfrak{J}_J)_n} \tau$$

and also

$$\chi\left(\frac{1}{|W|} \sum_{w \in W} \det w . w\tau\right) = \mathrm{tr}_{\hat{\mathfrak{J}}_n} \tau.$$

It follows that

$$\sum_J (-1)^{o_J} \mathrm{tr}_{(\mathfrak{J}_J)_n} \tau = \mathrm{tr}_{\hat{\mathfrak{J}}_n} \tau$$

as required. We observe that $\mathrm{tr}_{\hat{\mathfrak{J}}_n} \tau = \mathrm{tr}_{\mathfrak{J}_{n-N}} \tau$ since by 9.4.6 we have

$$\hat{\mathfrak{J}}_n = 0 \text{ for } n < N,$$

$$\hat{\mathfrak{J}}_n = \left(\prod_{r \in \Phi^+} H_r\right) . \hat{\mathfrak{J}}_{n-N} \text{ for } n \geqslant N$$

and

$$\tau\left(\prod_{r \in \Phi^+} H_r\right) = \prod_{r \in \Phi^+} H_r.$$

(e) The coefficient of t^n in

$$\frac{1}{\prod\limits_{i=1}^{l} (1 - \eta_i t) . \bar{P}_{W^1}(t)} = \prod_{i=1}^{l} \frac{1}{1 - \epsilon_i t^{d_i}}$$

is the trace of τ on $\mathfrak{J}_n$, as in 9.3.3. Thus the coefficient of t^n in

$$\prod_{i=1}^{l} \frac{t^N}{(1 - \eta_i t) . \bar{P}_{W^1}(t)}$$

is $\mathrm{tr}_{\mathfrak{J}_{n-N}} \tau = \mathrm{tr}_{\hat{\mathfrak{J}}_n} \tau$.

Now consider the coefficient of t^n in

$$\frac{1}{\prod\limits_{i=1}^{l} (1 - \eta_i t) . \bar{P}_{W_J^1}(t)} = \frac{1}{(1 - \epsilon_{J1} t^{d_{J1}}) \ldots (1 - \epsilon_{Jk} t^{d_{Jk}})(1 - \eta_{k+1}) \ldots (1 - \eta_l)},$$

where

$$\bar{P}_{W_J^1}(t) = \prod_{i=1}^{k} \left(\frac{1 - \epsilon_{Ji} t^{d_{Ji}}}{1 - \eta_i} \right).$$

Since $\mathfrak{V} = \mathfrak{V}_J \oplus \mathfrak{V}_J^\perp$, where W_J operates trivially on $\mathfrak{V}_J^\perp$, the basic invariants of W_J on $\mathfrak{V}$ may be taken as the basic invariants of W_J on $\mathfrak{V}_J$ together with $l-k$ additional elements satisfying $\tau(x_i) = \eta_i x_i$, where $\eta_{k+1}, \ldots, \eta_l$ are the eigenvalues of τ on $\mathfrak{V}_J^\perp$. Thus the coefficients of t^n in

$$\frac{1}{(1 - \epsilon_{J1} t^{d_{J1}}) \ldots (1 - \epsilon_{Jk} t^{d_{Jk}})(1 - \eta_{k+1} t^0) \ldots (1 - \eta_l t^0)}$$

is the trace of τ on $(\mathfrak{I}_J)_n$, again as in 9.3.3. By (d) it follows that

$$\sum_J (-1)^{0_J} \frac{1}{\bar{P}_{W_J^1}(t)} = \frac{t^N}{\bar{P}_{W^1}(t)}.$$

(f) We have now proved the identities

$$\sum_J (-1)^{0_J} \frac{P_{W^1}(t)}{P_{W_J^1}(t)} = t^N,$$

$$\sum_J (-1)^{0_J} \frac{\bar{P}_{W^1}(t)}{\bar{P}_{W_J^1}(t)} = t^N,$$

and so may deduce that $P_{W^1}(t) = \bar{P}_{W^1}(t)$ by induction on $|J|$. This completes the proof. ∎

14.3 The Orders of the Finite Twisted Groups

THEOREM 14.3.1. *The orders of the finite twisted groups are given by*

$$|G^1| = \frac{1}{d} q^N (q^{d_1} - \epsilon_1)(q^{d_2} - \epsilon_2) \ldots (q^{d_l} - \epsilon_l),$$

where d is as in section 14.1 and d_i, ϵ_i as in section 14.2.

PROOF. We have shown in 14.2.1 that

$$\sum_{w \in W^1} t^{l(w)} = \prod_{i=1}^{l} \left(\frac{1 - \epsilon_i t^{d_i}}{1 - \eta_i t} \right) = t^N \prod_{i=1}^{l} \left(\frac{t^{-d_i} - \epsilon_i}{t^{-1} - \eta_i} \right),$$

since $d_1 + \ldots + d_l = N + l$ by 9.3.4. Replacing t by t^{-1} we have

$$\sum_{w \in W^1} t^{N-l(w)} = \prod_{l=1}^{l} \left(\frac{t^{d_i} - \epsilon_i}{t - \eta_i} \right).$$

However, $w_0 \in W^1$, and for each w we have

$$l(w_0 w) = N - l(w).$$

It follows that

$$\sum_{w \in W^1} t^{N-l(w)} = \sum_{w \in W^1} t^{l(w)}.$$

Thus

$$\prod_{i=1}^{l} (t - \eta_i) \cdot \sum_{w \in W^1} t^{l(w)} = \prod_{i=1}^{l} (t^{d_i} - \epsilon_i).$$

The result now follows from 14.1.3 on replacing t by q.   ■

We shall now determine the numbers $\epsilon_1, \ldots, \epsilon_l$ for the individual twisted groups. If we put $t = 1$ in the identity

$$\sum_{w \in W^1} t^{l(w)} = \prod_{i=1}^{l} \left(\frac{1 - \epsilon_i t^{d_i}}{1 - \eta_i t} \right),$$

we see that the number of ϵ_i equal to 1 is the same as the number of η_i equal to 1. If the isometry τ of $\mathfrak{H}$ has order 2, then $\epsilon_i = \pm 1$ and $\eta_i = \pm 1$ for all i. Thus $\epsilon_1, \ldots, \epsilon_l$ is a permutation of $\eta_1, \ldots, \eta_l$. If τ has order 3 then $\mathfrak{L} = D_4$ and $\eta_1, \ldots, \eta_4$ are 1, 1, ω, ω^2, where $\omega = e^{2\pi i/3}$. Thus two of $\epsilon_1, \ldots, \epsilon_4$ are 1 and the other two must be ω, ω^2 since τ is a real transformation of order 3. Thus in all cases $\epsilon_1, \ldots, \epsilon_l$ is a permutation of $\eta_1, \ldots, \eta_l$.

The numbers $d_1, \ldots, d_l$ and $\epsilon_1, \ldots, \epsilon_l$ are now known; however we still have to determine which ϵ_i is associated with which d_i. If $\mathfrak{L}$ has type A_l, D_l (l odd), or E_6, then $\tau = -w_0$. (This follows easily from the fact that $w_0 \neq -1$ in these cases.) Thus τ operates on each invariant in the same way as -1. Hence $\tau(I_i) = (-1)^{d_i} I_i$ and $\epsilon_i = (-1)^{d_i}$.

If $\mathfrak{L}$ has type D_l we may take for the fundamental roots

$$p_1 = e_1 - e_2, \ p_2 = e_2 - e_3, \ \ldots, \ p_{l-1} = e_{l-1} - e_l, \ p_l = e_{l-1} + e_l,$$

as in section 3.6. Then τ operates on the orthonormal basis (e_i) of $\mathfrak{H}$ by

$$\tau(e_i) = e_i, \quad i = 1, \ldots, l-1, \quad \tau(e_l) = -e_l.$$

Each $x \in \mathfrak{V}$ may be expressed in the form

$$x = x_1 e_1 + \ldots + x_l e_l$$

and the basic invariants may be taken as the first $l-1$ elementary symmetric functions in $x_1^2, x_2^2, \ldots, x_l^2$, together with $x_1 x_2 \ldots x_l$. Thus the d_i and ϵ_i have values

$$d_1 = 2, \ d_2 = 4, \ldots, \ d_{l-1} = 2(l-1), \ d_l = l,$$

$$\epsilon_1 = 1, \ \epsilon_2 = 1, \ldots, \ \epsilon_{l-1} = 1, \ \epsilon_l = -1.$$

If $\mathfrak{L} = B_2$ we have $d_1 = 2, \ d_2 = 4$ and the ϵ's are $1, \ -1$. Now the invariant

$$\sum_i x_i^2$$

of degree 2 has $\epsilon = 1$, since τ is an isometry of $\mathfrak{V}$. Thus $\epsilon_1 = 1, \ \epsilon_2 = -1$ for B_2. If $\mathfrak{L} = G_2$ we have $d_1 = 2, \ d_2 = 6$ and the ϵ's are $1, \ -1$. Thus we have $\epsilon_1 = 1, \ \epsilon_2 = -1$ as before.

Finally suppose $\mathfrak{L} = F_4$. Then

$$d_1 = 2, \qquad d_2 = 6, \qquad d_3 = 8, \qquad d_4 = 12$$

and the ϵ's are $1, \ 1, \ -1, \ -1$, where $\epsilon_1 = 1$. We show that the other basic invariant with $\epsilon = 1$ is the one of degree 8. The roots of F_4 may be written with respect to an orthonormal basis $e_1, \ e_2, \ e_3, \ e_4$ in the form

$$\pm e_i \pm e_j, \qquad\qquad i \neq j,$$

$$\pm e_i,$$

$$\tfrac{1}{2}(\pm e_1 \pm e_2 \pm e_3 \pm e_4).$$

Let $x \in \mathfrak{V}$ be written as

$$x = x_1 e_1 + x_2 e_2 + x_3 e_3 + x_4 e_4$$

and define the polynomial I by the formula

$$I = \sum_{i \neq j} (\pm x_i \pm x_j)^8 + \sum_i (\sqrt{2}(\pm x_i))^8 + \sum \left(\frac{\sqrt{2}}{2} (\pm x_1 \pm x_2 \pm x_3 \pm x_4) \right)^8.$$

There is one term in I for each root, and the terms corresponding to the short roots have an additional factor $\sqrt{2}$. Since the elements of W permute the long roots and the short roots, I is a polynomial invariant of W. Since τ maps long roots into multiples of short roots by a factor of $\sqrt{2}$ we have $\tau(I) = I$. Also, I is not a multiple of the quadratic invariant $x_1^2 + x_2^2 + x_3^2 + x_4^2$. For substituting $(x_1, x_2, x_3, x_4) = (1, i, 0, 0)$ in the formula for I gives a non-zero value. Thus I may be taken as the basic invariant of degree 8. Hence $\epsilon_1 = 1$, $\epsilon_2 = -1$, $\epsilon_3 = 1$, $\epsilon_4 = -1$.

We have now obtained specific formulae for the orders of the finite twisted groups. They are as follows.

THEOREM 14.3.2.

$$| \, {}^2A_l(q^2) \, | = \frac{1}{(l+1, \, q+1)} \, q^{l(l+1)/2}(q^2-1)(q^3+1)$$
$$\times (q^4-1) \ldots (q^{l+1}+(-1)^l),$$

$$| \, {}^2D_l(q^2) \, | = \frac{1}{(4, \, q^l+1)} \, q^{l(l-1)}(q^2-1)(q^4-1)$$
$$\times (q^6-1) \ldots (q^{2l-2}-1)(q^l+1)$$

$$| \, {}^2E_6(q^2) \, | = \frac{1}{(3, \, q+1)} \, q^{36}(q^2-1)(q^5+1)(q^6-1)(q^8-1)(q^9+1)(q^{12}-1),$$

$$| \, {}^3D_4(q^3) \, | = q^{12}(q^2-1)(q^6-1)(q^8+q^4+1),$$

$$| \, {}^2B_2(q^2) \, | = q^4(q^2-1)(q^4+1), \qquad\qquad q^2 = 2^{2m+1},$$

$$| \, {}^2G_2(q^2) \, | = q^6(q^2-1)(q^6+1) \qquad\qquad q^2 = 3^{2m+1},$$

$$| \, {}^2F_4(q^2) \, | = q^{24}(q^2-1)(q^6+1)(q^8-1)(q^{12}+1) \qquad q^2 = 2^{2m+1}.$$

14.4 The Simplicity of the Twisted Groups

We shall now show that the twisted groups are all simple, with a few exceptions over very small fields.

THEOREM 14.4.1. *The twisted groups are all simple, except for* ${}^2A_2(2^2)$, ${}^2B_2(2)$, ${}^2G_2(3)$, ${}^2F_4(2)$.

PROOF. (a) By 13.5.4 each twisted group G^1 contains B^1, N^1 as a (B, N)-pair. We shall therefore use the criterion for simplicity given in 11.1.1. Three out of the four conditions given in 11.1.1 are easily checked. B^1 is soluble since it is a subgroup of the soluble group $B = UH$. Let G_1^1 be a normal subgroup of G^1 contained in B^1. Then G_1^1 is contained in U^1H^1 and its conjugate V^1H^1, so G_1^1 is contained in H^1. As in the proof of 11.1.2 we see that every element of G_1^1 commutes with every element of U^1 and every element of V^1. Let $h(\chi) \in G_1^1$. Since

$$h(\chi) \, x_r(t) \, h(\chi)^{-1} = x_r(\chi(r) \, t),$$

it follows from 13.6.3 that $\chi(r)=1$ for each root $r \in \Phi$. This is because each root r occurs in some equivalence class S, and so there is an element in X_S^1 involving r. Hence $G_1^1 = 1$ and

$$\bigcap_{g \in G^1} gB^1g^{-1} = 1.$$

The Weyl group of the (B, N)-pair B^1, N^1 is W^1, and the set I of distinguished generators of W^1 is the set of elements w_0^J as J runs over the ρ-orbits of Π. We have seen in section 13.3 that these elements either form the set of fundamental reflections in some indecomposable Weyl group, or $|I| = 2$ and the two elements of I generate a dihedral group of order 16. In either case it is impossible to decompose I into non-empty complimentary subsets which commute with one another. It is therefore sufficient by 11.1.1 to prove that $G^1 = (G^1)'$. We shall show in fact that $(G^1)'$ contains X_S^1 for each equivalence class S of Φ.

(b) Suppose all the roots have the same length. Then the equivalence classes S of Φ are of type A_1, $A_1 \times A_1$, $A_1 \times A_1 \times A_1$, or A_2. Let $h(\chi)$ be an element of H^1. We use the following relations:

$$h(\chi) \, x_r(t) \, h(\chi)^{-1} = x_r(\chi(r) \, t)$$

if $S = \{r\}$ has type A_1.

$$h(\chi) \, x_r(t) \, x_{\bar{r}}(\bar{t}) \, h(\chi)^{-1} = x_r(\chi(r) \, t) \, x_{\bar{r}}(\chi(\bar{r}) \, \bar{t})$$

if $S = \{r, \bar{r}\}$ has type $A_1 \times A_1$.

$$h(\chi) \, x_r(t) \, x_{\bar{r}}(\bar{t}) \, x_{\bar{\bar{r}}}(\bar{\bar{t}}) \, h(\chi)^{-1} = x_r(\chi(r) \, t) \, x_{\bar{r}}(\chi(\bar{r}) \, \bar{t}) \, x_{\bar{\bar{r}}}(\chi(\bar{\bar{r}}) \, \bar{\bar{t}})$$

if $S = \{r, \bar{r}, \bar{\bar{r}}\}$ has type $A_1 \times A_1 \times A_1$.

$$h(\chi) \, x_r(t) \, x_{\bar{r}}(\bar{t}) \, x_{r+\bar{r}}(u) \, h(\chi)^{-1} = x_r(\chi(r) \, t) \, x_{\bar{r}}(\chi(\bar{r}) \, \bar{t}) \, x_{r+\bar{r}}(\chi(r+\bar{r}) \, u)$$

if $S = \{r, \bar{r}, r+\bar{r}\}$ has type A_2 and $u + \bar{u} = - N_{r\bar{r}}t\bar{t}$.

Now we can find, given any $r \in \Phi$ and $t \neq 0 \in K$, a K-character χ of Q such that $\chi(r) = t^2$ (see 11.1.3). In fact, since the exceptions A_1 and C_l are not among the cases being considered here, we can find a K-character χ of Q such that $\chi(r) = t$. If $r = \bar{r}$ we can find, given any $t \neq 0$ in the fixed field K_0 of the automorphism $t \to \bar{t}$, a self-conjugate K-character of Q such that $\chi(r) = t$. If $r \neq \bar{r}$ we can find for each $t \neq 0$ in K a self-conjugate K-character of Q such that $\chi(r) = t$. [†]

If $S = \{r\}$ has type A_1 and $K_0 \neq GF(2)$ we can choose a self-conjugate character χ of Q with $\chi(r) \neq 1$. Then $x_r((\chi(r) - 1) \, t) \in (G^1)'$ for all $t \in K_0$.

[†] B. Hartley has pointed out that some of the statements in this paragraph are incorrect. In order to complete the proof, Hartley's Lemma on p. 313 must be used.

Let u be any non-zero element of K_0 and define

$$t = \frac{u}{\chi(r) - 1}.$$

Then $x_r(u) \in (G^1)'$, and so $X_S^1 \subseteq (G^1)'$.

If S has type $A_1 \times A_1$ or $A_1 \times A_1 \times A_1$, we see similarly that $X_S^1 \subseteq (G^1)'$, even if $K_0 = GF(2)$.

If S has type A_2 the above relations show that

$$x_r((\chi(r) - 1)\, t)\, x_{\bar{r}}((\chi(\bar{r}) - 1)\, \bar{t})\, x_{r+\bar{r}}((\chi(r + \bar{r}) - 1)\, u + N_{r\bar{r}}(\chi(\bar{r}) - 1)\, t\bar{t})$$

lies in $(G^1)'$. If $K_0 \neq GF(2)$ we can choose a self-conjugate K-character of Q such that $\chi(r + \bar{r}) \neq 1$. For if $\chi(r) = \lambda$ then $\chi(\bar{r}) = \bar{\lambda}$ and $\chi(r + \bar{r}) = \lambda\bar{\lambda}$. Suppose $\chi(r + \bar{r}) = 1$ for all λ. Then $\lambda\bar{\lambda} = 1$ for all $\lambda \neq 0$ in K and so $\lambda^2 = 1$ for all $\lambda \neq 0$ in K_0. Thus K_0 is a finite field. Let $K_0 = GF(q)$. Then $\bar{\lambda} = \lambda^q$ and $\lambda^{q+1} = 1$ for all $\lambda \neq 0$ in K. Thus $\mu = 1$ for all $\mu \neq 0$ in K_0, so that $K_0 = GF(2)$, the excluded case.

Let t_1, u_1 be elements of K satisfying $u_1 + \bar{u}_1 = -N_{r\bar{r}}t_1\bar{t}_1$. Choose a self-conjugate character χ of Q such that $\chi(r + \bar{r}) \neq 1$ and define t, $u \in K$ by

$$t = \frac{t_1}{\chi(r) - 1}, \qquad u = \frac{u_1}{\chi(r + \bar{r}) - 1} - \frac{N_{r\bar{r}}(\chi(\bar{r}) - 1)\, t\bar{t}}{\chi(r + \bar{r}) - 1}.$$

Then $u + \bar{u} = -N_{r\bar{r}}t\bar{t}$ and $x_r(t_1)\, x_{\bar{r}}(\bar{t}_1)\, x_{r+\bar{r}}(u_1)$ is in $(G^1)'$. Thus $X_S^1 \subseteq (G^1)'$, as required.

(c) Suppose all the roots have the same length and $K_0 = GF(2)$. For each equivalence class S of Φ we define elements $x_S(t)$ or $x_S(t, u)$ of X_S^1 as in 13.6.4. We recall that it has been shown in (b) that $x_S(t) \in (G^1)'$ if S has type $A_1 \times A_1$ or $A_1 \times A_1 \times A_1$.

Now the equivalence classes in Φ are in $1-1$ correspondence with the elements of a root system with Weyl group W^1. We have shown in section 13.3 that this root system has type:

$$
\begin{array}{lll}
C_k & \text{if} & G^1 = {}^2A_{2k-1}, \\
B_k & \text{if} & G^1 = {}^2A_{2k}, \\
B_{l-1} & \text{if} & G^1 = {}^2D_l, \\
F_4 & \text{if} & G^1 = {}^2E_6, \\
G_2 & \text{if} & G^1 = {}^3D_4.
\end{array}
$$

In each case, except $G^1 = {}^2A_2$, the root system with Weyl group W^1 has roots of two different lengths, and two equivalence classes correspond to roots of the same length if and only if they have the same type. Now W^1 operates transitively on roots of a given length, and since

$$n_w X^1_S n_w^{-1} = X^1_{w(S)}, \qquad w \in W^1,$$

we see that G^1 operates transitively by conjugation on the subgroups X^1_S of a given type. It is therefore sufficient to show that one such subgroup of each type lies in $(G^1)'$.

Suppose the Dynkin diagram of W^1 has a double bond. Let S_1, S_2 be two equivalence classes corresponding to fundamental roots joined by this double bond. If G^1 is not of type $^2A_{2k}$ these classes are of type $A_1 \times A_1$ and A_1. Let $S_1 = \{r, \bar{r}\}$ and $S_2 = \{s\}$. Then the commutator relations show that

$$[x_{S_2}(u), x_{S_1}(t)] = x_{S_3}(tu)\, x_{S_4}(t\bar{t}u),$$

where $S_3 = \{r+s, \bar{r}+s\}$ and $S_4 = \{r+\bar{r}+s\}$. (We have used the fact that K has characteristic 2.) Thus $x_{S_3}(tu)\, x_{S_4}(t\bar{t}u) \in (G^1)'$ for all $t \in K$, $u \in K_0$. However, $x_{S_3}(tu) \in (G^1)'$ since S_3 has type $A_1 \times A_1$. Thus $x_{S_4}(1) \in (G^1)'$ and so $X^1_{S_4} \subseteq (G^1)'$. Hence $(G^1)'$ contains a subgroup X^1_S of type A_1 as well as one of type $A_1 \times A_1$, thus $(G^1)'$ contains X^1_S for all S.

If G^1 is of type $^2A_{2k}$, the equivalence classes S_1, S_2 are of type $A_1 \times A_1$ and A_2. Let $S_1 = \{r, \bar{r}\}$ and $S_2 = \{s, \bar{s}, s+\bar{s}\}$. These fundamental roots $r, \bar{r}, s, \bar{s}$ of $\mathfrak{L}$ may be chosen so they are related in the manner shown in the Dynkin diagram of $\mathfrak{L}$.

The commutator relations show that

$$[x_{S_2}(u, v), x_{S_1}(t)] = x_{S_3}(t\bar{v})\, x_{S_4}(tu, t\bar{t}v),$$

where $S_3 = \{r+s+\bar{s}, \bar{r}+s+\bar{s}\}$ and $S_4 = \{r+s, \bar{r}+\bar{s}, r+\bar{r}+s+\bar{s}\}$. Thus $x_{S_3}(t\bar{v})\, x_{S_4}(tu, t\bar{t}v) \in (G^1)'$. However, $x_{S_3}(t\bar{v}) \in (G^1)'$ since S_3 has type $A_1 \times A_1$, thus $x_{S_4}(tu, t\bar{t}v) \in (G^1)'$ also. Putting $t = 1$ we have $x_{S_4}(u, \bar{v}) \in (G^1)'$ and so $X^1_{S_4} \subseteq (G^1)'$. It now follows as before that $X^1_S \subseteq (G^1)'$ for all S.

If G^1 has type $^3D_4$ the equivalence classes correspond to a root system of type G_2. Let S_1, S_2 be classes corresponding to short roots of G_2 whose sum is short. S_1, S_2 both have type A_1. Let $S_1 = \{r\}$, $S_2 = \{s\}$. Then the commutator relations show that

$$[x_{S_2}(u), x_{S_1}(t)] = x_{S_3}(tu),$$

where $S_3 = \{r+s\}$. Thus $x_{S_3}(1) \in (G^1)'$ and so $X^1_{S_3} \in (G^1)'$. Hence $X^1_S \subseteq (G^1)'$ for all S as before.

(d) Now suppose that Φ contains roots of different lengths. Then the equivalence classes S of Φ are of type $A_1 \times A_1$, B_2 or G_2. Suppose that $G^1 = {}^2B_2(K)$ or ${}^2F_4(K)$. Then the equivalence classes have type $A_1 \times A_1$ or B_2. Let $h(\chi)$ be an element of H^1. Then $\chi(\bar{r}) = \chi(r)^{\lambda(\bar{r})\theta}$ by 13.7.4. Let $S = \{r, \bar{r}\}$ be an equivalence class of type $A_1 \times A_1$, where r is short and $\bar{r}$ long. Then

$$h(\chi)\, x_S(t)\, h(\chi)^{-1} = x_S(\chi(\bar{r})\, t)$$

and so

$$h(\chi)\, x_S(t)\, h(\chi)^{-1}\, x_S(t)^{-1} = x_S((\chi(\bar{r}) - 1)\, t)$$

by 13.6.4 (v). Now by 13.7.4 we can find an element $h(\chi) \in H^1$ for which $\chi(\bar{r})$ takes any prescribed non-zero value in K. Thus if $K \neq GF(2)$ we may choose χ so that $\chi(\bar{r}) \neq 1$. It follows that $X_S^1 \subseteq (G^1)'$.

Now let $S = \{a, b, a+b, 2a+b\}$ be an equivalence class of type B_2. Suppose that $\alpha(t)$, $\beta(u)$ are defined as in 13.6.4 (vi). Then

$$h(\chi)\, \beta(u)\, h(\chi)^{-1} = \beta(\chi(a+b)\, u).$$

Thus we have

$$h(\chi)\, \beta(u)\, h(\chi)^{-1}\, \beta(u)^{-1} = \beta((\chi(a+b) - 1)\, u).$$

By 13.7.4 we can choose $h(\chi) \in H^1$ so that $\chi(a+b)$ takes any non-zero value in K. Thus $\beta(u) \in (G^1)'$ for all u, provided $K \neq GF(2)$. We also have

$$h(\chi)\, \alpha(t)\, h(\chi)^{-1} = \alpha(\chi(b)\, t)$$

and

$$h(\chi)\, \alpha(t)\, h(\chi)^{-1}\, \alpha(t)^{-1} = \alpha(\chi(b)\, t)\, \alpha(-t)\, \beta(t^{\theta+1})$$
$$= \alpha((\chi(b) - 1)\, t)\, \beta((\chi(b)^\theta - 1)\, t^{\theta+1})$$

by 13.6.4 (vi). Since $\beta(u) \in (G^1)'$ for all u we have $\alpha((\chi(b) - 1)\, t) \in (G^1)'$ for all t. As before we may choose χ with $\chi(b) \neq 1$ provided $K \neq GF(2)$. Thus $\alpha(t) \in (G^1)'$ for all t. Hence

$$x_S(t, u) = \alpha(t)\, \beta(u) \in (G^1)'$$

and $X_S^1 \subseteq (G^1)'$, as required.

(e) We suppose finally that $G^1 = {}^2G_2(K)$. In the proof of 13.7.4 we showed that H^1 contains the element

$$h(\chi) = h_a((-\lambda^{-2})^\theta)\, h_b(-\lambda^{-2})$$

for all $\lambda \in K$. Since

$$\chi = \chi a, \; (-\lambda^{-2})^\theta \chi \; b, \; -\lambda^{-2},$$

we have

$$\chi(a) = (-\lambda^{-2})^{\theta \cdot A_{aa}} \cdot (-\lambda^{-2})^{A_{ba}} = -\lambda^{-4\theta+2},$$

$$\chi(b) = (-\lambda^{-2})^{\theta \cdot A_{ab}} \cdot (-\lambda^{-2})^{A_{bb}} = -\lambda^{6\theta-4}.$$

Let $\alpha(t)$, $\beta(u)$, $\gamma(v)$ be defined as in 13.6.4 (vii). Then

$$h(\chi) \, \gamma(v) \, h(\chi)^{-1} = \gamma(\chi(3a+2b) \, v)$$

and

$$h(\chi) \, \gamma(v) \, h(\chi)^{-1} \, \gamma(v)^{-1} = \gamma((\chi(3a+2b)-1) \, v)$$

by 13.6.4 (vii). Now $\chi(3a+2b) = -\lambda^{-2}$. Suppose $\chi(3a+2b) = 1$ for all λ. Then $-\lambda^{-2} = 1$ for all $\lambda \neq 0$ in K, hence $\lambda^4 = 1$ for all $\lambda \neq 0$ in K. Now K is a field of characteristic 3, so if $K \neq GF(3)$ we can choose χ so that $\chi(3a+2b) \neq 1$. It follows that $\gamma(v) \in (G^1)'$ for all $v \in K$.

Now consider the elements $\beta(u)$. We have

$$h(\chi) \, \beta(u) \, h(\chi)^{-1} = \beta(\chi(3a+b) \, u)$$

and so

$$h(\chi) \, \beta(u) \, h(\chi)^{-1} \, \beta(u)^{-1} = \beta((\chi(3a+b)-1) \, u)$$

by 13.6.4 (vii). Now $\chi(3a+b) = \lambda^{-6\theta+2}$. Suppose $\chi(3a+b) = 1$ for all λ. Then $\lambda^{-6\theta+2} = 1$ and so $\lambda^{-2+2\theta} = 1$ since $6\theta^2 = 2$. It follows that $\lambda^{-6+6\theta} = 1$ and so $\lambda^{-4} = 1$. Thus $\lambda^4 = 1$ for all $\lambda \neq 0$ in K. If $K \neq GF(3)$ we can therefore choose χ so that $\chi(3a+b) \neq 1$. Then $\beta(u) \in (G^1)'$ for all $u \in K$.

Now consider the elements $\alpha(t)$. We have

$$h(\chi) \, \alpha(t) \, h(\chi)^{-1} = \alpha(\chi(b) \, t)$$

and so, by 13.6.4 (vii), we have

$$h(\chi) \, \alpha(t) \, h(\chi)^{-1} \, \alpha(t)^{-1} = \alpha((\chi(b)-1) \, t) \, \beta(u) \, \gamma(v)$$

for certain elements $u, v \in K$. However, $\beta(u)$ and $\gamma(v)$ belong to $(G^1)'$ and therefore $\alpha((\chi(b)-1) \, t)$ is in $(G^1)'$ also. Now $\chi(b) = -\lambda^{6\theta-4}$. Suppose $\chi(b) = 1$ for all λ. Then $-\lambda^{6\theta-4} = 1$ and so $-\lambda^{2-4\theta} = 1$ since $6\theta^2 = 2$. It follows that $\lambda^{12\theta-8} = 1$ and $-\lambda^{6-12\theta} = 1$, whence $-\lambda^{-2} = 1$. Therefore $\lambda^4 = 1$ for all $\lambda \neq 0$ in K. If $K \neq GF(3)$ we can choose χ so that $\chi(b) \neq 1$. Thus $\alpha(t) \in (G^1)'$ for all $t \in K$.

We have now shown that

$$x_S(t, u, v) = \alpha(t)\,\beta(u)\,\gamma(v) \in (G^1)'$$

for all t, u, $v \in K$, hence $X_S^1 \subseteq (G^1)'$ as required.

(f) We have now shown that $X_S^1 \subseteq (G^1)'$ for all equivalence classes S of Φ, except when $G^1 = {}^2A_2(2^2)$, ${}^2B_2(2)$, ${}^2G_2(3)$, ${}^2F_4(2)$. Since the subgroups X_S^1 generate G^1 we have $G^1 = (G^1)'$ except in these cases. Thus G^1 is simple by 11.1.1.   ∎

Note. The four twisted groups which have not been proved to be simple are all in fact not simple. ${}^2A_2(2^2)$ is a soluble group of order 72, ${}^2B_2(2)$ is a soluble group of order 20, ${}^2G_2(3)$ is a group of order 1512 which has as its commutator subgroup the simple group $A_1(8)$ of order 504, and ${}^2F_4(2)$ has as its commutator subgroup a subgroup of index 2 which is simple of order $2^{11}.3^3.5^2.13$. $({}^2F_4(2))'$ can be proved to be simple by an argument along the lines of the preceding discussion (cf. Tits [12]). This group is not isomorphic to any of the other simple groups of Lie type which we have discussed.

14.5 Identification with some Classical Groups

We show now that the twisted groups ${}^2A_l(K)$ and ${}^2D_l(K)$ are isomorphic to certain classical groups.

THEOREM 14.5.1. ${}^2A_l$ *is isomorphic to the unitary group* $PSU_{l+1}(K, f)$ *leaving invariant the Hermitian form*

$$f = \epsilon(\bar{x}_0 x_l - \bar{x}_1 x_{l-1} + \bar{x}_2 x_{l-2} - \dots)$$

with matrix

$$A = \epsilon \begin{pmatrix} & & & & & 1 \\ & & & & -1 & \\ & & & 1 & & \\ & & -1 & & & \\ & \cdot & & & & \\ \cdot & & & & & \end{pmatrix}.$$

Here ϵ *is defined to be* 1 *if* l *is even and an element of* K *satisfying* $\epsilon + \bar{\epsilon} = 0$ *if* l *is odd.*

PROOF.   By 3.5.2 there is an automorphism of the Lie algebra $\mathfrak{L}$ such that

$$e_r \to e_{\bar{r}}, \qquad h_r \to h_{\bar{r}}$$

for $r \in \Pi$ or $-r \in \Pi$. If we transfer to the field K we obtain a corresponding automorphism of the Lie algebra $\mathfrak{L}_K$. If we then combine this automorphism with the field automorphism $t \to \bar{t}$ of K we obtain a 'semi-automorphism' ψ of $\mathfrak{L}_K$ which satisfies the conditions

$$\psi(h_r) = h_{\bar{r}}, \qquad \psi(e_r) = e_{\bar{r}}, \qquad r \in \pm \Pi,$$

$$\psi(\lambda x + \mu y) = \bar{\lambda}\psi(x) + \bar{\mu}\psi(y), \qquad \lambda, \mu \in K; \ x, y \in \mathfrak{L}_K.$$

We now consider the map

$$\theta \to \psi\theta\psi^{-1},$$

where θ is an element of the Chevalley group $G = \mathfrak{L}(K)$. The transformation $\psi\theta\psi^{-1}$ is easily seen to be an automorphism of $\mathfrak{L}_K$ and we have

$$\psi x_r(t) \, \psi^{-1} = \psi \exp (\text{ad } te_r) \, \psi^{-1}$$

$$= \exp \text{ad } (\psi . te_r) = \exp (\bar{t}e_{\bar{r}}) = x_{\bar{r}}(\bar{t})$$

for $r \in \pm \Pi$. Thus $\psi x_r(t) \, \psi^{-1} \in G$, and since the elements $x_r(t)$ for $r \in \pm \Pi$, $t \in K$, generate G it follows that the map $\theta \to \psi\theta\psi^{-1}$ transforms G into itself. It is in fact an automorphism of G, for it is invertible and satisfies

$$\psi\theta_1\theta_2\psi^{-1} = \psi\theta_1\psi^{-1} . \psi\theta_2\psi^{-1}.$$

However, this map coincides with the automorphism σ of G on the generators $x_r(t)$ for $r \in \pm \Pi$, so is equal to σ. Thus the elements of G invariant under σ are those which commute with the semi-automorphism ψ.

In the present case $\mathfrak{L}_K$ may be taken as the Lie algebra of $(l+1) \times (l+1)$ matrices of trace 0. The fundamental root vectors may be taken as $e_{i, \, i+1}$ for $i = 0, 1, \ldots, l-1$ and the fundamental co-roots are then

$$e_{ii} - e_{i+1, \, i+1}, \qquad i = 0, 1, \ldots, l-1.$$

The automorphism of $\mathfrak{L}_K$ given by the symmetry of the Dynkin diagram acts as follows:

$$e_{i, \, i+1} \to e_{l-i-1, \, l-i},$$

$$e_{ii} - e_{i+1, \, i+1} \to e_{l-i-1, \, l-i-1} - e_{l-i, \, l-i}.$$

K

Now it is easily verified that the maps $M \to -M'$ (where M' is the transpose of M) and $M \to A^{-1}MA$ are both automorphisms of $\mathfrak{L}_K$. Combining them we obtain an automorphism $M \to -A^{-1}M'A$. Under this automorphism we have

$$e_{ij} \to (-1)^{i+j+1} e_{l-j,\, l-i}.$$

Thus this automorphism behaves in the same way as the automorphism induced by the symmetry of the Dynkin diagram when operating on $e_{i,\, i+1}$ and $e_{ii} - e_{i+1,\, i+1}$. Thus these two automorphisms coincide. The semi-automorphism ψ defined above is therefore given by

$$M \overset{\psi}{\to} -A^{-1}\bar{M}'A.$$

Now we have seen that G consists of the automorphisms of $\mathfrak{L}_K$ given by

$$M \to TMT^{-1}, \qquad T \in SL_{l+1}(K).$$

We consider which of these automorphisms commute with ψ. In order for this to be so, T must satisfy the condition

$$-A^{-1}(\bar{T}^{-1})'\bar{M}'\bar{T}'A = -TA^{-1}\bar{M}'AT^{-1},$$

which implies

$$\bar{M}'\bar{T}'ATA^{-1} = \bar{T}'ATA^{-1}\bar{M}'.$$

As this holds for all $M \in \mathfrak{L}_K$ we must have

$$\bar{T}'ATA^{-1} = \lambda I$$

for some $\lambda \in K$. Thus $\bar{T}'AT = \lambda A$.

Suppose T is an upper unitriangular matrix. Then by comparing the $(0, l)$-coefficients on each side we have $\lambda = 1$. Similarly, if T is lower unitriangular a comparison of the $(l, 0)$-coefficients shows that $\lambda = 1$. Thus the matrices T giving rise to the elements of U^1 and V^1 are precisely the upper and lower unitriangular matrices of the group $SU_{l+1}(K, f)$. However, $SU_{l+1}(K, f)$ is generated by its upper and lower unitriangular matrices. Thus G^1, the group generated by U^1 and V^1, consists of all transformations $M \to TMT^{-1}$, where $T \in SU_{l+1}(K, f)$. Therefore G^1 is isomorphic to $PSU_{l+1}(K, f)$. ∎

We note that the index $\nu(f)$ of the Hermitian form f is $(l+1)/2$ if l is odd and $l/2$ if l is even. Thus $\nu(f)$ is as large as it can be in the light of Witt's theorem.

THEOREM 14.5.2. $^2D_l(K)$ *is isomorphic to the orthogonal group* $P\Omega_{2l}(K_0, f)$, *where* K_0 *is the fixed field of* K *under the field automorphism used to define* $^2D_l(K)$ *and* f *is the quadratic form*

$$x_1x_{-1} + x_2x_{-2} + \ldots + x_{l-1}x_{-(l-1)} + (x_l - \alpha x_{-l})(x_l - \bar\alpha x_{-l}),$$

where α *is a generator of* K *over* K_0.

(Note that the quadratic form f is defined over K_0. It has index $l-1$ regarded as a form over K_0 and l regarded as a form over K.)

PROOF. It was shown in 11.3.2 that the Chevalley group $G = D_l(K)$ is isomorphic to the orthogonal group $P\Omega_{2l}(K, f_D)$, where f_D is the quadratic form

$$y_1y_{-1} + y_2y_{-2} + \ldots + y_ly_{-l}.$$

Let

$$A = \begin{pmatrix} 0 & I_l \\ I_l & 0 \end{pmatrix}$$

be the matrix of f_D. The group $\Omega_{2l}(K, f_D)$ is the commutator subgroup of $O_{2l}(K, f_D)$. It is generated by matrices exp (te_r), where $r \in \Pi$ or $-\Pi$ and $t \in K$, and all its matrices T satisfy $T'AT = A$. If we use the matrix representation given in 11.2.3, the matrices exp (te_r) for $r \in \Pi$ are

$$I + t(e_{12} - e_{-2, -1}),\ I + t(e_{23} - e_{-3, -2}),\ \ldots,\ I + t(e_{l-1, l} - e_{-l, -(l-1)}),$$

$$I + t(e_{l-1, -l} - e_{l, -(l-1)}).$$

These matrices correspond to the nodes 1, 2, ..., l respectively in the Dynkin diagram

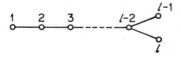

Now consider the map

$$\exp(te_r) \to \exp(te_{\bar r}) \qquad r \in \pm\Pi.$$

The operation of transformation by the matrix B below induces this map.

$$
B = \begin{pmatrix}
1 & & & & & & & & \\
& \cdot & & & & & & & \\
& & \cdot & & & & & & \\
& & & \cdot & & & & & \\
& & & & 1 & & & & \\
& & & & & 0 & & & 1 \\
& & & & & & 1 & & \\
& & & & & & & \cdot & \\
& & & & & & & & \cdot 1 \\
& & & & & 1 & & & 0
\end{pmatrix}
\begin{matrix}
1 \\ 2 \\ \cdot \\ \cdot \\ l \\ -1 \\ \cdot \\ \cdot \\ -l
\end{matrix}
$$

Transformation by B leaves the first $l-2$ matrices exp (te_r) invariant and interchanges the last two. It is convenient to write the rows and columns in the order $1, 2, \ldots, l-1, -1, -2, \ldots, -(l-1), l, -l$. Then

$$
B = \begin{pmatrix} I_{2l-2} & 0 \\ 0 & B_0 \end{pmatrix},
$$

where

$$
B_0 = \begin{pmatrix} 0 & 1 \\ 1 & 0 \end{pmatrix}.
$$

Since the map exp $(te_r) \to$ exp $(\bar{t}e_{\bar{r}})$ is induced by the transformation

$$
T \to B^{-1}\bar{T}B,
$$

we shall need to consider the matrices fixed under this transformation. These are the matrices $T \in \Omega_{2l}(K, f_D)$ which satisfy $BT = \bar{T}B$.

We now change the basis of the underlying $2l$-dimensional space over K so that the point which originally had coordinates

$$
y_1, \ldots, y_l, y_{-1}, \ldots, y_{-l}
$$

now has coordinates $x_1, \ldots, x_l, x_{-1}, \ldots, x_{-l}$, where

$$
y_i = x_i, \qquad i = 1, \ldots, l-1, -1, \ldots, -(l-1),
$$

$$
y_l = x_l - \alpha x_{-l},
$$

$$
y_{-l} = x_l - \bar{\alpha} x_{-l}
$$

(α is a generator of K over K_0). Then the matrix of the form f_D with respect to $x_1, \ldots, x_l, x_{-1}, \ldots, x_{-l}$ is $S'AS$, where

$$S = \begin{pmatrix} I_{2l-2} & 0 \\ 0 & S_0 \end{pmatrix},$$

$$A = \begin{pmatrix} 0 & I_{l-1} & & \\ & & & 0 \\ I_{l-1} & 0 & & \\ \hline & & 0 & 1 \\ & 0 & 1 & 0 \end{pmatrix},$$

$$S_0 = \begin{pmatrix} 1 & -\alpha \\ 1 & -\bar{\alpha} \end{pmatrix}.$$

We note that a matrix M satisfies

$$M'(S'AS)\,M = S'AS$$

if and only if $T = SMS^{-1}$ satisfies

$$T'AT = A.$$

We consider the conjugate subgroup $S^{-1}\Omega_{2l}(K, f_D)S$ and investigate which matrices M in this subgroup correspond to matrices $T \in \Omega_{2l}(K, f_D)$ such that $BT = \bar{T}B$. Let

$$M = \begin{pmatrix} M_{11} & M_{12} \\ M_{21} & M_{22} \end{pmatrix} \begin{matrix} 2l-2 \\ 2 \end{matrix}$$

$$\phantom{M = \begin{pmatrix} M_{11} \end{pmatrix}} 2l-2 \quad\; 2$$

Then, using the fact that $B_0 S_0 = \bar{S}_0$, we see by matrix multiplication that $BT = \bar{T}B$ if and only if

$$M_{11} = \bar{M}_{11}, \qquad M_{12} = \bar{M}_{12}, \qquad M_{21} = \bar{M}_{21}, \qquad M_{22} = \bar{M}_{22}.$$

Thus the subgroup of $\Omega_{2l}(K, f_D)$ of elements T invariant under σ is $S\Omega_{2l}(K_0, f)\,S^{-1}$. Since G is isomorphic to $P\Omega_{2l}(K, f_D)$ it follows that G^1 is isomorphic to $P\Omega_{2l}(K_0, f)$.

CHAPTER 15

Associated Geometrical Structures

Certain geometrical structures, called buildings, on which groups with a (B, N)-pair operate as groups of automorphisms, have recently been introduced by J. Tits. We shall describe these structures here and demonstrate the connection with groups with a (B, N)-pair. Closely connected to these geometries are somewhat simpler structures on which Weyl groups (or more generally Coxeter groups) operate as groups of automorphisms. These structures are the Coxeter complexes, which we have described in earlier chapters. However, in order to describe the buildings we first need an axiomatic system which picks out the essential features of the Coxeter complex. A geometry satisfying these axioms will be called an *abstract Coxeter complex*. We shall introduce these ideas by means of a series of definitions of increasing complexity.

15.1 Chamber Complexes

We consider a set Θ endowed with a relation $\subseteq$ of partial ordering. Θ is called a *simplex* if Θ is isomorphic to the set of all subsets of some set, partially ordered by inclusion. Θ is called a *complex* if:

(a) For each $A \in \Theta$ the set of elements $B \in \Theta$ such that $B \subseteq A$ forms a simplex.

(b) Each pair of elements $A, B \in \Theta$ have a greatest lower bound $A \cap B$.

It follows from (a), (b) that a complex Θ contains a unique minimal element, which will be called 0.

We define the rank of each element of a complex Θ. Rank A is the number of elements B such that B is minimal with respect to the properties $B \subseteq A$, $B \neq 0$. Thus the set of elements B with $B \subseteq A$ is isomorphic to the set of subsets of a set with cardinality rank A.

A subset Θ' of Θ with the induced partial ordering is called a *subcomplex* if, for all $A \in \Theta'$, $B \in \Theta$ with $B \subseteq A$, we have $B \in \Theta'$. There exist subsets of a complex Θ which are complexes, although not subcomplexes of Θ. For example, let $A \in \Theta$ and define St A (the star of A) by:

$$\text{St } A = \{B \in \Theta; \, B \supseteq A\}.$$

274

Then St A is a complex contained in Θ but not a subcomplex of Θ (unless $A = 0$).

For any two elements A, $B \in \Theta$ with $A \subseteq B$ we define the codimension of A in B by

$$\operatorname{codim}_B A = \operatorname{rank}_{\operatorname{St} A} B$$

In particular $\operatorname{codim}_B A = 1$ if and only if $A \neq B$ and there is no element X with $A \subset X \subset B$. Also $\operatorname{codim}_B A = 0$ if and only if $A = B$.

A complex Θ is called a *chamber complex* if:

(a) Every element of Θ is contained in a maximal element.

(b) Given any two maximal elements C, C' of Θ there exists a finite sequence

$$C = C_0, C_1, C_2, \ldots, C_m = C'$$

of elements of Θ such that

$$\operatorname{codim}_{C_{i-1}} (C_{i-1} \cap C_i) = \operatorname{codim}_{C_i} (C_{i-1} \cap C_i) \leqslant 1$$

for $i = 1, \ldots, m$.

The maximal elements of a chamber complex will be called chambers. The above condition on the codimension means that C_{i-1} is either equal to C_i or that $C_{i-1} \neq C_i$ and the intersection has codimension 1 in each.

LEMMA 15.1.1. *An element of a chamber complex has the same codimension in all the chambers containing it.*

PROOF. Let $A \in \Theta$ be contained in two chambers C, C'. Then there exists a sequence

$$C = C_0, C_1, \ldots, C_m = C'$$

as above. Let $B = C_0 \cap C_1 \cap \ldots \cap C_m \cap A$. Then

$$\operatorname{codim}_{C_{i-1}} B = \operatorname{codim}_{C_i} B \quad \text{for} \quad i = 1, \ldots, m$$

and so

$$\operatorname{codim}_C B = \operatorname{codim}_{C'} B.$$

Now it is readily verified that $\operatorname{codim}_C (C_0 \cap C_1 \cap \ldots \cap C_m)$ is finite and it follows that $\operatorname{codim}_A B$ is finite also. Hence

$$\operatorname{codim}_C A = \operatorname{codim}_C B - \operatorname{codim}_A B$$

$$= \operatorname{codim}_{C'} B - \operatorname{codim}_A B = \operatorname{codim}_{C'} A. \qquad \blacksquare$$

It follows from 15.1.1 that the terms of a sequence $C_0, C_1, \ldots, C_m$ of the type described above are all chambers. A sequence of chambers of this type will be called a *gallery*. Two chambers C, C' are said to be *adjacent* if codim $(C \cap C') = 1$. Thus a gallery is a sequence of chambers in which every pair of consecutive chambers are either identical or adjacent.

We now consider maps from one chamber complex to another. Let Θ, Θ' be chamber complexes. A map $\alpha : \Theta \to \Theta'$ is called a *morphism* of chamber complexes if:

(a) $\alpha(C)$ is a chamber of Θ' for each chamber $C \in \Theta$.

(b) For each chamber $C \in \Theta$, α induces an isomorphism between the simplexes determined by C, $\alpha(C)$.

It is clear that a morphism of chamber complexes preserves the partial ordering ($B \subseteq A$ implies $\alpha(B) \subseteq \alpha(A)$) and leaves invariant the rank of each element. A morphism of Θ into itself is called an endomorphism, and an endomorphism which is invertible is called an automorphism.

A chamber complex is said to be *thin* if every element of codimension 1 is contained in exactly two chambers and *thick* if every element of co-dimension 1 is contained in at least three chambers.

LEMMA 15.1.2. *Let Θ, Θ' be two chamber complexes in which each element of codimension 1 is contained in at most two chambers, and let α, β be two morphisms of Θ into Θ' injective on the set of chambers. Suppose there exists a chamber $C \in \Theta$ such that $\alpha(A) = \beta(A)$ for all $A \subseteq C$. Then $\alpha = \beta$.*

PROOF. The elements A satisfying $A \subseteq C$ will be called the *faces* of C. Suppose there is a chamber $C' \in \Theta$ such that α, β do not coincide on the faces of C'. Let Γ be a gallery of minimal length joining C to C'. Then

$$\Gamma = \{C = C_0, C_1, \ldots, C_m = C'\}.$$

Now there exists an integer i such that α, β coincide on the faces of C_{i-1} but not on all faces of C_i. We have

$$\beta(C_i) \supset \beta(C_{i-1} \cap C_i) = \alpha(C_{i-1} \cap C_i).$$

But $\alpha(C_{i-1} \cap C_i)$ is contained in only two chambers, which are $\alpha(C_{i-1})$ and $\alpha(C_i)$ since α is injective on chambers. Thus $\beta(C_i) = \alpha(C_{i-1})$ or $\beta(C_i) = \alpha(C_i)$.

Suppose $\beta(C_i) = \alpha(C_{i-1})$. Then $\beta(C_i) = \beta(C_{i-1})$, contradicting the fact that β is injective on chambers. (C_{i-1}, C_i are distinct since Γ is a gallery of minimal length joining C to C'.) Thus $\beta(C_i) = \alpha(C_i)$. Hence

α, β coincide on C_i and on all faces of the element $C_{i-1} \cap C_i$ of codimension 1 in C_i. Since the faces of C_i form a simplex and α, β are isomorphisms between the simplexes determined by C_i and $\alpha(C_i)$ it follows that α, β coincide on all faces of C_i, and we have a contradiction. ∎

15.2 Foldings

Let Θ be a thin chamber complex. A folding of Θ is an endomorphism α of Θ satisfying:

(a) $\alpha^2 = \alpha$ (α is idempotent).

(b) Each chamber in $\alpha(\Theta)$ is the image under α of exactly two chambers of Θ.

The idea of a folding is the key to the definition of an abstract Coxeter complex. We shall now elucidate some properties of foldings.

LEMMA 15.2.1. *Let α be a folding of Θ. Then there exist adjacent chambers C, C' of Θ such that $C \in \alpha(\Theta)$, $C' \notin \alpha(\Theta)$. If C, C' are any two such chambers we have $\alpha(C') = C$.*

PROOF. Since α is idempotent there exist chambers C_1, C_2 with $C_1 \in \alpha(\Theta)$, $C_2 \notin \alpha(\Theta)$. Since C_1, C_2 can be joined by a gallery there exists a pair of adjacent chambers C, C' in such a gallery with $C \in \alpha(\Theta)$, $C' \notin \alpha(\Theta)$.

Now all the faces of C are in $\alpha(\Theta)$ and so α acts as the identity on them. In particular $\alpha(C \cap C') = C \cap C'$. Thus $\alpha(C') \supset C \cap C'$. Since Θ is thin this implies that $\alpha(C') = C$ or $\alpha(C') = C'$. But $C' \notin \alpha(\Theta)$, thus $\alpha(C') = C$. ∎

A set of chambers in a chamber complex is called *convex* if every gallery of minimal length joining two chambers in the set has all its terms in the set.

LEMMA 15.2.2. *Let α be a folding of Θ. Then the chambers in $\alpha(\Theta)$ form a convex set in Θ.*

PROOF. Let C, C' be two chambers in $\alpha(\Theta)$ and

$$\Gamma = \{C = C_0, C_1, \ldots, C_m = C'\}$$

be a gallery of minimal length joining C, C'. Suppose Γ contains some term $C_i \notin \alpha(\Theta)$. Then we can find two consecutive terms C_{j-1}, C_j in Γ such that $C_{j-1} \in \alpha(\Theta)$, $C_j \notin \alpha(\Theta)$. Now $\alpha(\Gamma) = \{\alpha(C_0), \ldots, \alpha(C_m)\}$ is also a gallery joining C to C'. By 15.2.1 we have $\alpha(C_j) = C_{j-1}$ and so

$$\alpha(C_{j-1}) = \alpha(C_j).$$

Thus two consecutive terms of $\alpha(\Gamma)$ are the same, and by omitting one we could obtain a gallery joining C to C' of shorter length than Γ, a contradiction. ∎

LEMMA 15.2.3. *Let α be a folding of Θ. Then the chambers not in $\alpha(\Theta)$ also form a convex set in Θ.*

PROOF. To establish this result we introduce a map $\bar{\alpha}$ on the set of chambers of Θ. If C is a chamber not in $\alpha(\Theta)$ we define $\bar{\alpha}(C) = C$; while if $C \in \alpha(\Theta)$, C is the image under α of exactly two chambers. One of these is C and the other is defined as $\bar{\alpha}(C)$.

Let C, C' be two chambers not in $\alpha(\Theta)$ and

$$\Gamma = \{C = C_0, C_1, \ldots, C_m = C'\}$$

be a gallery of minimal length joining C, C'. We show that $\bar{\alpha}(\Gamma)$ is also a gallery. To do this we must prove that the images under $\bar{\alpha}$ of two adjacent chambers D, D' are either adjacent or identical. This is clear if neither D, D' is in $\alpha(\Theta)$. So suppose $D \in \alpha(\Theta)$. Then $\alpha(\bar{\alpha}(D)) = D$. Let A be the face of $\bar{\alpha}(D)$ such that $\alpha(A) = D \cap D'$. Let D'' be the chamber containing A other than $\bar{\alpha}(D)$.

If $D'' \in \alpha(\Theta)$, we have $A \in \alpha(\Theta)$ and so $\alpha(A) = A$. Then $\bar{\alpha}(D) = D'$ since Θ is thin and $\bar{\alpha}(D) \notin \alpha(\Theta)$. Thus $\bar{\alpha}(D') = D'$ and $\bar{\alpha}(D) = \bar{\alpha}(D')$.

If $D'' \notin \alpha(\Theta)$, we have $\alpha(D'') \neq D$ (since $D'' \neq \bar{\alpha}(D)$). Thus $\alpha(D'') = D'$ and $D'' = \bar{\alpha}(D')$. Hence $\bar{\alpha}(D)$, $\bar{\alpha}(D')$ are adjacent.

Thus $\bar{\alpha}(D)$, $\bar{\alpha}(D')$ are either adjacent or identical and so

$$\bar{\alpha}(\Gamma) = \{\bar{\alpha}(C_0), \bar{\alpha}(C_1), \ldots, \bar{\alpha}(C_m)\}$$

is also a gallery joining C, C'. Suppose some term of Γ is in $\alpha(\Theta)$. Then we can find two consecutive terms C_{i-1}, C_i of Γ with $C_{i-1} \notin \alpha(\Theta)$, $C_i \in \alpha(\Theta)$.

Then $\alpha(C_{i-1}) = C_i$ by 15.2.1 and so $\bar{\alpha}(C_i) = C_{i-1}$. Thus $\bar{\alpha}(C_i) = \bar{\alpha}(C_{i-1})$ and the gallery $\bar{\alpha}(\Gamma)$ has two consecutive terms which are equal. By omitting one of these we would obtain a gallery joining C, C' which is shorter than Γ, a contradiction. ∎

Let C, C' be chambers in a chamber complex Θ. The distance between C, C' is defined as the shortest length of a gallery joining C, C'. Thus if $\Gamma = \{C = C_0, C_1, \ldots, C_m = C'\}$ is a gallery for which m is minimal we define

$$\text{dist } CC' = m.$$

In particular, dist $CC' = 0$ if and only if $C = C'$ and dist $CC' = 1$ if and only if C, C' are adjacent.

LEMMA 15.2.4. *Let α be a folding of the thin chamber complex Θ and C, C' be adjacent chambers of Θ such that $C \in \alpha(\Theta)$, $C' \notin \alpha(\Theta)$. Then for any chamber D of Θ we have:*

$$\text{dist } C'D = \text{dist } CD + 1 \text{ if } D \in \alpha(\Theta),$$
$$\text{dist } C'D = \text{dist } CD - 1 \text{ if } D \notin \alpha(\Theta).$$

PROOF. Since C, C' are adjacent it is clear that

$$\text{dist } CD - 1 \leqslant \text{dist } C'D \leqslant \text{dist } CD + 1.$$

Suppose $D \in \alpha(\Theta)$ and Γ is a gallery of minimal length joining D to C'. Then Γ contains two consecutive terms, one in $\alpha(\Theta)$ and the other not. By 15.2.1 $\alpha(\Gamma)$ is a gallery joining D to C which contains two consecutive terms which are equal. Hence dist $CD \leqslant$ dist $C'D - 1$, and we must have equality.

If $D \notin \alpha(\Theta)$ we take a gallery Γ of minimal length joining D to C. By the proof of 15.2.3 $\bar{\alpha}(\Gamma)$ is a gallery joining D to C' which contains two consecutive terms which are equal. Hence dist $C'D \leqslant$ dist $CD - 1$, and we again have equality. ∎

PROPOSITION 15.2.5. *Let α be a folding of Θ and C, C' be adjacent chambers of Θ such that $C \in \alpha(\Theta)$, $C' \notin \alpha(\Theta)$. Then α is the only folding such that $\alpha(C') = C$.*

PROOF. We showed in 15.2.1 that $\alpha(C') = C$. Let β be any folding of Θ such that $\beta(C') = C$. Then 15.2.4 implies that, for any chamber

$D \in \Theta$, D belongs to $\alpha(\Theta)$ if and only if D belongs to $\beta(\Theta)$. Since $\alpha(\Theta)$, $\beta(\Theta)$ are chamber complexes we have $\alpha(\Theta) = \beta(\Theta)$. α and β, being idempotent, both operate as the identity on $\alpha(\Theta)$.

Let $\bar{\alpha}(\Theta)$ be the subcomplex of Θ consisting of all faces of all chambers of the form $\bar{\alpha}(D)$ for $D \in \Theta$. The chambers of form $\bar{\alpha}(D)$ for $D \in \Theta$ are those which are not in $\alpha(\Theta)$, and so form a convex set by 15.2.3. Hence $\bar{\alpha}(\Theta)$ is a chamber complex. Similarly $\bar{\beta}(\Theta)$ is a chamber complex. By 15.2.4 a chamber belongs to $\bar{\alpha}(\Theta)$ if and only if it belongs to $\bar{\beta}(\Theta)$, hence $\bar{\alpha}(\Theta) = \bar{\beta}(\Theta)$.

Now consider the two morphisms:

$$\bar{\alpha}(\Theta) \xrightarrow{\alpha} \Theta, \qquad \bar{\alpha}(\Theta) \xrightarrow{\beta} \Theta.$$

Since α, β are foldings, they are injective on chambers when restricted to $\bar{\alpha}(\Theta)$. Moreover, α, β coincide on the chamber $C' \in \bar{\alpha}(\Theta)$ and on all the faces of the element $C \cap C'$. Thus α, β coincide on all faces of C'. By 15.1.2 it follows that α, β coincide on $\bar{\alpha}(\Theta)$. But $\alpha(\Theta) \cup \bar{\alpha}(\Theta) = \Theta$, thus α, β coincide on Θ. ∎

PROPOSITION 15.2.6. *Let α be a folding of Θ and C, C' be adjacent chambers of Θ such that $\alpha(C') = C$. Suppose there is a folding β such that $\beta(C) = C'$. Then β has the same property for any other pair of adjacent chambers of this type, viz., if D, D' are adjacent and $\alpha(D') = D$, then $\beta(D) = D'$.*

PROOF. By 15.2.4 a chamber is in $\alpha(\Theta)$ if and only if it is not in $\beta(\Theta)$. Thus $D' \in \beta(\Theta)$, $D \notin \beta(\Theta)$, and so $\beta(D) = D'$ by 15.2.1. ∎

The folding β of 15.2.6 which, when it exists, is uniquely determined by α, is called the *opposite folding* of α.

PROPOSITION 15.2.7. *Let α, β be opposite foldings of Θ. Then there exists an automorphism ρ of Θ which coincides with β on $\alpha(\Theta)$ and with α on $\beta(\Theta)$. Also, ρ^2 is the identity.*

PROOF. α, β both operate as the identity on $\alpha(\Theta) \cap \beta(\Theta)$, so ρ is well-defined. It is clearly an endomorphism of Θ. Now ρ is injective on chambers. For if $\rho(C_1) = \rho(C_2) \in \alpha(\Theta)$ we have C_1, $C_2 \in \beta(\Theta)$ and

$$\alpha(C_1) = \alpha(C_2),$$

whence $C_1 = C_2$. The same applies if $\rho(C_1) = \rho(C_2) \in \beta(\Theta)$. It follows that

ρ^2 is injective on chambers also. Let C, C' be a pair of adjacent chambers such that $\alpha(C')=C$ and $\beta(C)=C'$. Then ρ^2 fixes C and also fixes each face of $C \cap C'$, hence ρ^2 fixes each face of C. By 15.1.2 ρ^2 is the identity. In particular ρ is invertible, so is an automorphism. ∎

15.3 Abstract Coxeter Complexes

An abstract Coxeter complex is a thin chamber complex Σ such that, given any pair C, C' of adjacent chambers, there exists a folding α of Σ with $\alpha(C')=C$.

LEMMA 15.3.1. *Let Σ be an abstract Coxeter complex and A be an element of Σ. Then* $\mathrm{St}_\Sigma A$ *is also an abstract Coxeter complex.*

PROOF. $\mathrm{St}_\Sigma A$ is certainly a complex. To show it is a chamber complex, let C, C' be chambers containing A. There exists a gallery

$$\Gamma = \{C = C_0,\, C_1,\, \ldots,\, C_m = C'\}$$

in Σ joining C, C', where $m = \mathrm{dist}\ CC'$. Let α be the folding of Σ such that $\alpha(C) = C_1$. Then

$$\mathrm{dist}\ C_1 C' = \mathrm{dist}\ CC' - 1,$$

hence $C' \in \alpha(\Sigma)$ by 15.2.4. Thus all faces of $C \cap C'$ are in $\alpha(\Sigma)$. Consider the minimal non-zero faces of C. These are all faces of $C \cap C_1$ except one, which we call V. Now $V \subseteq C$ and so $\alpha(V) \subseteq C_1$, hence $\alpha(V) \neq V$. Thus $V \notin \alpha(\Sigma)$, and so V cannot be a face of $C \cap C'$. It follows that $C \cap C' \subseteq C \cap C_1$. Hence A is contained in C_1. Using induction we see that A is contained in each C_i. Thus St A is a chamber complex. $\mathrm{St}_\Sigma A$ is certainly thin, so it remains only to check the existence of all possible foldings.

Let C, C' be adjacent chambers both containing A. There is a folding α of Σ with $\alpha(C') = C$. α fixes A, so induces an idempotent endomorphism of $\mathrm{St}_\Sigma A$. The same applies to the opposite folding β of α. Let D be any chamber in $\alpha(\Sigma)$ containing A. Then $\beta(D)$ also contains A, and

$$\alpha\beta(D) = \rho^2(D) = D$$

by 15.2.7. Thus the two chambers D and $\beta(D)$ which α transforms into D both contain A. Hence α induces a folding of $\mathrm{St}_\Sigma A$. ∎

We shall now prove some general properties of abstract Coxeter complexes, showing first that such a complex can be 'folded down' into a single chamber.

PROPOSITION 15.3.2. *Let* Σ *be an abstract Coxeter complex,* C *a chamber in* Σ *and* $S(C)$ *the simplex of all faces of* C. *Then there exists a unique idempotent morphism of* Σ *onto* $S(C)$.

Note. An idempotent morphism will be called a retraction.

PROOF. Let $\Gamma = \{C = C_0, C_1, \ldots, C_m\}$ be any gallery beginning with C. We show there is an endomorphism γ of Σ leaving invariant all faces of C such that $\gamma(\Gamma) = \{C, C, \ldots, C\}$. We use induction on the length m of Γ, the result being clear if $m = 0$. If $C_1 = C$ the result is clear by induction, so we assume $C_1 \neq C$. There is a folding α of Σ with $\alpha(C_1) = C$. Thus $\alpha(\Gamma) = \{C, C, \ldots, \alpha(C_m)\}$. By induction there exists an endomorphism δ leaving invariant all faces of C such that $\delta\alpha(\Gamma) = \{C, C, \ldots, C\}$. Then $\gamma = \delta\alpha$ has the required properties.

It follows that, given any finite set C_i of chambers of Σ, there exists an endomorphism γ of Σ leaving invariant all faces of C such that $\gamma(C_i) = C$ for each i. For there exists a gallery beginning with C and containing all the C_i.

Now for each $A \in \Sigma$ there exists an endomorphism γ leaving invariant all faces of C such that $\gamma(A) \in S(C)$. For if we choose a chamber C' containing A there is an endomorphism γ fixing all faces of C with $\gamma(C') = C$. We now show that the element $\gamma(A) \in S(C)$ is uniquely determined, i.e. that if γ, δ are endomorphisms of Σ leaving invariant all faces of C such that $\gamma(A) \in S(C)$ and $\delta(A) \in S(C)$, then $\gamma(A) = \delta(A)$.

Let C' be a chamber containing A and $\Gamma = \{C = C_0, C_1, \ldots, C_m = C'\}$ be a gallery joining C to C'. Consider the galleries $\gamma(\Gamma)$ and $\delta(\Gamma)$. There is an endomorphism ϵ of Σ which leaves invariant all faces of C and maps each chamber in $\gamma(\Gamma)$ and in $\delta(\Gamma)$ to C. Consider the endomorphisms $\epsilon\gamma, \epsilon\delta$ of Σ. They agree on all faces of $C = C_0$. Suppose by induction that they agree on all faces of C_{i-1}. Then they agree on C_i and on all faces of $C_{i-1} \cap C_i$, thus (by the usual argument) they agree on all faces of C_i. In particular $\epsilon\gamma, \epsilon\delta$ agree on all faces of $C_m = C'$. Thus $\epsilon\gamma(A) = \epsilon\delta(A)$. Since $\gamma(A), \delta(A)$ are in $S(C)$ this gives $\gamma(A) = \delta(A)$.

We now define $\rho_C(A)$ to be the common value $\gamma(A)$ for all endomorphisms γ of Σ leaving invariant all faces of C and such that $\gamma(A) \in S(C)$. It is clear that ρ_C is an endomorphism from Σ into $S(C)$ and that ρ_C

leaves invariant all faces of C. It is equally clear that ρ_C is the only endomorphism with this property. ∎

The map ρ_C is called the retraction of Σ on to the simplex of faces of C.

We now introduce an equivalence relation on Σ. Given A, $B \in \Sigma$ we write $A \sim B$ if $\rho_C(A) = \rho_C(B)$.

LEMMA 15.3.3. *The equivalence relation defined on Σ is independent of the chamber C.*

PROOF. Let C' be another chamber of Σ. We show that the equivalence relations for C, C' are the same. Since any two chambers can be joined by a gallery we may assume C, C' are adjacent. By 15.2.7 there exists an automorphism δ of Σ such that $\delta(C) = C'$. Also δ maps each face of $C \cap C'$ into itself. Thus δ coincides with $\rho_{C'}$ on each face of $C \cap C'$. Since $\delta(C) = \rho_{C'}(C)$, δ coincides with $\rho_{C'}$ on each face of C. Similarly δ^{-1} coincides with ρ_C on each face of C'. Thus $\delta\rho_C$ is an endomorphism of Σ into $S(C')$:

$$\Sigma \overset{\rho_C}{\rightarrow} S(C) \overset{\delta}{\rightarrow} S(C')$$

and $\delta\rho_C$ leaves invariant each face of C'. Hence $\delta\rho_C = \rho_{C'}$. It follows that $\rho_C(A) = \rho_C(B)$ if and only if $\rho_{C'}(A) = \rho_{C'}(B)$. ∎

We shall say that equivalent elements of Σ have the same type. It is clear that each chamber has exactly one face of each type.

LEMMA 15.3.4. *Let Σ be an abstract Coxeter complex and C be a chamber of Σ. Let γ be an endomorphism of Σ leaving invariant the type of each face of C. Then γ preserves the type of each element of Σ.*

PROOF. Let C' be another chamber of Σ. We show that γ leaves invariant the type of each face of C'. Since any two chambers may be joined by a gallery we may assume that C, C' are adjacent. Now γ leaves invariant the type of each face of $C \cap C'$ and it also leaves invariant the type of C'. (All chambers have the same type.) Thus, since $\gamma(C')$ has just one face of each type, γ leaves invariant the type of each face of C'. ∎

Endomorphisms and automorphisms of the kind discussed in 15.3.4 will be called type-preserving.

LEMMA 15.3.5. *The automorphisms ρ defined in 15.2.7 are type-preserving.*

PROOF. Let α, β be a pair of opposite foldings of Σ with respect to which ρ is defined. Let C, C' be a pair of adjacent chambers such that $C \in \alpha(\Sigma)$, $C' \in \beta(\Sigma)$. Then ρ leaves invariant each face of $C \cap C'$ and interchanges C, C'. Thus ρ preserves the type of C and of each face of $C \cap C'$, hence ρ preserves the type of each face of C. By 15.3.4 ρ is type-preserving.   ∎

Let $W(\Sigma)$ be the group generated by all the automorphisms of Σ of the kind described in 15.2.7.

PROPOSITION 15.3.6. *Let Σ be an abstract Coxeter complex. Then $W(\Sigma)$ is the group of all type-preserving automorphisms of Σ.*

PROOF. By 15.3.5 each element of $W(\Sigma)$ is a type-preserving automorphism of Σ. Conversely, let δ be any type-preserving automorphism of Σ. Let C be a chamber of Σ. Since any two chambers can be joined by a gallery it follows from 15.2.7 and the fact that Σ is a Coxeter complex that there exists $\gamma \in W(\Sigma)$ such that $\gamma(C) = \delta(C)$. Thus $\gamma^{-1}\delta$ is a type-preserving automorphism of Σ which fixes C. But a chamber C has just one face of each type. Thus $\gamma^{-1}\delta$ fixes each face of C, so must be the identity, by 15.1.2. Hence $\delta \in W(\Sigma)$.   ∎

Let C be a chamber of the abstract Coxeter complex Σ. Each chamber adjacent to C intersects C in a face of codimension 1. Now the number of faces of codimension 1 in C is rank C (which may be finite or infinite), and each such face is contained in just one chamber other than C. Thus there are rank C chambers of Σ adjacent to C. Each of these chambers gives rise to an involutary type-preserving automorphism of Σ as in 15.2.7. These automorphisms will be called the reflections in the faces of codimension 1 in C.

THEOREM 15.3.7. *Let Σ be an abstract Coxeter complex and C be a chamber of Σ. Then the reflections in the faces of codimension 1 in C generate the group $W(\Sigma)$. Moreover $W(\Sigma)$ is a Coxeter group with respect to this set of generators.* (cf. 2.4.2.)

PROOF. Let the reflections in the faces of codimension 1 in C be denoted by w_i, $i \in I$; and let H be the subgroup of $W(\Sigma)$ generated by the w_i. We show that for each chamber D of Σ there exists $w \in H$ such that $D = w(C)$. We do this by induction on dist CD, the result being clear if D is identical with C or adjacent to it. Otherwise there exists a chamber E adjacent to D such that

$$\text{dist } CE = \text{dist } CD - 1.$$

By induction $E = w'(C)$ for some $w' \in H$. Now $w'^{-1}(D)$ is adjacent to C, so has form $w_i(C)$ for some $i \in I$. Thus $w'^{-1}(D) = w_i(C)$ and so $D = w'w_i(C)$, where $w'w_i \in H$ as required.

Let w be any element of $W(\Sigma)$. Then there exists $w' \in H$ such that $w'(C) = w(C)$. Thus $w^{-1}w'$ is a type-preserving automorphism of Σ fixing C, so must be the identity by 15.1.2. Thus $w \in H$ and $H = W(\Sigma)$.

We now show that the reflections w_i generate $W(\Sigma)$ as a Coxeter group. Each element $w \in W(\Sigma)$ can be expressed as a product

$$w = w_{i_1} w_{i_2} \ldots w_{i_k}, \qquad i_\alpha \in I.$$

Let $l(w)$ be the minimal length of any expression of w in this form. We show that
$$l(w) = \text{dist } (C, w(C)).$$

Suppose $l(w) = k$ and $w = w_{i_1} w_{i_2} \ldots w_{i_k}$ with $i_\alpha \in I$. Then

$$\{C, w_{i_1}(C), w_{i_1} w_{i_2}(C), \ldots, w_{i_1} w_{i_2} \ldots w_{i_k}(C)\}$$

is a gallery joining C to $w(C)$ of length k. Thus dist $(C, w(C)) \leqslant k$ and so dist $(C, w(C)) \leqslant l(w)$. Suppose conversely that dist $(C, w(C)) = k$. Then there is a gallery

$$\Gamma = \{C = C_0, C_1, \ldots, C_k = w(C)\}$$

joining C to $w(C)$. Since C, C_1 are adjacent we have $C_1 = w_{i_1}(C)$ for some $i_1 \in I$. Since C_1, C_2 are adjacent we have $C_2 = w_{i_1}(C')$, where C, C' are adjacent. Hence $C_2 = w_{i_1} w_{i_2}(C)$ for some $i_2 \in I$. Arguing in a similar way we see that $C_k = w_{i_1} w_{i_2} \ldots w_{i_k}(C)$ with each $i_\alpha \in I$. Hence

$$w(C) = w_{i_1} w_{i_2} \ldots w_{i_k}(C),$$

and since w and $w_{i_1} w_{i_2} \ldots w_{i_k}$ are both type-preserving we have

$$w = w_{i_1} w_{i_2} \ldots w_{i_k}.$$

Thus $l(w) \leqslant \text{dist } (C, w(C))$, and so we have $l(w) = \text{dist } (C, w(C))$.

We show next that if $l(w_{i_1} w_{i_2} \ldots w_{i_k}) < k$ and $l(w_{i_2} \ldots w_{i_k}) = k - 1$, then there is some $j \geqslant 2$ such that

$$w_{i_2} \ldots w_{i_j} = w_{i_1} \ldots w_{i_{j-1}}.$$

This will be sufficient, using a theorem of Matsumoto, to show that $W(\Sigma)$ is a Coxeter group. Let $w = w_{i_1}w_{i_2} \ldots w_{i_k}$ and write

$$C_r = w_{i_1}w_{i_2} \ldots w_{i_r}(C)$$

for $r = 1, 2, \ldots, k$. Then

$$\begin{aligned}
\text{dist} (C_1, w(C)) &= \text{dist} (w_{i_1}(C), w_{i_1} \ldots w_{i_k}(C)) \\
&= \text{dist} (C, w_{i_2} \ldots w_{i_k}(C)) \\
&= l(w_{i_2} \ldots w_{i_k}) \\
&= k - 1.
\end{aligned}$$

Also

$$\text{dist} (C, w(C)) = l(w) \leqslant k - 1.$$

Thus we have

$$\text{dist} (C, w(C)) \leqslant \text{dist} (C_1, w(C)).$$

Let α be the folding with $\alpha(C_1) = C$. By 15.2.4 we have $w(C) \in \alpha(\Sigma)$. Since $C_1 \notin \alpha(\Sigma)$, there exists $j \geqslant 2$ such that $C_{j-1} \notin \alpha(\Sigma)$ but $C_j \in \alpha(\Sigma)$. Then $\alpha(C_{j-1}) = C_j$. By definition of α we have $\alpha(C_{j-1}) = w_{i_1}(C_{j-1})$, hence $w_{i_1}(C_{j-1}) = C_j$. Thus we have

$$w_{i_1} \cdot w_{i_1}w_{i_2} \ldots w_{i_{j-1}}(C) = w_{i_1}w_{i_2} \ldots w_{i_j}(C).$$

Since all the automorphisms involved are type-preserving it follows that

$$w_{i_2} \ldots w_{i_{j-1}} = w_{i_1}w_{i_2} \ldots w_{i_j},$$

whence

$$w_{i_1} \ldots w_{i_{j-1}} = w_{i_2} \ldots w_{i_j}.$$

The proof of 15.3.7 is completed by establishing the following result, due to Matsumoto.

THEOREM 15.3.8. *Let W be a group generated by a set of involutions w_i. For each $w \in W$ let $l(w)$ be the shortest length of any expression of w as a product of the involutary generators. A product $w_{i_1} \ldots w_{i_k}$ is called reduced if $l(w_{i_1} \ldots w_{i_k}) = k$. Suppose W satisfies the condition that, whenever $w_{i_2} \ldots w_{i_k}$ is reduced but $w_{i_1}w_{i_2} \ldots w_{i_k}$ is not reduced, there exists an integer j with $2 \leqslant j \leqslant k$ such that $w_{i_2} \ldots w_{i_j} = w_{i_1} \ldots w_{i_{j-1}}$. Then W is generated by the w_i as a Coxeter group.*

(The group $W(\Sigma)$ considered above satisfies the hypotheses of this theorem.)

PROOF. Any relation between the generators w_i can be expressed in the form

$$w_{i_1} \ldots w_{i_r} = 1.$$

We show first that any such relation is a consequence of relations of form $w_i^2 = 1$ and

$$w_{j_1} \ldots w_{j_s} = w_{k_1} \ldots w_{k_s},$$

where both expressions are reduced. We prove this by induction on r Since $w_{i_1} \ldots w_{i_r}$ is not reduced there exists α such that $w_{i_{\alpha+1}} \ldots w_{i_r}$ is reduced but $w_{i_\alpha} \ldots w_{i_r}$ is not. Thus there exists an integer $j \leqslant r$ such that $w_{i_\alpha} \ldots w_{i_{j-1}} = w_{i_{\alpha+1}} \ldots w_{i_j}$. Both expressions in this equation are reduced. Using this relation and $w_{i_\alpha}^2 = 1$ we deduce

$$w_{i_1} \ldots w_{i_r} = w_{i_1} \ldots w_{i_{\alpha-1}} w_{i_{\alpha+1}} \ldots w_{i_{j-1}} w_{i_{j+1}} \ldots w_{i_r}.$$

Since the expression on the right has fewer than r terms the result follows by induction.

We show next that each relation

$$w_{j_1} \ldots w_{j_s} = w_{k_1} \ldots w_{k_s},$$

where both expressions are reduced, is a consequence of the Coxeter relations $(w_i w_j)^{m_{ij}} = 1$. We again use induction on s. Since $w_{j_1} \ldots w_{j_s}$ is reduced but $w_{k_1} w_{j_1} \ldots w_{j_s}$ is not reduced, there exists $\alpha \geqslant 2$ such that

$$w_{k_1} w_{j_1} \ldots w_{j_{\alpha-1}} = w_{j_1} \ldots w_{j_\alpha}.$$

It follows that

$$w_{k_1} w_{j_1} \ldots w_{j_{\alpha-1}} w_{j_{\alpha+1}} \ldots w_{j_s} = w_{k_1} \ldots w_{k_s}.$$

Both expressions in this equation are reduced, and by cancelling w_{k_1} we see by induction that this relation can be deduced from the Coxeter relations.

We now distinguish two cases. Suppose $\alpha < s$. Then

$$w_{k_1} w_{j_1} \ldots w_{j_{\alpha-1}} = w_{j_1} \ldots w_{j_\alpha}$$

can be deduced from the Coxeter relations, by induction. Thus

$$w_{j_1} \ldots w_{j_s} = w_{k_1} w_{j_1} \ldots w_{j_{\alpha-1}} w_{j_{\alpha+1}} \ldots w_{j_s}$$

can be deduced from the Coxeter relations, and therefore so can

$$w_{j_1} \ldots w_{j_s} = w_{k_1} \ldots w_{k_s}.$$

Suppose $\alpha = s$. Then we have

$$w_{j_1} \ldots w_{j_s} = w_{k_1} w_{j_1} \ldots w_{j_{s-1}} = w_{k_1} \ldots w_{k_s}.$$

By induction, the relation

$$w_{k_1} w_{j_1} \ldots w_{j_{s-1}} = w_{k_1} \ldots w_{k_s}$$

can be deduced from the Coxeter relations, so the required relation

$$w_{j_1} \ldots w_{j_s} = w_{k_1} \ldots w_{k_s}$$

can be deduced from the Coxeter relations, provided that

$$w_{j_1} \ldots w_{j_s} = w_{k_1} w_{j_1} \ldots w_{j_{s-1}}$$

can be deduced from these relations.

We now repeat the argument. The relation

$$w_{j_1} \ldots w_{j_s} = w_{k_1} w_{j_1} \ldots w_{j_{s-1}}$$

can be deduced from the Coxeter relations provided that

$$w_{k_1} w_{j_1} \ldots w_{j_{s-1}} = w_{j_1} w_{k_1} w_{j_1} \ldots w_{j_{s-2}}$$

can be deduced from these relations. Repeating the argument a number of times, we eventually have to show that

$$w_{j_1} w_{k_1} w_{j_1} \ldots = w_{k_1} w_{j_1} w_{k_1} \ldots$$

is a consequence of the Coxeter relations, which is obvious. ∎

15.4 The Complex $\Sigma(W, \Pi)$

We shall now show, as is to be expected, that the Coxeter complex as defined in section 2.6 is an abstract Coxeter complex.

Let Φ be a root system and W be the Weyl group of Φ. Let Π be a fundamental system of roots in Φ.

THEOREM 15.4.1. *The following two partially ordered sets are abstract Coxeter complexes which are isomorphic to one another:*

(i) *The Coxeter complex Σ of W (defined as in section 2.6), with partial order $\leqslant$ defined by $K_1 \leqslant K_2$ if and only if K_1 is contained in the closure $\bar{K}_2$ of K_2.*

(ii) *The set $\Sigma(W, \Pi)$ of all cosets wW_J for all $w \in W$ and all subsets J of Π, with partial ordering $\leqslant$ defined by $K_1 \leqslant K_2$ if and only if K_2 is a subset of K_1.*

PROOF. We show that Σ and $\Sigma(W, \Pi)$ are isomorphic as partially ordered sets, and then show that $\Sigma(W, \Pi)$ is an abstract Coxeter complex.

We recall some facts about Σ. Let C be the chamber corresponding to the fundamental system Π. Then the elements of Σ contained in the closure $\bar{C}$ are the elements C_J defined in section 2.6. The stablizer of C_J under the action of W on Σ is W_J, by 2.6.1. Moreover, each element of Σ can be expressed in the form $w(C_J)$ for exactly one subset J of Π, by 2.6.3. Let w_1, w_2 be two elements of W. Then $w_1(C_J) = w_2(C_J)$ if and only if $w_1 W_J = w_2 W_J$. Thus the map

$$w(C_J) \to w W_J$$

is a bijection between Σ and $\Sigma(W, \Pi)$. We show that this bijection preserves the partial order relations.

Suppose $w W_J \geqslant w' W_K$. Then $w W_J$ is a subset of $w' W_K$. It follows that $w'^{-1} w W_J$ is a subset of W_K, and so $w'^{-1} w \in W_K$ and W_J is contained in W_K. This implies that J is a subset of K, since the fundamental roots are linearly independent. Hence $w W_J \geqslant w' W_K$ if and only if $w'^{-1} w \in W_K$ and J is a subset of K.

Now suppose $w(C_J) \geqslant w'(C_K)$. This means that $w'(C_K)$ is in the closure of $w(C_J)$, and so $w^{-1} w'(C_K)$ is in $\bar{C}_J$. In particular $w^{-1} w'(C_K)$ is in $\bar{C}$. However, the only elements of the complex contained in $\bar{C}$ are those of the form C_L, for some subset L of Π, and if $w^{-1} w'(C_K) = C_L$ then we have $L = K$ by our earlier remarks. Thus $w^{-1} w' \in W_K$ and C_K is contained in $\bar{C}_J$. It is clear from the definitions of C_J, C_K that this implies that J is a subset of K. Thus $w(C_J) \geqslant w'(C_K)$ if and only if $w^{-1} w' \in W_K$ and J is a subset of K.

Thus we have shown that the bijection between Σ and $\Sigma(W, \Pi)$ preserves the partial orderings $\leqslant$. We must now verify that $\Sigma(W, \Pi)$ (and hence Σ also) is an abstract Coxeter complex. We observe that the maximal elements of $\Sigma(W, \Pi)$ under the ordering $\leqslant$ are the elements of W. Two elements w_1, w_2 of W are adjacent if and only if $w_2 = w_1 w_r$ for some $r \in \Pi$. Since the fundamental reflections w_r generate W, it follows that $\Sigma(W, \Pi)$ is a chamber complex. $\Sigma(W, \Pi)$ is thin, because the elements of codimension 1 are sets of the form $\{w, w w_r\}$ for some $w \in W$, $r \in \Pi$. Thus each element of codimension 1 is contained in two chambers. To show that $\Sigma(W, \Pi)$ is an abstract Coxeter complex it remains to prove the existence of all possible foldings.

Let w' and $w' w_r$ $(r \in \Pi)$ be two adjacent chambers of $\Sigma(W, \Pi)$. We show there exists a folding α such that $\alpha(w' w_r) = w'$. Let $s = w'(r)$ and define

$$W_1 = \{w'' \in W; \ w''^{-1}(s) \in \Phi^+\},$$

$$W_2 = \{w'' \in W; \ w''^{-1}(s) \in \Phi^-\}.$$

Thus $W = W_1 \cup W_2$ and $W_1 \cap W_2$ is empty. Let α be the map of $\Sigma(W, \Pi)$

into itself given by

$$\alpha(wW_J) = \begin{cases} wW_J & \text{if } wW_J \nsubseteq W_2, \\ w_s wW_J & \text{if } wW_J \subseteq W_2. \end{cases}$$

α clearly maps chambers into chambers. If $w \in W_1$ we have $\alpha(wW_J) = wW_J$ for all subsets J of Π. Thus α operates as the identity on all the faces of the chamber w. Now suppose $w \in W_2$. Then $\alpha(w) = w_s w$, and

$$\alpha(wW_J) = w_s wW_J$$

whenever $wW_J \subseteq W_2$. We show that $\alpha(wW_J) = w_s wW_J$ even if $wW_J \nsubseteq W_2$ and this will prove that α induces an isomorphism between the faces of w and the faces of $w_s w$. Now $w \in W_2$ and so $w^{-1}(s) \in \Phi^-$. Also $wW_J \nsubseteq W_2$, so there is some element $w_J \in W_J$ such that $ww_J \in W_1$. Then $w_J^{-1} w^{-1}(s) \in \Phi^+$, whence $w^{-1}(s) \in W_J(\Phi^+)$. Thus we have

$$w^{-1}(s) \in \Phi^- \cap W_J(\Phi^+) = \Phi_J^-$$

by 9.4.1. It follows that

$$w_{w^{-1}(s)} = w^{-1} w_s w \in W_J.$$

Therefore

$$\alpha(wW_J) = wW_J = w_s wW_J.$$

We have now shown that α is an endomorphism of $\Sigma(W, \Pi)$.

If $w \in W_1$, α operates as the identity on all faces of w. If $w \in W_2$ then $\alpha(w) = w_s w$ is in W_1. Thus α operates as the identity on all faces of $\alpha(w)$. Hence $\alpha^2 = \alpha$ and α is a folding. Finally $w'w_r \in W_2$ and so

$$\alpha(w'w_r) = w_s w'w_r = w'.$$

Thus α is a folding which maps $w'w_r$ to w'.   ∎

We determine next the group of type-preserving automorphisms of the abstract Coxeter complex $\Sigma(W, \Pi)$.

PROPOSITION 15.4.2. *The group of type-preserving automorphisms of* $\Sigma(W, \Pi)$ *is isomorphic to* W.

PROOF. Let α be the folding of $\Sigma(W, \Pi)$ which maps $w'w_r$ to w'. The effect of α on the chambers is given by:

$$\alpha(w) = \begin{cases} w & \text{if } w \in W_1, \\ w_s w & \text{if } w \in W_2. \end{cases}$$

(The notation is as in 15.4.1.) By replacing w' by $w'w_r$ we may determine the opposite folding β of α. β operates on the chambers by:

$$\beta(w) = \begin{cases} w_s w & \text{if } w \in W_1, \\ w & \text{if } w \in W_2. \end{cases}$$

Let ρ be the involutary automorphism of $\Sigma(W, \Pi)$ determined by α and β. Then

$$\rho(w) = w_s w$$

for all $w \in W$. Thus ρ operates on the elements of W by left-multiplication by the reflection w_s. Now the automorphisms of this kind generate the whole group of type-preserving automorphisms of $\Sigma(W, \Pi)$, by 15.3.6. Since the Weyl group W is generated by its reflections it follows that the group of type-preserving automorphisms is isomorphic to W. ∎

We now show that $\Sigma(W, \Pi)$ is, to within isomorphism, the only abstract Coxeter complex whose group of type-preserving automorphisms is isomorphic to W.

THEOREM 15.4.3. *Let W be a Weyl group and Σ be an abstract Coxeter complex whose group of type-preserving automorphisms $W(\Sigma)$, when regarded as a Coxeter group as in 15.3.7, has the same relations as W when generated by a system of fundamental reflections. Then Σ is isomorphic to $\Sigma(W, \Pi)$.*

PROOF. Let C be a chamber of Σ. Then by 15.3.7 $W(\Sigma)$ is generated as a Coxeter group by the reflections in the faces of codimension 1 in C. In the present instance the number of such reflections is $l = \text{rank } W$. Let these reflections be $\rho_1, \rho_2, \ldots, \rho_l$. Then

$$W \cong W(\Sigma) = \langle \rho_1, \rho_2, \ldots, \rho_l \rangle.$$

Now rank $C = l$ and C has faces $A_1, A_2, \ldots, A_l$ of codimension 1 in natural 1–1 correspondence with $\rho_1, \rho_2, \ldots, \rho_l$. Let J be any subset of $\{1, 2, \ldots, l\}$ and

$$C_J = \bigcap_{i \in J} A_i.$$

Then the elements C_J are all faces of C, and every element of Σ is expressible in the form $w(C_J)$ for some $w \in W(\Sigma)$ and some J.

Consider the complex St C_J. This is also an abstract Coxeter complex by 15.3.1. The group of type-preserving automorphisms of St C_J is isomorphic to the subgroup W_J of $W(\Sigma)$ generated by the elements ρ_i for $i \in J$, also by 15.3.7. In particular, W_J operates transitively on the chambers of Σ containing C_J, and $| W_J |$ is the number of such chambers. Now W_J stabilizes C_J since each of its generators has this property. Suppose the stabilizer H of C_J in W were greater than W_J. Then H would operate on the chambers containing C_J, and so the chambers $w(C)$ for $w \in H$ could not all be distinct. However, this contradicts 15.1.2. Thus W_J is the stabilizer of C_J in W.

Let $w_1, w_2 \in W$. Then $w_1(C_J) = w_2(C_J)$ if and only if $w_1 W_J = w_2 W_J$. Thus the map

$$\Sigma \to \Sigma(W, \Pi),$$

$$w(C_J) \to w W_J.$$

is a bijection. This bijection transforms the chamber $w(C)$ in Σ into the chamber w in $\Sigma(W, \Pi)$. Moreover, it induces an isomorphism between the faces $w(C_J)$ of $w(C)$ and the faces $w W_J$ of w. Thus it is an isomorphism between Σ and $\Sigma(W, \Pi)$. ∎

Note. We have shown that there is just one abstract Coxeter complex Σ (to within isomorphism) whose group $W(\Sigma)$ of type-preserving auto-morphisms is isomorphic to W, if W is a Weyl group. In general $W(\Sigma)$ is a Coxeter group. Now there are Coxeter groups which are not Weyl groups. (A classification of the finite Coxeter groups is given in Bourbaki [1].) Although such groups will not be discussed in the present volume it is possible to extend the result mentioned above to all Coxeter groups. Given any Coxeter group W there is, to within isomorphism, a unique abstract Coxeter complex whose group of type-preserving automorphisms is isomorphic to W.

15.5 Buildings

A building is a pair $(\Omega, \mathscr{A})$ where Ω is a chamber complex and $\mathscr{A}$ is a set of subcomplexes, called apartments, satisfying the following axioms:

    B1. Ω is a thick chamber complex.

    B2. The apartments of Ω are thin chamber complexes.

    B3. Given any two chambers C, C' in Ω there exists an apartment $\Sigma \in \mathscr{A}$ such that $C \in \Sigma$ and $C' \in \Sigma$.

    B4. If A, A' are elements of Ω which are both contained in each of the apartments Σ, $\Sigma' \in \mathscr{A}$, there exists an isomorphism between Σ, Σ' leaving invariant A, A' and all their faces.

It follows from these axioms that any two apartments of a building Ω are isomorphic. For let Σ, Σ' be apartments and C, C' be chambers with $C \in \Sigma$ and $C' \in \Sigma'$. Let $\Gamma = \{C = C_0, C_1, \ldots, C_m = C'\}$ be a gallery joining C, C'. There exist apartments Σ_1 containing C_0, C_1; Σ_2 containing C_1, C_2; etc. By $B4$ we have

$$\Sigma \cong \Sigma_1 \cong \Sigma_2 \cong \ldots \cong \Sigma_m \cong \Sigma'.$$

Example 15.5.1.   *The building* $\Omega(G; B, N)$.

Let G be a group with a (B, N)-pair. Then there exist two subgroups B, N of G such that:

BN 1. G is generated by B and N.

BN 2. $B \cap N$ is a normal subgroup of N.

BN 3.   The group $W = N/B \cap N$ is generated by a set w_i of involutions $(i \in I)$.

BN 4.   If $n_i \in N$ maps to w_i under the natural homomorphism, and if n is an element of N, then

$$B n_i B . B n B \subseteq B n_i n B \cup B n B.$$

BN 5. $n_i B n_i \neq B$, $i \in I$.

The subgroups of G containing B are in 1–1 correspondence with the subsets J of I, by 8.3.2, and have the form $P_J = \langle B, n_i; i \in J \rangle$ by 8.3.1. Let Ω be the set of left cosets gP_J for all $g \in G$ and all subsets J of I. We introduce a partial ordering on Ω which is the reverse of set-theoretical inclusion. Let Σ_0 be the subset of Ω which consists of the elements nP_J for all $n \in N$ and all J. Then $g\Sigma_0$ is the set of cosets gnP_J for all $n \in N$ and all J, and we define $\mathscr{A}$ to be the family of subsets $g\Sigma_0$ of Ω for all $g \in G$. We shall show that Ω is a building and that $\mathscr{A}$ is a set of apartments in Ω. The building constructed in this way will be called $\Omega(G; B, N)$.

We show first that Ω is a chamber complex. The maximal elements (chambers) of Ω are the elements gB and the elements of codimension 1 are those of form gP_J, where J is a 1-element subset of I. Thus B and n_iB are adjacent chambers if $i \in I$. It follows that the chambers B, nB can be joined by a gallery for all $n \in N$. Let g be any element of G. Since $G = BNB$ we have $g = bnb'$ with b, $b' \in B$ and $n \in N$. Thus $gB = bnB$. Now B, nB can be joined by a gallery and so bB, bnB can be joined by a gallery also. Thus B, gB can be joined by a gallery. Let g_1, g_2 be arbitrary elements of G and put $g = g_2^{-1} g_1$. Since B, gB can be joined by a gallery it follows that $g_1 B$, $g_2 B$ can be joined also. Thus Ω is a chamber complex.

We now show that the chamber complex Ω is thick. Consider the element $P_J = \langle B, n_i \rangle$ of codimension 1 in Ω, where $J = \{i\}$. It is clear that B and $n_i B$ are chambers containing P_J. However they are not the only chambers containing P_J. For if they were, B would be a subgroup of index 2 in P_J and hence normal in P_J, and we would have $n_i B n_i = B$, which contradicts $BN\,5$. It follows that gB and $g n_i B$ are not the only chambers containing $g P_J$. Thus Ω is thick.

Now $P_J = B \cup B n_i B$ and the only elements of N contained in P_J are 1, n_i by 8.3.1 and 8.3.4. Thus B and $n_i B$ are the only chambers containing P_J of the form nB for $n \in N$. It follows that nB and $n n_i B$ are the only chambers in Σ_0 containing $n P_J$. Thus Σ_0 is thin, and hence all the other apartments of Ω are thin also.

We verify next that any two chambers of Ω are contained in some apartment. Let $g_1 B$, $g_2 B$ be two chambers and let $g = g_1^{-1} g_2$. Since $G = BNB$ we have $g = bnb'$ for b, $b' \in B$ and $n \in N$. Thus $gB = bnB$. Now B and nB are in Σ_0, and so B and gB are in $b\Sigma_0$. It follows that $g_1 B$ and $g_2 B$ are in the apartment $g_1 b \Sigma_0$.

Finally, suppose we have two elements of Ω and two apartments which each contain both these elements. By multiplying on the left by a suitable element of G we may assume one of the apartments is Σ_0 and that the chambers are P_J and $n P_K$, where $n \in N$. Let the other apartment be $g \Sigma_0$. We show there is an isomorphism between Σ_0 and $g \Sigma_0$ which leaves invariant P_J, $n P_K$ and all their faces. Since $P_J \in g \Sigma_0$ we have $g^{-1} P_J \in \Sigma_0$ and so $g^{-1} P_J = n' P_J$ for some $n' \in N$. Thus $g n' = p_J \in P_J$, and $g \Sigma_0 = p_J \Sigma_0$. Now $n P_K \in p_J \Sigma_0$ and so $p_J^{-1} n P_K = n' P_K$ for some $n' \in N$. It follows that

$$P_J n P_K = P_J n' P_K.$$

We now require the following lemma.

LEMMA 15.5.2. *Let G be a group with a (B, N)-pair. Then for each $n \in N$ we have*

$$P_J n P_K \cap N = N_J n N_K.$$

(The notation is as in 8.2.2.)

PROOF. Axiom $BN4$ applied repeatedly shows that, for any subset N_0 of N, we have

$$N_J B N_0 \subseteq B N_J N_0 B.$$

By inverting both sides we also obtain

$$N_0 B N_J \subseteq B N_0 N_J B$$

for any subset N_0 of N. By applying these formulae we have

$$P_J n P_K \cap N = B N_J B n B N_K B \cap N$$
$$\subseteq B N_J n B N_K B \cap N$$
$$\subseteq B N_J n N_K B \cap N$$
$$= N_J n N_K$$

by 8.2.3. The reverse inclusion is obvious. ∎

We can now complete the argument to show that $\Omega(G; B, N)$ is a building. We have $P_J n P_K = P_J n' P_K$ and so, by 15.5.2, we obtain $N_J n N_K = N_J n' N_K$. Thus

$$N_J n N_K n^{-1} = N_J n' N_K n^{-1}$$

and, in particular, we have

$$N_J \cap n' N_K n^{-1} \neq \phi.$$

Let $\bar{n} \in N_J \cap n' N_K n^{-1}$. We show that left multiplication by the element $p_J \bar{n}$ gives an isomorphism from Σ_0 to $g\Sigma_0$ with the required properties. We have

$$p_J \bar{n} \Sigma_0 = p_J \Sigma_0 = g\Sigma_0$$

and so we have an isomorphism from Σ_0 to $g\Sigma_0$.
Also

$$p_J \bar{n} P_J = p_J P_J = P_J$$

and

$$p_J \bar{n} . n P_K = p_J n' P_K = n P_K.$$

Thus this isomorphism fixes P_J and $n P_K$ and clearly fixes also all faces of these two elements, since the faces are larger subsets of G. Thus axiom $B4$ has been established, and so $\Omega(G; B, N)$ is a building.

15.6 Retractions onto an Apartment

We have shown above that there is a building associated to each group with a (B, N)-pair, and have therefore established the existence of a large number of buildings. We shall now prove some further general properties of buildings, concentrating on the relationship between a building and its apartments.

Let Ω be a building, Σ an apartment of Ω and C a chamber in Σ. For each element $A \in \Omega$ there exists an apartment Σ' containing A and C, by axiom $B3$. By $B4$ there is an isomorphism $\Sigma' \to \Sigma$ which leaves invariant all faces of C. By 15.1.2 there is only one such isomorphism. The image of A under this isomorphism is an element of Σ which is independent of the choice of Σ', by $B4$. This image will be called $\mathrm{retr}_{\Sigma, \, c}(A)$.

LEMMA 15.6.1.   *The map*

$$\Omega \to \Sigma,$$

$$A \to \mathrm{retr}_{\Sigma, \, c}(A)$$

is a retraction from Ω onto Σ.

PROOF.   The map is clearly a morphism, and since it acts as the identity on Σ it is idempotent.   ∎

We use retractions of this kind to prove the following important result.

THEOREM 15.6.2.   *The apartments of a building are abstract Coxeter complexes.*

PROOF.   Let Σ be an apartment of a building Ω. We know that Σ is a thin chamber complex and so must prove the existence of all possible foldings. Let C, C' be adjacent chambers of Σ and let $A = C \cap C'$. Since Ω is thick there is a third chamber C'' containing A. Let Σ' be an apartment containing C, C''. Let α be the map of Σ into itself given by

$$\alpha = \mathrm{retr}_{\Sigma, \, c'} . \mathrm{retr}_{\Sigma', \, c}.$$

The retractions are restricted to the apartments being considered, which are as shown below.

$$\Sigma \xrightarrow{\mathrm{retr}_{\Sigma', \, c}} \Sigma' \xrightarrow{\mathrm{retr}_{\Sigma, \, c''}} \Sigma$$

Then α is an endomorphism of Σ, and we have

$$\alpha(C) = \mathrm{retr}_{\Sigma, \, c'}(C) = C,$$

$$\alpha(C') = \mathrm{retr}_{\Sigma, \, c'}(C'') = C.$$

We may define similarly another endomorphism β of Σ by interchanging the rôles of C, C'. Thus $\beta(C') = C'$ and $\beta(C) = C'$. Furthermore α and β both leave invariant all faces of A. We shall show that α, β are a pair of opposite foldings of Σ, thus proving that Σ is an abstract Coxeter complex.

We define a set $\Gamma_\Sigma(A)$ of galleries of Σ. A gallery Γ of Σ lies in $\Gamma_\Sigma(A)$ if:

(a) The first term of Γ contains A.

(b) There is no gallery of Σ shorter than Γ whose first term contains A and last term coincides with the last term of Γ.

Thus $\Gamma_\Sigma(A)$ is the set of galleries of Σ which start from a chamber containing A and reach their destination as quickly as possible. We shall show that if $\Gamma \in \Gamma_\Sigma(A)$ then $\alpha(\Gamma) \in \Gamma_\Sigma(A)$ also.

Now
$$\alpha(\Gamma) = \mathrm{retr}_{\Sigma, \, C'} \, \mathrm{retr}_{\Sigma', \, C}(\Gamma).$$

The map
$$\Sigma \xrightarrow{\mathrm{retr}_{\Sigma', \, C}} \Sigma'$$

is the unique isomorphism from Σ to Σ' which leaves invariant all the faces of C. Since $\Gamma \in \Gamma_\Sigma(A)$ it follows that $\mathrm{retr}_{\Sigma', \, C}(\Gamma) \in \Gamma_{\Sigma'}(A)$. Similarly the map
$$\Sigma' \xrightarrow{\mathrm{retr}_{\Sigma, \, C'}} \Sigma$$

is the unique isomorphism from Σ' to Σ which leaves invariant all faces of C'. Since $\mathrm{retr}_{\Sigma', \, C}(\Gamma) \in \Gamma_{\Sigma'}(A)$ it follows that

$$\mathrm{retr}_{\Sigma, \, C'} \, \mathrm{retr}_{\Sigma', \, C}(\Gamma) \in \Gamma_\Sigma(A).$$

Thus α maps $\Gamma_\Sigma(A)$ into itself. Similarly $\Gamma_\Sigma(A)$ is mapped into itself by β.

Let $\Gamma = \{C = C_0, \, C_1, \ldots, \, C_m\}$ be a gallery in $\Gamma_\Sigma(A)$. We show by induction on m that all faces of C_m are invariant under α and under $\alpha\beta$, but that C_m is not fixed by β. These facts are clear if $m = 0$, since $\alpha(C) = C$, $\beta(C) = C'$ and $\alpha\beta(C) = \alpha(C') = C$. Suppose therefore that $m > 0$ and write $\gamma = \alpha$ or $\alpha\beta$. By induction γ fixes all faces of C_{m-1}. Now $\gamma(C_m)$ contains $C_{m-1} \cap C_m$, and so must be either C_{m-1} or C_m. However, $\gamma(C_m) = C_{m-1}$ would imply that

$$\gamma(\Gamma) = \{C = \gamma(C_0), \, \gamma(C_1), \ldots, \, C_{m-1}, \, C_{m-1}\}$$

is not in $\Gamma_\Sigma(A)$, contrary to the fact that α, β transform $\Gamma_\Sigma(A)$ into itself. Hence γ fixes C_m. It also fixes all faces of $C_{m-1} \cap C_m$, and so it fixes all faces of C_m.

Now suppose by way of contradiction that $\beta(C_m) = C_m$. Then we have

$$\beta(\Gamma) = \{C' = \beta(C_0), \ldots, C_m\} \in \Gamma_\Sigma(A)$$

and so β fixes all the faces of C_m. In particular β fixes $C_{m-1} \cap C_m$. Hence $\beta(C_{m-1})$ is either C_{m-1} or C_m. However, $\beta(C_{m-1}) = C_{m-1}$ is false by induction, and $\beta(C_{m-1}) = C_m$ contradicts the fact that $\beta(\Gamma) \in \Gamma_\Sigma(A)$. Thus we have a contradiction, and hence $\beta(C_m) \neq C_m$.

We are now able to show that α, β are a pair of opposite foldings of Σ. We show first that α is idempotent. Let D be a chamber in Σ and let $\Gamma \in \Gamma_\Sigma(A)$ be a gallery whose last term is D. Then $\alpha(\Gamma)$ is also a gallery in $\Gamma_\Sigma(A)$ and has first term C and last term $\alpha(D)$. Thus α fixes all the faces of $\alpha(D)$, as shown above. Hence $\alpha^2 = \alpha$.

We now show that each chamber in $\alpha(\Sigma)$ is the image of just two chambers in Σ. Let D be a chamber in $\alpha(\Sigma)$ and let $D = \alpha(E)$, where $E \in \Sigma$. Now we have shown above that for each chamber $E \in \Sigma$, either $\alpha(E) = E$ and $\beta(E) \neq E$, or $\alpha(E) \neq E$ and $\beta(E) = E$. If $\alpha(E) = E$ then $E = D$. If $\alpha(E) \neq E$ then $\beta(E) = E$ and $\beta\alpha(E) = E$, as shown above. Thus

$$E = \beta(\alpha(E)) = \beta(D).$$

Hence there are just two chambers in Σ such that $\alpha(E) = D$, viz., D and $\beta(D)$. Therefore α is a folding. It follows by symmetry that β is also a folding, and by definition that α, β are opposite foldings. Thus Σ is an abstract Coxeter complex. ∎

Definition 15.6.3.   Let A, A' be two elements of a building Ω. Then A, A' are said to have the same type in Ω if they have the same type in any apartment containing A, A'.

We observe that this condition is independent of the apartment chosen. For let Σ_1, Σ_2 be two apartments containing A, A'. Let C, C' be chambers of Σ_1 containing A, A' respectively. Let γ be the retraction of Σ_1 onto the simplex $S(C)$ defined in 15.3.2. Then A, A' have the same type in Σ_1 if and only if $\gamma(A') = A$. Now by $B4$ there exists an isomorphism $\delta : \Sigma_1 \to \Sigma_2$ which leaves A and A' invariant. Thus the retraction of Σ_2 onto the simplex $S(\delta(C))$ is given by

$$\delta(X) \to \delta\gamma(X), \qquad X \in \Sigma_1.$$

A and A' have the same type in Σ_2 if and only if A' is mapped to A under this retraction, which holds if and only if $\gamma(A') = A$.

15.7 Groups of Type-Preserving Automorphisms

We shall be concerned with groups of type-preserving automorphisms of a building Ω, and first give an example in the building $\Omega(G; B, N)$ constructed from a group G with a (B, N)-pair.

PROPOSITION 15.7.1. *In the building $\Omega(G; B, N)$, the group G operates by left multiplication as a group of type-preserving automorphisms which is transitive on the pairs (C, Σ), where C is a chamber and Σ is an apartment containing C.*

PROOF. We use the notation of 15.5.1. The elements of Ω are cosets of the form gP_J. Consider the apartment Σ_0 of Ω. The chambers of Σ_0 have form nP_J, where $n \in N$, and the map

$$nP_J \to P_J$$

is an idempotent morphism from Σ_0 to the simplex of faces of the chamber B. It is therefore the retraction described in 15.3.2. Thus n_1P_J and n_2P_K have the same type if and only if $J = K$. Similarly in the apartment $\Sigma = g\Sigma_0$ of Ω, the elements gn_1P_J and gn_2P_K have the same type if and only if $J = K$. We now consider any two elements g_1P_J and g_2P_K of Ω. They have the same type in Ω if and only if they have the same type in some apartment containing both. Thus g_1P_J and g_2P_K have the same type if and only if $J = K$. It is now clear that the map of Ω into itself given by

$$gP_J \to xgP_J, \qquad x \in G,$$

is a type-preserving automorphism of Ω.

Let C be a chamber and Σ an apartment of Ω containing C. Then $\Sigma = g\Sigma_0$ for some $g \in G$, and $C = gnB$ for some $n \in N$. Let $x = gn$. Then $C = xB$ and $\Sigma = x\Sigma_0$. Thus the element x of G transforms the pair (B, Σ_0) into the pair (C, Σ). It follows that G operates transitively on the pairs (C, Σ) with $C \in \Sigma$. ∎

We shall now prove a converse of this result, namely that a group of type-preserving automorphisms of a building which is transitive on the pairs (C, Σ) with $C \in \Sigma$ is a group with a (B, N)-pair.

THEOREM 15.7.2. *Let $(\Omega, \mathscr{A})$ be a building and G be a group of type-preserving automorphisms of Ω which is transitive on the pairs (C, Σ) with*

$C \in \Sigma$, where C is a chamber and Σ an apartment of Ω. Let C_0, Σ_0 be a fixed chamber and apartment with $C_0 \in \Sigma_0$, let B be the stabilizer of C_0 in G and N be the stabilizer of Σ_0. Then the subgroups B, N form a (B, N)-pair in G. Moreover, $W = N/B \cap N$ is isomorphic to the group $W(\Sigma)$ of type-preserving automorphisms of each apartment Σ of Ω.

PROOF. We verify the last assertion first. N operates on Σ_0 as a group of type-preserving automorphisms. The transitivity hypothesis shows that N operates transitively on the chambers of Σ_0. Thus N induces on Σ_0 the full group $W(\Sigma_0)$ of type-preserving automorphisms, by 15.1.2. Thus we have an epimorphism $N \to W(\Sigma_0)$ with kernel $B \cap N$. For a type-preserving automorphism of Σ_0 which fixes C_0 must be the identity, again by 15.1.2. Thus $N/B \cap N$ is isomorphic to $W(\Sigma_0)$. This is isomorphic to $W(\Sigma)$ for any apartment Σ, since any two apartments are isomorphic.

We now show that the subgroups B, N satisfy the axioms for a (B, N)-pair. It has already been verified that $B \cap N$ is normal in N and that $N/B \cap N$ is generated by a set of involutions. We choose for the set I of generating involutions the reflections of Σ_0 in the faces of codimension 1 in C_0 (see 15.3.7).

Let $g \in G$ and let Σ be an apartment containing C_0 and $g(C_0)$. Then C_0 is contained in Σ and $g^{-1}(\Sigma)$, so the transitivity condition shows there exists $b_1 \in B$ such that $g^{-1}(\Sigma) = b_1(\Sigma)$. Also, since C_0 is in Σ_0 and Σ, there exists $b_2 \in B$ such that $b_2(\Sigma_0) = \Sigma$. Thus $g^{-1}b_2(\Sigma_0) = b_1b_2(\Sigma_0)$ and it follows that $b_2^{-1}gb_1b_2 \in N$. Hence $g \in BNB$ and we have $G = BNB$.

Let n_i be an element of N which induces on Σ_0 a generating involution w_i of W ($i \in I$). We show $n_iBn_i \neq B$. Suppose this is false, so that $n_iB = Bn_i$. Then we have

$$B . n_i(C_0) = n_i . B(C_0) = n_i(C_0).$$

Thus B stabilizes both C_0 and $n_i(C_0)$. Now C_0 and $n_i(C_0)$ are adjacent chambers and, since Ω is thick, there is a third chamber C' containing $C_0 \cap n_i(C_0)$. Let Σ be an apartment containing C_0 and C'. By transitivity there exists $b \in B$ such that $b(\Sigma_0) = \Sigma$. Since B stabilizes both C_0 and $n_i(C_0)$ we see that C_0, $n_i(C_0)$, C' are all in Σ. This contradicts the fact that Σ is thin.

Finally we check the axiom

$$Bn_iB . BnB \subseteq Bn_inB \cup BnB.$$

Given elements n_i, $n \in N$ and $b \in B$ we consider the adjacent chambers C_0 and $n_i(C_0)$. Let $A = C_0 \cap n_i(C_0)$. The element b fixes C_0 and is type-preserving, so fixes all faces of C_0. In particular $bA = A$ and bn_iC_0 contains A. Thus the chambers nC_0, nn_iC_0 and nbn_iC_0 all contain nA.

Consider the set of galleries whose first term is C_0 and last term is a chamber containing nA. Let Γ be such a gallery with as few terms as possible and let

$$\Gamma = \{C_0, C_1, \ldots, C_m\}.$$

Let Σ be an apartment containing C_0 and nbn_iC_0. Then there exists $b_1 \in B$ such that $b_1\Sigma = \Sigma_0$.

Now Σ contains C_0 and nA, and we show that Σ contains each term of the gallery Γ. Suppose this is false, and let C_i be the first term in Γ not in Σ. Let D be the chamber of Σ other than C_{i-1} containing $C_{i-1} \cap C_i$. Then

$$\text{retr}_{\Sigma, D}(C_{i-1}) = C_{i-1},$$

$$\text{retr}_{\Sigma, D}(C_i) = C_{i-1}$$

and $\text{retr}_{\Sigma, D}(\Gamma)$ is a gallery whose first term is C_0, whose last term contains nA, and which has two consecutive terms identical. This contradicts the fact that the length of Γ is minimal. Thus Σ contains each term of Γ. Similarly Σ_0 contains C_0 and nA, so contains each term of Γ.

We shall show by induction on i that $b_1C_i = C_i$. This is clear if $i = 0$. Assume inductively that $b_1C_{i-1} = C_{i-1}$. Then $b_1(C_{i-1} \cap C_i) = C_{i-1} \cap C_i$ since b_1 is type-preserving. Thus b_1C_i contains $C_{i-1} \cap C_i$. Now $C_i \in \Sigma$ and $b_1C_i \in \Sigma_0$. The two chambers of Σ_0 containing $C_{i-1} \cap C_i$ are C_{i-1} and C_i. Now $b_1C_i \neq C_{i-1}$ since b_1 is bijective, hence $b_1C_i = C_i$. In particular $b_1C_m = C_m$. Since b_1 is type-preserving we have also $b_1nA = nA$.

Now nbn_iC_0 is a chamber in Σ which contains nA and so $b_1nbn_iC_0$ is a chamber in Σ_0 which contains nA. Thus $b_1nbn_iC_0$ is either nC_0 or nn_iC_0. If $b_1nbn_iC_0 = nC_0$ we have $nbn_i \in BnB$, and if $b_1nbn_iC_0 = nn_iC_0$ we have $nbn_i \in Bnn_iB$. Thus

$$nBn_i \subseteq Bnn_iB \cup BnB.$$

By taking inverses it follows that

$$n_iBn \subseteq Bn_inB \cup BnB.$$

Thus G has a (B, N)-pair. ∎

Note. The above proof shows clearly the geometrical meaning of the axiom *BN* 4. It was shown that some element of B transforms the chamber nbn_iC_0 into a chamber of Σ_0 containing $nC_0 \cap nn_iC_0$, which must therefore be either nC_0 or nn_iC_0.

The relation between the algebraic and geometric structures discussed in this chapter may be summarized in the following scheme.

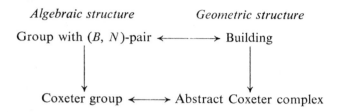

Algebraic structure   *Geometric structure*

Group with (B, N)-pair ⟷ Building

Coxeter group ⟷ Abstract Coxeter complex

J. Tits has recently carried out a detailed investigation of the buildings of the various different types. In particular he has shown that a finite building whose associated Coxeter group is an indecomposable Weyl group of rank at least 3 must be a building $\Omega(G;\ B,\ N)$, where G is a finite Chevalley group or twisted group. Tits is able to deduce from this that the only finite simple groups with a $(B,\ N)$-pair of rank at least 3 are the finite Chevalley groups and twisted groups which have this property (see Tits [22]).

CHAPTER 16

Sporadic Simple Groups

The simple groups of Lie type which we have described in the earlier chapters include almost all the finite simple groups at present known. In addition to the cyclic groups, alternating groups and simple groups of Lie type over finite fields, there are 26 further finite simple groups. This was established when the classification of the finite simple groups was completed in 1981. In order to complete the picture regarding the finite simple groups we give a description in this final chapter of these 26 so called 'sporadic' simple groups. This description will be exceedingly brief in comparison with the amount of information available about these groups. The best source of information is the 'Atlas of Finite Groups' published by the Clarendon Press, Oxford. The Atlas was compiled by J.H. Conway, R.T. Curtis, S.P. Norton, R.A. Parker and R.A. Wilson. We shall follow the notation of the Atlas in describing the sporadic groups.

16.1 The Mathieu Groups

Five sporadic simple groups, which we denote by M_{11}, M_{12}, M_{22}, M_{23}, M_{24} were described by Mathieu in papers in 1861 and 1873 [1,2]. They can be described most easily as groups of automorphisms of Steiner systems.

A Steiner system of type (r, s, t) is a set A together with a family of subsets B_i of A, called blocks, such that $|A| = t$, $|B_i| = s$, and each subset of A with r elements is contained in exactly one block B_i. It is known that there is, to within isomorphism, exactly one Steiner system of each of the types $(5, 6, 12)$ and $(5, 8, 24)$ (see Witt [1, 2]). These Steiner systems will be denoted by $S(5, 6, 12)$ and $S(5, 8, 24)$.

An automorphism of a Steiner system A is a permutation of A which maps each block into a block. It can be shown that the groups of automorphisms of $S(5, 6, 12)$ and $S(5, 8, 24)$ are simple and we write

$$M_{12} = \text{Aut } S(5, 6, 12),$$
$$M_{24} = \text{Aut } S(5, 8, 24).$$

The groups M_{12} and M_{24} have the property of being quintuply transitive permutation groups. They are in fact the only known quintuply transitive permutation groups other than the symmetric and alternating groups. In M_{12}, the only element which fixes five symbols is the identity, but in M_{24} the elements stabilizing each of five given symbols form a subgroup of order 48. Thus the orders of these groups are given by

$$| M_{12} | = 12.11.10.9.8,$$

$$| M_{24} | = 24.23.22.21.20.48.$$

Given a Steiner system of type (r, s, t) we can obtain a Steiner system of type $(r-1, s-1, t-1)$ by taking all the blocks B_i of A which contain a fixed element $a \in A$, and then taking the sets $B_i - \{a\}$ as blocks in the set $A - \{a\}$. Steiner systems $S(4, 7, 23)$ and $S(3, 6, 22)$ can be obtained from $S(5, 8, 24)$ by this process, and a Steiner system $S(4, 5, 11)$ from $S(5, 6, 12)$.

We define M_{23} to be the subgroup of M_{24} stabilizing one symbol $a \in A$. Thus M_{23} is a permutation group on twenty-three symbols, and is a group of automorphisms of $S(4, 7, 23)$. Define M_{22} to be the stabilizer of one symbol in M_{23} and M_{11} to be the stabilizer of one symbol in M_{12}. Then M_{22} is a group of automorphisms of $S(3, 6, 22)$ and M_{11} is a group of automorphisms of $S(4, 5, 11)$. It is known that M_{23}, M_{11} are the full automorphism groups of $S(4, 7, 23)$, $S(4, 5, 11)$ respectively, and that M_{22} is a subgroup of index 2 in Aut $S(3, 6, 22)$.

Let M_{10}, M_{21} be the stabilizers of one symbol in M_{11}, M_{22} respectively. Then M_{10} has a subgroup of index 2 isomorphic to $PSL_2(9)$ and operates as a permutation group on the ten points of the projective line over $GF(9)$. By 12.5.1 Aut $PSL_2(9)$ is generated by the inner automorphisms and by a diagonal automorphism d of order 2 and a field automorphism f of order 2. Thus

$$| \text{Aut } PSL_2(9) : PSL_2(9) | = 4.$$

M_{10} is the subgroup of Aut $PSL_2(9)$ generated by $PSL_2(9)$ and the automorphism df.

M_{21} turns out to be isomorphic to $PSL_3(4)$ and operates as a permutation group on the twenty-one points of the projective plane over $GF(4)$.

The Mathieu groups may also be defined as transitive extensions. A transitive extension of a permutation group G on a set A is a transitive permutation group $\bar{G}$ on a set $\bar{A} = A \cup \{\bar{a}\}$ such that the stabilizer of $\bar{a}$ in $\bar{G}$, when restricted to A, is G. It can be shown that M_{11}, M_{12}, M_{22}, M_{23}, M_{24} are the unique transitive extensions of M_{10}, M_{11}, M_{21}, M_{22}, M_{23} respectively, and that M_{12}, M_{24} do not possess transitive extensions.

Thus the Mathieu groups are uniquely determined in terms of the classical groups M_{10} and M_{21}.

16.2 Sporadic Simple Groups Characterized by the Centralizer of an Involution

Apart from the Mathieu groups, all the other known sporadic simple groups have been discovered within the last ten years. The Feit–Thompson theorem [1] shows that every non-cyclic finite simple group contains an element of order 2, i.e. an involution. It was shown by Brauer and Fowler [1] that if C is any finite group, there are only finitely many finite simple groups G which have an involution j whose centralizer $C(j)$ is isomorphic to C. Brauer initiated the programme of trying to characterize all finite simple groups in which the centralizer of an involution is a given group, and very many such characterizations have been obtained. This programme has in fact led to the discovery of several sporadic simple groups. A typical result of this type asserts that if G is a simple group which has an involution j whose centralizer $C(j)$ is isomorphic to a fixed group C, then G is either one of the previously known simple groups or (say) a group determined to within isomorphism whose order and character table are both known. The existence or non-existence of this hypothetical new simple group then has to be decided. In most of the cases which have arisen, the existence of the new group has been checked with the aid of a computer.

An involution in G is called a central involution if it is contained in the centre of some Sylow 2-subgroup of G. Every non-cyclic simple group contains some central involution, and the characterizations of simple groups are often carried out in terms of centralizers of central involutions.

Six sporadic simple groups, denoted by J_1, J_2, J_3, He, Ly, J_4 have been discovered by these methods. Janko [1] characterized simple groups with a central involution j such that $C(j)$ is isomorphic to $C_2 \times A_5$, and showed that there is at most one such simple group and that its order must be 175560. He also succeeded in constructing such a group as a subgroup of the orthogonal group $O_7(11)$. This group is denoted by J_1. It is known that J_1 is a subgroup of the Chevalley group $G_2(11)$ (Janko [1]).

Janko then considered simple groups with a central involution j such that $C(j)$ is an extension of a group of order 2^5 by A_5. He showed that there are at most two such groups, of orders 604,800 and 50,232,960 but did not prove their existence. The existence of the smaller group, denoted by J_2, was proved by M. Hall using a computer, although it can now be described by other methods. The existence of the larger group, called J_3, was proved by G. Higman and MacKay, also using a computer.

The existence of a further sporadic simple group was predicted by Held. Let $G_0 = PSL_5(2)$ and j_0 be the involution $x_r(1)$ in G_0, where r is some root. Let C be the centralizer of j_0 in G_0. Held investigated simple groups G with an involution j such that $C(j)$ is isomorphic to C, and was able to show that there are at most three such groups. The first is $PSL_5(2)$, the second is M_{24}, and the third would be a sporadic group of order 4,030,387,200. Held did not prove the existence of this third group, denoted by He, but this was later checked by G. Higman and MacKay by computer.

Lyons has recently investigated simple groups in which there is a central involution j whose centralizer $C(j)$ is a non-split central extension of C_2 by A_{11}. There is at most one such group, of order

$$51,765,179,004,000,000,$$

and its existence has recently been proved by Sims using a computer. This group is denoted by Ly.

The existence of yet another sporadic simple group was predicted by Janko using this method. This group has an involution j in which $C(j)$ has the following structure. The maximal normal 2-subgroup $O_2(C(j))$ is an extraspecial group of order 2^{13}. The factor group $C(j)/O_2(C(j))$ has a normal subgroup $N/O_2(C(j))$ of order 3. $C(j)/N$ is an extension of the Mathieu group M_{22} of degree 2, and in fact is the automorphism group of M_{22}. Janko showed that a group of this type, and satisfying certain additional conditions, must have order $2^{21}.3^3.5.7.11^3.23.29.31.37.43$. The existence of such a sporadic simple group J_4 was subsequently proved by Norton by computational techniques, using the fact that J_4 has a representation of degree 112 over $GF(2)$.

16.3 Towers of Permutation Groups

A further method of constructing sporadic simple groups was suggested by a property of the group J_2. This group, whose existence had been conjectured by Janko, was exhibited by M. Hall as a transitive permutation group of degree 100. The stabilizer of one symbol is a simple subgroup isomorphic to $PSU_3(3^2)$, and this subgroup has three orbits in the given representation of degrees 1,36,63. Thus J_2 can be obtained as a permutation group by taking two permutational representations of $PSU_3(3^2)$, adjoining one additional point, and enlarging the group to a transitive extensive on the resulting set of 100 elements.

The question then arose as to whether a similar procedure could be adopted beginning with two permutational representations of some other classical group. If G is a group operating transitively on two sets A_1, A_2, we introduce an additional element $\bar{a}$ and seek to construct a transitive extension $\bar{G}$ of G on $\bar{A} = A_1 \cup A_2 \cup \{\bar{a}\}$. Such a transitive extension is usually found, when it exists, as a group of automorphisms of a graph with the elements of $\bar{A}$ as vertices. The problem thus becomes to construct a graph with vertex set $\bar{A}$ on which G operates as a group of automorphisms, such that the full automorphism group of the graph is transitive on $\bar{A}$. This can be done in a number of special cases to give further sporadic simple groups.

The first new group to be discovered by this method was obtained by D. Higman and C. Sims. They started with the Mathieu group M_{22}, taking the two transitive permutational representations of degrees 22, 77 obtained from the 22 points and 77 blocks of the Steiner system $S(3, 6, 22)$. Adding one additional point, they showed that M_{22} could be extended to a group HS operating transitively on these 100 symbols. HS is a simple group of order 44,352,000.

A similar construction was carried out by McLaughlin starting from the group $PSU_4(3^2)$. He took two permutational representations of this group of degrees 112 and 162. The representation of degree 112 is obtained from the operation of $PSU_4(3^2)$ on the 112 isotropic lines on the 'Hermitian quadric' in 3-dimensional projective space over $GF(3^2)$ left invariant by $PSU_r(3^2)$. The representation of degree 162 is obtained from the fact that $PSU_4(3^2)$ contains a subgroup $PSL_3(4)$ of index 162. Taking these two representations together and adding an extra point we obtain an intransitive representation of $PSU_4(3^2)$ of degree 275. It is possible, by constructing a suitable graph with these vertices, to find a transitive extension McL. McL turns out to be a simple group of order 898,128,000.

The next group of this type was obtained by Suzuki, beginning with $G_2(4)$. This group has permutational representations of degree 416 and 1365. The stabilizer of a symbol in the first representation is isomorphic to HaJ and the stabilizer of a symbol in the second representation is the centralizer of the centre of a Sylow 2-subgroup of $G_2(4)$. Adding one extra point we obtain an intransitive representation of degree 1782, and by constructing a suitable graph with this vertex set the existence of a transitive extension can be proved. This is a simple group, denoted by Suz, of order 448,345,497,600.

The existence of a further group of this type was predicted by A. Rudvalis. Rudvalis started from the Ree group $^2F_4(2)$ which, as we mentioned in §14.4, is not simple but has a simple subgroup of index 2. $^2F_4(2)$ has order $2^{12}.3^3.5^2.13$. $^2F_4(2)$ has two permutation representations of degrees 1755 and 2304. These are obtained as follows. $^2F_4(2)$ has a conjugacy class of subgroups isomorphic to $PSL_2(25)$. They have index 2304 in $^2F_4(2)$, so giving a permutation representation of this degree. The other permutation representation is obtained by considering the central involutions in $^2F_4(2)$. These are all conjugate and their centralizers have order $2^{12}.5$. They have index 1755 in $^2F_4(2)$. Rudvalis conjectured that by taking these two permutation representations of $^2F_4(2)$ together and adding an extra point it should be possible to obtain a transitive extension. In this way he conjectured the existence of a sporadic simple group Ru of order $2^{14}.3^3.5^3.7.13.29$. The existence of the Rudvalis group Ru was subsequently established by Conway and Wales by computational methods. They used the fact that the Rudvalis group has a double cover which has a representation of degree 28 over $\mathbb{C}$.

16.4 The Leech Lattice

We mention next three sporadic groups discovered by Conway. These can be described as groups of isometries of positive definite quadratic forms over the integers. Let $\mathbb{Z}^n$ be a free Abelian Group of rank n. A positive definite quadratic form of degree n over $\mathbb{Z}$ is a map $Q: \mathbb{Z}^n \to \mathbb{Z}$ given by

$$Q(x_1, \ldots, x_n) = \sum_{i,j} a_{ij} x_i x_j,$$

where $a_{ij} \in \mathbb{Z}$ and $a_{ji} = a_{ij}$, such that $Q(x) \geqslant 0$ for all $x \in \mathbb{Z}^n$ and $Q(x) = 0$ only if $x = 0$. The group $\mathbb{Z}^n$ together with the map $Q: \mathbb{Z}^n \to \mathbb{Z}$ is called a lattice.

We restrict attention to positive definite integral quadratic forms with two further properties. We assume Q is unimodular, i.e. $\det(a_{ij}) = 1$, and that Q is even, i.e. $Q(x)$ is an even integer for all x. It is known that there is an even unimodular, positive definite $\mathbb{Z}$-form of degree n if and only if n is divisible by 8 (cf. O'Meara [1]). If $n = 8$ there is, to within isomorphism, just one such form. If $n = 16$ there are two such forms. If $n = 24$ the forms of this type have recently been classified by Niemeier [1]. There are twenty-four such forms. One of them is distinguished from all the others by the fact that there is no x for which $Q(x) = 2$. The corresponding lattice was discovered independently by Leech, as the set of centres of spheres in a sphere-packing of greatest density in twenty-four dimensions

[1]. It is called the Leech lattice. Let G be the group of integral unimodular matrices which are isometries of this form. The centre Z of G consists of the two elements $\pm I$, and G/Z can be shown to be a simple group. It was first investigated by Conway and its simplicity was proved by Thompson. It is denoted by Co_1 and its order is

$$4,157,776,806,543,360,000.$$

Since G is represented geometrically as the group of automorphisms of the Leech lattice, subgroups of G can readily be obtained as stabilizers of sublattices. Consider the set of points $x \in \mathbb{Z}^{24}$ such that $Q(x) = 4$. These points are permuted transitively by G. The stabilizer of such a point in G can also be shown to be a simple group. This group is denoted by Co_2 and its order is $42,305,421,312,000$.

We next consider points $x \in \mathbb{Z}^{24}$ such that $Q(x) = 6$. These points are also permuted transitively by G, and the stabilizer of one of them is also a simple group. It is denoted by Co_3 and its order is $495,766,656,000$.

In addition to the three new sporadic simple groups Co_1, Co_2, Co_3 contained in G, several of the other sporadic groups occur in G as stabilizers of sublattices or as composition factors of such stabilizers. The groups HiS and McL and all five Mathieu groups can be obtained in this way. The group Suz also occurs inside G, although not (so far as is known) as the stabilizer of a sublattice.

16.5 Groups Generated by a Class of 3-Transpositions

We next describe three further sporadic simple groups discovered by B. Fischer [1]. Fischer investigated groups generated by a class of conjugate involutions such that the product of two distinct involutions in the class has order 2 or 3. Such a conjugacy class of involutions is called by Fischer a class of 3-transpositions. For example, each symmetric group is generated in this way by its class of transpositions. More generally, the Weyl groups of type A_l, D_l, E_6, E_7, E_8 are generated in this way by their classes of reflections. In addition certain classical groups over very small fields are also generated in this way. The symplectic groups $Sp_{2l}(2)$ are generated in this way by the class of symplectic transvections, the orthogonal groups $O_{2l}^+(2)$, $O_{2l}^-(2)$ by orthogonal transvections and the unitary groups $PSU_n(2^2)$ by unitary transvections. Moreover, in the orthogonal groups $O_{2l}^+(3)$, $O_{2l}^-(3)$, $O_{2l+1}(3)$ the reflections corresponding to the elements a satisfying $(a, a) = 1$ form a class of 3-transpositions in the subgroup they generate, and so do the reflections corresponding to the elements satisfying $(a, a) = -1$.

Fischer considered finite groups G generated by a class of 3-transpositions which satisfy the further conditions:

(i) Every normal subgroup of G of order a power of 2 or a power of 3 lies in the centre of G.

(ii) The derived group G' of G is its own derived group, viz., $G' = G''$.

He showed that if G is such a group with centre $Z(G)$ then $G/Z(G)$ is isomorphic to either a symmetric group, or one of the classical groups mentioned above (factored by the centre in the case of the groups over $GF(3)$), or one of three further groups which we shall denote by Fi_{22}, Fi_{23}, Fi_{24}. Moreover, there is just one class of 3-transpositions generating each of these groups, except in the group $S_6 \cong Sp_4(2) \cong PO_4^-(3)$ which has two such classes. We observe that the Weyl groups of type E_6, E_7, E_8 are included in this classification in view of the isomorphisms

$$
\begin{aligned}
W(E_6) &\cong O_6^-(2), \\
W(E_7)/Z &\cong Sp_6(2), \\
W(E_8)/Z &\cong O_8(2).
\end{aligned}
$$

If d is an element in the class of 3-transpositions of G it is shown in Fischer [1] that the set D_d of elements of this class distinct from d but commuting with it forms a conjugacy class of 3-transpositions in the group it generates. Thus the classification of groups generated by 3-transpositions proceeds by induction. In particular Fischer is able to show that if the group $\langle D_d \rangle / Z \langle D_d \rangle$ is a symmetric group or a classical group not isomorphic to $PSU_6(2^2)$ then G is either a symmetric group or on of the classical groups described above. If $\langle D_d \rangle / Z \langle D_d \rangle$ is isomorphic to $PSU_6(2^2)$ then the class D has 3510 elements and G is shown to be a simple group Fi_{22} of order $2^{17}.3^9.5^2.7.11.13$. A maximal commuting subset L of D has twenty-two elements and the group $\mathfrak{N}_G(L)/\langle L \rangle$ of permutations induced by G on L is isomorphic to the Mathieu group M_{22}.

If $\langle D_d \rangle / Z \langle D_d \rangle$ is isomorphic to Fi_{22} then the class D has 31,671 elements and G is a simple group Fi_{23} of order $2^{18}.3^{13}.5^2.7.11.13.17.23$. A maximal commuting subset L of D has twenty-three elements and the group $\mathfrak{N}_G(L)/\langle L \rangle$ of permutations induced by G on L is isomorphic to the Mathieu group M_{23}.

If $\langle D_d \rangle / Z \langle D_d \rangle$ is isomorphic to Fi_{23} then the class D has 306,936 elements and G is a group Fi_{24} of order $2^{22}.3^{16}.5^2.7^3.11.13.17.23.29$. Fi_{24} is not simple, but contains a simple subgroup Fi_{24}' of index 2. A maximal commuting subset L of D has twenty-four elements and the group

$\mathfrak{N}_G(L)/\langle L\rangle$ of permutations induced by G on L is isomorphic to the Mathieu group M_{24}.

Finally it can be shown that there can be no group G in which $\langle D_d\rangle/Z\langle D_d\rangle$ is isomorphic to Fi_{24}.

16.6 The O'Nan Group

The existence of a further sporadic simple group was predicted by O'Nan by investigating a situation studied by Alperin. Alperin considered simple groups containing an elementary abelian subgroup E of order 8 such that $C(E)$ has 2-rank 3 and $\mathfrak{N}(E)/C(E)$ is isomorphic to $PSL_3(2)$. It was shown by Alperin that a Sylow 2-subgroup V of $C(E)$ is isomorphic to $C_{2^n} \times C_{2^n} \times C_{2^n}$ for some $n \geq 1$. Moreover a Sylow 2-subgroup P of $\mathfrak{N}(E)$ is specified by precise generators and relations. (There are two possibilities.) Such a 2-group with $n \geq 2$ is called a group of Alperin type. The Higman-Sims group HS has Sylow 2-subgroups of Alperin type. O'Nan was able to show that a simple group with Sylow 2-subgroups of Alperin type must be either the Higman-Sims group or a new group of order $2^9.3^4.5.7^3.11.19.31$. This sporadic simple group $O'N$ was proved to exist by Sims by computational methods.

$O'N$ has a covering group of degree 3 which has a representation of degree 45 over $GF(7)$.

16.7 The Monster and its Subgroups

Generalising his previous ideas on groups generated by 3-transpositions Fischer subsequently investigated groups generated by $\{3,4\}$-transpositions. In such a group there is a class of conjugate involutions such that the product of any two involutions in the class has order 2,3 or 4. Fischer predicted the existence of such a group B in which an involution j in the class has $C(j)/\langle j\rangle$ equal to an extension of $^2E_6(2^2)$ of degree 2. Such a group B would have order $2^{41}.3^{13}.5^6.7^2.11.13.17.19.23.31.47$. Its existence was proved subsequently by Leon and Sims by computational methods.

Fischer also conjectured the existence of a further group M containing an involution j such that $C(j)/\langle j\rangle$ is isomorphic to B. M would have order $2^{46}.3^{20}.5^9.7^6.11^2.13^3.17.19.23.29.31.41.47.59.71$ and so be larger than any of the previously discovered sporadic simple groups. M is called the monster and B the baby monster. The existence of the monster was predicted independently by Griess. It was eventually proved to exist by

Griess as a group of automorphisms of a certain algebra of dimension 196883. Remarkably, no computational work was needed in Griess' proof despite the size of the group involved!

Before the existence of the monster had been proved it was realised that further sporadic simple groups would be contained in it. It was shown by Thompson that the monster would contain an element a of order 3 such that $C(a)/\langle a \rangle$ is a new sporadic simple group of order $2^{15}.3^{10}.5^3.7^2.13.19.31$. This group Th was proved to exist subsequently by Thompson. He showed it had a 248-dimensional representation over $\mathbb{R}$ which, on reduction mod 3, give a 248-dimensional representation over GF(3). Thompson obtained the group as a subgroup of $E_8(3)$.

The monster also contains an element a of order 5 such that $C(a)/\langle a \rangle$ is a new sporadic simple group of order $2^{14}.3^6.5^6.7.11.19$. This group was investigated by Harada before the existence of the monster had been proved. The existence of Harada's group was proved by computational methods by Norton and P.E. Smith. This group is known as the Harada-Norton group HN.

The classification of the finite simple groups, completed in 1981, showed that the sporadic simple groups we have mentioned are the only ones. Information about the classification can be found in a number of works by D. Gorenstein, including a survey of the classification in the book 'Finite Simple Groups — An introduction to their classification', Plenum Press, New York, 1982.

We conclude by giving a list of the sporadic simple groups together with their orders.

M_{11}	$2^4.3^2.5.11$	Mathieu
M_{12}	$2^6.3^3.5.11$	Mathieu
M_{22}	$2^7.3^2.5.7.11$	Mathieu
M_{23}	$2^7.3^2.5.7.11.23$	Mathieu
M_{24}	$2^{10}.3^3.5.7.11.23$	Mathieu
J_1	$2^3.3.5.7.11.19$	Janko
J_2	$2^7.3^3.5^2.7$	Hall, Janko
J_3	$2^7.3^5.5.17.19$	Janko, Higman, McKay
HS	$2^9.3^2.5^3.7.11$	Higman, Sims
McL	$2^7.3^6.5^3.7.11$	McLaughlin
Suz	$2^{13}.3^7.5^2.7.11.13$	Suzuki
He	$2^{10}.3^3.5^2.7^3.17$	Held, Higman, McKay
Ru	$2^{14}.3^3.5^3.7.13.29$	Rudvalis, Conway, Wales
Co_1	$2^{21}.3^9.5^4.7^2.11.13.23$	Conway, Leech
Co_2	$2^{18}.3^6.5^3.7.11.23$	Conway
Co_3	$2^{10}.3^7.5^3.7.11.23$	Conway

Fi_{22}	$2^{17}.3^9.5^2.7.11.23$	Fischer
Fi_{23}	$2^{18}.3^{13}.5^2.7.11.13.17.23$	Fischer
Fi'_{24}	$2^{21}.3^{16}.5^2.7^3.11.13.17.23.29$	Fischer
$O'N$	$2^9.3^4.5.7^3.11.19.31$	O'Nans, Sims
Ly	$2^8.3^7.5^6.7.11.31.37.67$	Lyons, Sims
J_4	$2^{21}.3^3.5.7.11^3.23.29.31.37.43$	Janko, Norton
HN	$2^{14}.3^6.5^6.7.11.19$	Harada, Norton, Smith
Th	$2^{15}.3^{10}.5^3.7^2.13.19.31$	Thompson, Smith
B	$2^{41}.3^{13}.5^6.7^2.11.13.17.19.23.31.47$	Fischer, Sims, Leon
M	$2^{46}.3^{20}.5^9.7^6.11^2.13^3.17.19.23.29$ $.31.41.47.59.71$	Fischer, Griess

Supplement
Hartley's Lemma

(See page 263).

(a) Let G^1 be a finite twisted group of type $^2A_\ell$, $^2D_\ell$, $^2E_6$ or $^3D_4$ and let $r\epsilon\Phi$.

Suppose $r=\bar{r}$. Then given any $t\epsilon K^*$ with $t=\bar{t}$ there exists $h(\chi)\epsilon H^1$ with $\chi(r)=t$.

Suppose $r\neq\bar{r}$. Then given any $t\epsilon K^*$ there exists $h(\chi)\epsilon H^1$ with $\chi(r)=t$ except when $G^1=^2A_3$ or $^2D_\ell$ or when $G^1=^2A_2$ and $q\equiv-1$ mod 3. If $G^1=^2A_3$ or $^2D_\ell$ there exists $h(\chi)\epsilon H^1$ with $\chi(r)=t^2$. If $G^1=^2A_2$ and $q\equiv-1$ mod 3 there exists $h(\chi)\epsilon H^1$ with $\chi(r)=t^3$.

(b) Let G^1 be a finite twisted group of type $^2B_2$, $^2F_4$ or $^2G_2$ and let $r\epsilon\Phi$. Then given any $t\epsilon K^*$ there exists $h(\chi)\epsilon H^1$ with $\chi(r)=t$ except when $G^1=^2G_2$ and $r=a+b$ or $3a+b$. In these cases there exists $h(\chi)\epsilon H^1$ with $\chi(r)=t^2$.

Bibliography

E. Abe
1. On the groups of C. Chevalley, *J. Math. Soc. Japan*, **11** (1959), 15–41.
2. Geometry in certain simple groups, *Tôhoku Math. J.*, **14** (1962), 64–72.
3. On simple groups associated with the real simple Lie algebras, *Tôhoku Math. J.*, **14** (1962), 244–262.
4. Finite groups admitting Bruhat decompositions of type A_n, *Tôhoku Math. J.*, **16** (1964), 130–141.

J. F. Adams
1. *Lectures on Lie Groups*, Benjamin, New York (1969).

K. Aomoto
1. On some double coset decompositions of complex semi-simple Lie groups, *J. Math. Soc. Japan*, **18** (1966), 1–44.

E. Artin
1. *Geometric Algebra*, Interscience Publishers, New York (1957).
2. The orders of the linear groups, *Comm. Pure Appl. Math.*, **8** (1955), 355–365.
3. The orders of the classical simple groups, *Comm. Pure Appl. Math.*, **8** (1955), 455–472.

H. Asano
1. A remark on the Coxeter–Killing transformations of finite reflection groups, *Yokohama Math. J.*, **15** (1967), 45–49.

A. Borel
1. *Linear Algebraic Groups*, Benjamin, New York (1969).
2. Sur la cohomologie des espaces fibres principaux et des espaces homogènes de groupes de Lie compacts, *Ann. of Math.*, **57** (1953), 115–207.
3. Groupes linéaires algébriques, *Ann. of Math.*, **64** (1956), 20–80.

A. Borel and T. A. Springer
1. Rationality properties of linear algebraic groups, (I) *Proceedings of Symposia in Pure Mathematics*, *Vol. 9, Algebraic groups and discontinuous subgroups*, A.M.S. (1966), 26–32; (II) *Tôhoku Math. J.*, **20** (1968), 443–497.

A. Borel and J. de Siebenthal
1. Les sous-groupes fermés de rang maximum des groupes de Lie clos, *Comment. Math. Helv.*, **23** (1949), 200–221.

A. Borel and J. Tits
1. Groupes réductifs, *I.H.E.S. Publ. Math.*, **27** (1965), 55–151.

A. Borel, R. Carter, C. W. Curtis, N. Iwahori, T. A. Springer and R. Steinberg
1. Seminar on algebraic groups and related finite groups, *Lecture Notes in Mathematics*, **131** (1970), Springer.

R. Bott
1. An application of the Morse theory to the topology of Lie groups, *Bull. Soc. Math. France*, **84** (1956), 251–282.

N. Bourbaki
1. *Groupes et algèbres de Lie*, IV, V, VI, Hermann, Paris (1968).
R. Brauer and K. A. Fowler
1. On groups of even order, *Ann. of Math.*, **62** (1955), 565–583.
F. Bruhat
1. Représentations induites des groupes de Lie semi-simples connexes, *C. R. Acad. Sci. Paris*, **238** (1954), 437–439.
2. Sous-groupes compacts maximaux des groupes semi-simples $\mathscr{P}$-adiques, *Seminaire Bourbaki*, (1963), Expose 271.
3. Sur une classe de sous-groupes compacts maximaux des groupes de Chevalley sur un corps $\mathscr{P}$-adique, *I.H.E.S. Publ. Math.*, **23** (1964), 45–74.
4. $\mathscr{P}$-adic groups, *Proceedings of Symposia in Pure Mathematics, Vol. 9, Algebraic groups and discontinuous subgroups*, A.M.S. (1966).
F. Bruhat and J. Tits
1. Groupes algébriques simples sur un corps local, *Proceedings of the Conference on Local Fields (Driebergen)*, 23–36, Springer (1967).
2. *C. R. Acad. Sci. Paris*, **263** (1966), 598–601, 766–768, 822–825, 867–869.
W. Burnside
1. *Theory of Groups of Finite Order*, 2nd edition, Dover, New York (1955).

E. Cartan
1. *Oeuvres Complètes*, Gauthier-Villars, Paris (1952).
R. W. Carter
1. Simple groups and simple Lie algebras, *J. London Math. Soc.*, **40** (1965), 193–240.
2. Weyl groups and finite Chevalley groups, *Proc. Cambridge Philos. Soc.*, **67** (1970), 269–276.
3. Conjugacy classes in the Weyl groups, To appear.
B. Chang
1. The conjugate classes of Chevalley groups of type (G_2), *J. Algebra*, **9** (1968), 190–211.
C. Chevalley
1. *Theory of Lie Groups*, Princeton University Press (1946).
2. *Théorie des groupes de Lie*, Hermann, Paris (1968).
3. *Séminaire Chevalley Vols. I, II; Classifications des groupes de Lie algébriques*, Paris (1956–8).
4. Sur certain groupes simples, *Tôhoku Math. J.*, **7** (1955), 14–66.
5. Invariants of finite groups generated by reflections, *Amer. J. Math.*, **77** (1955), 778–782.
6. Certain schémas de groupes semi-simples, *Seminaire Bourbaki*, **13** (1960). Exposé 219.
7. La théorie des groupes algébriques, *Proceedings of the International Congress of Mathematicians, Edinburgh*, Cambridge University Press (1960).
P. M. Cohn
1. *Lie Groups*, Cambridge University Press (1957).

A. J. Coleman
1. The Betti numbers of the simple Lie groups, *Canad. J. Math.*, **10** (1958), 349–356.

J. H. Conway
1. A group of order 8,315,553,613,086,720,000, *Bull. London Math. Soc.*, **1** (1969), 79–88.
2. A characterisation of Leech's lattice, *Invent. Math.*, **7** (1969), 137–142.

H. S. M. Coxeter
1. *Regular polytopes*, 2nd edition, Macmillan, New York (1963).
2. Discrete groups generated by reflections, *Ann. of Math.*, **35** (1934), 588–621.
3. The product of the generators of a finite group generated by reflections, *Duke Math. J.*, **18** (1951), 765–782.
4. Extreme forms, *Canad. J. Math.*, **3** (1951), 391–441.

H. S. M. Coxeter and W. Moser
1. Generators and relations for discrete groups, 2nd edition, *Ergebnisse der Mathematik und ihrer Grenzgebiete*, **14** (1965), Springer.

C. W. Curtis
1. Representations of Lie algebras of classical type with applications to linear groups, *J. Math. Mech.*, **9** (1960), 307–326.
2. On projective representations of certain finite groups, *Proc. Amer. Math. Soc.*, **11** (1960), 852–860.
3. An isomorphism theorem for certain finite groups, *Illinois J. Math.*, **7** (1963), 279–304.
4. Groups with a Bruhat decomposition, *Bull. Amer. Math. Soc.*, **70** (1964), 357–360.
5. Irreducible representations of finite groups of Lie type, *J. Reine Ang. Math.*, **219** (1965), 180–199.
6. Central extensions of groups of Lie type, *J. Reine Ang. Math.*, **220** (1965), 174–185.
7. The Steinberg character of a finite group with a (B, N)-pair, *J. Algebra*, **4** (1966), 433–441.
8. A character-theoretic criterion for a coset geometry to be a generalized polygon, *J. Algebra*, **7** (1967), 208–217.

C. W. Curtis and I. Reiner
1. *Representation Theory of Finite Groups and Associative Algebras*, Interscience Publishers, New York (1962).

Constance Davis
1. *A Bibliographical Survey of Simple Groups of Finite Order* 1900–1965, Courant Institute of Mathematical Sciences, New York (1969).

M. Demazure and A. Grothendieck
1. Schémas en Groupes, I, II, III, *Lecture Notes in Mathematics*, **151, 152, 153**, Springer.

L. E. Dickson
1. *Linear Groups with an Exposition of the Galois Field Theory*, Dover, New York (1958).
2. Linear groups in an arbitrary field, *Trans. Amer. Math. Soc.* (1901), 363–394.

3. A new system of simple groups, *Math. Ann.*, **60** (1905), 137–150.

J. Dieudonné

1. *Sur les groupes classiques*, Actualités scientifiques et industrielles 1040, Hermann, Paris (1948).
2. Le géométrie des groupes classiques, *Ergebnisse der Mathematik und ihrer Grenzgebiete*, **5** (1955), Springer.
3. Sur les générateurs des groupes classiques, *Summa Brasil. Math.*, **3** (1955), 149–179.
4. On simple groups of type B_n, *Amer. J. Math.*, **79** (1957), 922–923.
5. Les algèbres de Lie simples associées aux groupes simples algébriques sur un corps de caractéristique $p > 0$, *Rend. Circ. Mat. Palermo*, **6** (1957), 198–204.

Anne Duncan

1. An automorphism of the symplectic group $Sp_4(2^n)$, *Proc. Cambridge Philos. Soc.*, **64** (1968), 5–9.

E. B. Dynkin

1. The structure of semi-simple Lie algebras, *Amer. Math. Soc. Transl.* (1), **9** (1955), 328–469.
2. Semisimple subalgebras of semisimple Lie algebras, *Amer. Math. Soc. Transl.* (2), **6** (1957), 111–244.

M. Eichler

1. *Quadratische Formen und orthogonale Gruppen*, Springer (1952).

V. Ennola

1. On the characters of the finite unitary groups, *Ann. Acad. Sci. Fenn.*, **323** (1963).

W. Feit

1. *Characters of Finite Groups*, Benjamin, New York (1967).
2. *The Current Situation in the Theory of Finite Simple Groups*, Yale University (1971).

W. Feit and G. Higman

1. The non-existence of certain generalised polygons, *J. Algebra*, **1** (1964), 114–131.

W. Feit and J. G. Thompson

1. Solvability of groups of odd order, *Pacific J. Math.*, **13** (1963), 755–1029.

B. Fischer

1. Finite groups generated by 3-transpositions, *Invent. Math.*, to appear.

L. Flatto

1. Basic sets of invariants for finite reflections groups, *Bull. Amer. Math Soc.*, **74** (1968), 730–734.

L. Flatto and Margaret Weiner

1. Invariants of finite reflection groups and mean value problems, *Amer. J. Math.*, **91** (1969), 591–598.

H. Freudenthal

1. Zur Klassifikation der einfachen Lie-Gruppen, *Indag. Math.*, **20** (1958), 379–383.
2. Lie groups in the foundations of geometry, *Advances in Math.*, **1** (1964), 145–190.

H. Freudenthal and H. de Vries
1. *Linear Lie Groups*, Academic Press (1969).
G. Frobenius
1. Über Matrizen aus positiven Elementen, *Preuss. Akad. Wiss. Sitzungsber* (1908), 471–476.

I. M. Gel'fand and M. I. Graev
1. Construction of irreducible representations of simple algebraic groups over a finite field, *Dokl. Akad. Nauk. SSSR*, **147** (1962), 529–532.
J. A. Gibbs
1. Automorphisms of certain unipotent groups, *J. Algebra*, **14** (1970), 203–228.
D. Gorenstein
1. *Finite Groups*, Harper and Row, New York (1968).
J. A. Green
1. The characters of the finite general linear groups, *Trans. Amer. Math. Soc.*, **80** (1955), 402–447.
2. On the Steinberg characters of finite Chevalley groups, *Math. Z.*, **117** (1970), 272–288.

M. Hall
1. *Theory of Groups*, Macmillan, New York (1959).
M. Hall and D. Wales
1. The simple group of order 604,800, *J. Algebra*, **9** (1968), 417–450.
G. Harder
1. Über die Galoiskohomologie halbeinfacher Matrizengruppen I, *Math. Z.*, **90** (1965), 404–428.
Harish-Chandra
1. On a lemma of F. Bruhat, *J. Math. Pures Appl.*, **35** (1956), 203–210.
2. Automorphic forms on semisimple Lie groups, *Lecture Notes in Mathematics*, **62** (1968), Springer.
3. Eisenstein series over finite fields, *Functional Analysis and Related Fields*, edited by F. E. Browder, Springer (1970).
M. Hausner and J. T. Schwartz
1. *Lie Groups. Lie Algebras*, Nelson (1968).
D. Held
1. The simple group related to M_{24}, *J. Algebra*, **13** (1969), 253–296.
C. Hering, W. M. Kantor and G. M. Seitz
1. Finite groups with a split *BN*-pair of rank 1, to appear.
D. Hertzig
1. On simple algebraic groups. *Proceedings of the International Congress of Mathematicians, Edinburgh*, Cambridge University Press (1960).
2. Forms of algebraic groups, *Proc. Amer. Math. Soc.*, **12** (1961), 657–660.
D. G. Higman and C. C. Sims
1. A simple group of order 44,352,000, *Math. Z.*, **105** (1968), 110–113.
G. Higman
1. On the simple group of D. G. Higman and C. C. Sims, *Illinois Math. J.*, **13** (1969), 74–80.

G. Higman and J. McKay
 1. On Janko's simple group of order 50,232,960, *Bull. London Math. Soc.*, **1** (1969), 89–94.
H. Hopf
 1. Über die Topologie der Gruppen-Mannigfaltigkeiten und ihre Verallge-meinerungen, *Ann. of Math.*, **42** (1941), 22–52.
J. E. Humphreys
 1. Algebraic groups and modular Lie algebras, *Mem. Amer. Math. Soc.*, **71** (1967).
 2. On the automorphisms of infinite Chevalley groups, *Canad. J. Math.*, **21** (1969), 908–911.
B. Huppert
 1. *Endliche Gruppen I*, Springer (1967).

S. Ihara and T. Yokonuma
 1. On the second cohomology groups (Schur-multipliers) of finite reflection groups, *J. Fac. Sci. Univ. Tokyo*, **11** (1965), 155–171.
N. Iwahori
 1. On the structure of a Hecke ring of a Chevalley group over a finite field, *J. Fac. Sci. Univ. Tokyo*, **10** (1964), 215–236.
 2. Generalized Tits system (Bruhat decomposition) on $\mathscr{P}$-adic semisimple groups, *Proceedings of Symposia in Pure Mathematics, Vol. 9, Algebraic groups and discontinuous subgroups*, A.M.S. (1966), 71–83.
N. Iwahori and H. Matsumoto
 1. On some Bruhat decomposition, *I.H.E.S. Publ. Math.*, **25** (1965), 5–48.

N. Jacobson
 1. *Lie Algebras*, Interscience Publishers, New York (1962).
 2. *Lectures in abstract algebra III Theory of fields*, Van Nostrand (1964).
 3. Cayley numbers and simple Lie algebras of type G, *Duke Math. J.*, **5** (1939), 775–783.
 4. Some groups of transformations defined by Jordan algebras, I, II, III, *J. Reine Ang. Math.*, **201** (1959), 178–195, **204** (1960), 74–98, **207** (1961), 61–85.
 5. A note on automorphisms of Lie algebras, *Pacific J. Math.*, **12** (1962), 303–315.
Z. Janko
 1. A new finite simple group with abelian Sylow 2-subgroups, and its charac-terisation, *J. Algebra*, **3** (1966), 147–186.
 2. Some new simple groups of finite order, *Theory of finite groups (Harvard Symposium)*, 63–64. Benjamin (1969).
Z. Janko and J. G. Thompson
 1. On a class of finite simple groups of Ree, *J. Algebra*, **4** (1966), 274–292.
C. Jordan
 1. *Traité des substitutions et des équations algébriques*, Gauthier-Villars, Paris (1870).

W. Killing
1. Die Zusammensetzung der stetigen endlichen Transformationsgruppen, I–IV, *Math. Ann.*, **31** (1888), 252–290, **33** (1889), 1–48, **34** (1889), 57–122, **36** (1890), 161–189.

M. Kneser
1. Über die Ausnahme—Isomorphismen zwischen endlichen klassischen Gruppen, *Abh. Math. Sem. Hamburg*, **31** (1967), 136–140.
2. Semisimple algebraic groups, *Proceedings of the Instructional Conference (Brighton)* (1965), 250–265.

T. Kondo
1. The characters of the Weyl group of type F_4, *J. Fac. Sci. Univ. Tokyo*, **11** (1965), 145–153.

B. Kostant
1. The principal three-dimensional subgroup and the Betti numbers of a complex simple Lie Group, *Amer. Jour. Math.*, **81** (1959), 973–1032.
2. Lie group representations on polynomial rings, *Bull. Amer. Math. Soc.*, **69** (1963), 518–526.
3. Lie group representations on polynomial rings, *Amer. J. Math.*, **85** (1963), 327–404.
4. Groups over *Z*, *Proceedings of Symposia in Pure Mathematics, Vol. 9, Algebraic groups and discontinuous subgroups*, A.M.S. (1966).

S. Lang
1. Algebraic groups over finite fields, *Amer. J. Math.*, **78** (1956), 555–563.

J. Leech
1. Some sphere packings in higher space, *Canad. J. Math.*, **16** (1964), 657–682.

S. Lie
1. *Seminaire Sophus Lie. Theorie des algèbres de Lie. Topologie des groupes de Lie*, Paris (1955).

D. Livingstone
1. On a permutation representation of the Janko group, *J. Algebra*, **6** (1967), 43–55.

L. H. Loomis and S. Sternberg
1. *Advanced Calculus*, Addison-Wesley (1968).

B. Lou
1. The centralizer of a regular unipotent element in a semisimple algebraic group, *Bull. Amer. Math. Soc.*, **74** (1968), 1144–1146.

H. Lüneburg
1. Die Suzukigruppen und ihre Geometrien, *Lecture Notes in Mathematics*, **10** (1965), Springer.
2. Transitive Erweiterungen endlicher Permutationsgruppen, *Lecture Notes in Mathematics*, **84** (1969), Springer.
3. Some remarks concerning the Ree groups of type (G_2), *J. Algebra*, **3** (1966), 256–259.
4. Über die Gruppen von Mathieu, *J. Algebra*, **10** (1968), 194–210.

I. G. Macdonald
  1. Spherical functions on a $\mathscr{P}$-adic Chevalley group, *Bull. Amer. Math. Soc.*, **74** (1968), 520–525.
  2. The Hall algebra, *Jber. Deutsch. Math.-Verein*, to appear.
J. MacLaughlin
  1. Some groups generated by transvections, *Arch. Math.* (Basel), **18** (1967), 364–368.
  2. A simple group of order 898,128,000, *Theory of finite groups* (*Harvard Symposium*), Benjamin (1969).
E. Mathieu
  1. Mémoire sur l'étude des fonctions de plusieurs quantités, *J. Math. Pures Appl.*, **6** (1861), 241–243.
  2. Sur les fonctions cinq fois transitives de 24 quantités. *J. Math. Pures Appl.*, **18** (1873), 25–46.
H. Matsumoto
  1. Quelques remarques sur les groupes de Lie algébriques réels, *J. Math. Soc. Japan*, **16** (1964), 419–446.
  2. Générateurs et relations des groupes de Weyl généralisés, *C. R. Acad. Sci. Paris*, **258** (1964), 3419–3422.
  3. Un théorème de Sylow pour les groupes semi-simples $\mathscr{P}$-adiques, *C. R. Acad. Sci. Paris*, **262** (1966), 425–427.
W. H. Mills and G. B. Seligman
  1. Lie algebras of classical type, *J. Math. Mech.*, **6** (1957), 519–548.

H. V. Niemeyer
  1. *Definite quadratische Formen der Dimension* 24 *und Diskriminante* 1 (Thèse), Göttingen (1968).

O. T. O'Meara
  1. *Introduction to Quadratic Forms*, Springer (1963).
T. Ono
  1. Sur les groupes de Chevalley, *J. Math. Soc. Japan*, **10** (1958), 307–313.

R. Ree
  1. On some simple groups defined by C. Chevalley, *Trans. Amer. Math. Soc.*, **84** (1957), 392–400.
  2. A family of simple groups associated with the simple Lie algebra of type (F_4), *Amer. J. Math.*, **83** (1961), 401–420.
  3. A family of simple groups associated with the simple Lie algebra of type (G_2), *Amer. J. Math.*, **83** (1961).
  4. Construction of certain semisimple groups, *Canad. J. Math.*, **16** (1964), 490–508.
  5. Sur une famille de groupes de permutations doublement transitifs, *Canad. J. Math.*, **16** (1964), 797–820.
  6. Classification of involutions and centralizers of involutions in finite Chevalley groups of exceptional types, *Proceedings of the International Conference on the Theory of Groups* (*Canberra*). Gordon and Breach (1967).

R. Richardson
1. Conjugacy classes in Lie algebras and algebraic groups, *Ann. of Math.*, **86** (1967), 1–15.

F. Richen
1. Modular representations of split (B, N)-pairs, *Trans. Amer. Math. Soc.*, **140** (1969), 435–460.

B. A. Rozenfel'd
1. Forms of simplicity and semi-simplicity, *Trudy Sem. Vektor. Tenzor. Anal.*, **12** (1963), 269–285.

H. Samelson
1. *Notes on Lie Algebras*, Van Nostrand Reinhold (1969).

W. R. Scott
1. *Group Theory*, Prentice-Hall (1964).

G. B. Seligman
1. Modular Lie algebras, *Ergebni sse der Mathematik und ihrer Grenzgebiete*, **40** (1967), Springer.
2. Some remarks on classical Lie algebras, *J. Math. Mech.*, **6** (1957), 549–558.
3. On automorphisms of Lie algebras of classical type, I–III, *Trans. Amer. Math. Soc.*, **92** (1959), 430–448, **94** (1960), 452–481, **97** (1960), 286–316.

J-P. Serre
1. Cohomologie galoisienne, *Lecture Notes in Mathematics*, **5** (1964), Springer.
2. *Lie Algebras and Lie Groups*, Benjamin, New York (1965).
3. *Algèbras de Lie semi-simples complexes*, Benjamin, New York (1966).

G. C. Shephard
1. Regular complex polytopes, *Proc. London Math. Soc.*, **2** (1952), 82–97.
2. Some problems on finite reflection groups, *Enseignement Math.*, **2** (1956), 42–48.

D. Soda
1. Some groups of type D_4 defined by Jordan algebras. *J. Reine Ang. Math.*, **223** (1966), 150–163.

L. Solomon
1. Invariants of finite reflection groups, *Nagoya Math. J.*, **22** (1963), 57–64.
2. Invariants of Euclidean reflection groups, *Trans. Amer. Math. Soc.*, **113** (1964), 274–286.
3. A fixed-point formula for the classical groups over a finite field, *Trans. Amer. Math. Soc.*, **117** (1965), 423–440.
4. The orders of the finite Chevalley groups, *J. Algebra*, **3** (1966), 376–393.
5. A decomposition of the group algebra of a finite Coxeter group, *J. Algebra*, **9** (1968), 220–239.
6. The Steinberg character of a finite group with BN-pair, *Theory of Finite Groups (Harvard Symposium)*, 213–221, Benjamin (1969).

E. L. Spitznagel
1. Hall subgroups of certain families of finite groups, *Math. Z.*, **97** (1967), 259–290.

    2. Terminality of the maximal unipotent subgroups of Chevalley groups, *Math. Z.*, **103** (1968), 112–116.

T. A. Springer
    1. Some arithmetic results on semi-simple Lie algebras, *I.H.E.S. Publ. Math.*, **30** (1966), 115–141.
    2. A note on centralizers in semisimple groups, *Indag. Math.*, **28** (1966), 75–77.
    3. The unipotent variety of a semisimple group, *Proceedings of the Colloquium in Algebraic Geometry. (Tata Institute)*, (1969), 373–391.

B. Srinivasan
    1. The characters of the finite symplectic group $Sp(4, q)$, *Trans. Amer. Math. Soc.*, **131** (1968), 488–525.

R. Steinberg
    1. A geometric approach to the representations of the full linear group over a Galois field, *Trans. Amer. Math. Soc.*, **71** (1951), 274–282.
    2. Prime power representations of finite linear groups, I, II, *Canad. J. Math.*, **8** (1956), 580–591, **9** (1957), 347–351.
    3. Finite reflection groups, *Trans. Amer. Math. Soc.*, **91** (1959), 493–504.
    4. Variations on a theme of Chevalley, *Pacific J. Math.*, **9** (1959), 875–891.
    5. Invariants of finite reflection groups, *Canad. J. Math.*, **12** (1960), 616–618.
    6. Automorphisms of finite linear groups, *Canad. J. Math.*, **12** (1960), 606–615.
    7. The simplicity of certain groups, *Pacific J. Math.*, **10** (1960), 1039–1041.
    8. Automorphisms of classical Lie algebras, *Pacific J. Math.*, **11** (1961), 1119–1129.
    9. A closure property of a set of vectors, *Trans. Amer. Math. Soc.*, **105** (1962), 118–125.
  10. Generators for simple groups, *Canad. J. Math.*, **14** (1962), 277–283.
  11. Générateurs, relations et revêtements de groupes algébriques, *Colloquium sur la théorie des groupes algébriques (Bruxelles)*, (1962), 113–127.
  12. Representations of algebraic groups, *Nagoya Math. J.*, **22** (1963), 33–56.
  13. Differential equations invariant under finite reflection groups, *Trans. Amer. Math. Soc.*, **112** (1964), 392–400.
  14. Regular elements of semisimple algebraic groups, *I.H.E.S. Publ. Math.*, **25** (1965), 49–80.
  15. *Lectures on Chevalley Groups*, Yale University (1967).
  16. Endomorphisms of linear algebraic groups, *Mem. Amer. Math. Soc.*, **80** (1968).
  17. Algebraic groups and finite groups, *Illinois J. Math.*, **13** (1969), 81–86.

I. N. Stewart
    1. Lie algebras, *Lecture Notes in Mathematics*, **127** (1970), Springer.

M. Suzuki
    1. A new type of simple groups of finite order, *Proc. Nat. Acad. Sci. U.S.A.*, **46** (1960), 868–870.
    2. On a class of doubly transitive groups, *Ann. of Math.*, **75** (1962), 105–145.
    3. A simple group of order 448,345,497,600, *Theory of finite groups (Harvard Symposium)*, Benjamin (1969).

J. G. Thompson
1. Toward a characterisation of $E_2^*(q)$, *J. Algebra*, **7** (1967), 406–414.
2. Nonsolvable finite groups all of whose local subgroups are solvable, (I) *Bull. Amer. Math. Soc.*, **74** (1968), 383–437; (II) *Pacific J. Math.*, **33** (1970), 451–536.
3. Quadratic pairs, to appear.

J. Tits
1. Sur les analogue algébriques des groupes semi-simples complexes, *Colloquium d'algèbre supérieure (Bruxelles)*, (1956), 261–289.
2. Les groupes de Lie exceptionnels et leur interprétation géométrique, *Bull. Math. Soc. Belg.*, **8** (1956), 48–81.
3. Les 'formes réelles' des groupes de type E_6, *Seminaire Bourbaki* (1958), Exposé 162.
4. Sur la trialité et certains groupes qui s'en déduisent, *I.H.E.S. Publ. Math.*, **2** (1959), 14–60.
5. Sur la classification des groupes algébriques semi-simples, *C. R. Acad. Sci. Paris*, **249** (1959), 1438–1440.
6. Les groupes simples de Suzuki et de Ree, *Seminaire Bourbaki*, **13** (1960), Exposé 210.
7. Groupes algébriques semi-simples et géometries associées, *Proceedings of the colloquium on the algebraic and topological foundations of geometry (Utrecht)*, 175–192, Pergamon Press, Oxford (1962).
8. Groupes simples et géométries associées, *Proceedings of the International Congress of Mathematicians (Stockholm)*, (1962), 197–221.
9. Ovoïdes et groupes de Suzuki, *Arch. Math.*, **13** (1962), 187–198.
10. Groupes semi-simples isotropes, *Colloquium on the Theory of Algebraic Groups (Paris)*, 137–147, Gauthier-Villars (1962).
11. Théorème de Bruhat et sous-groupes paraboliques, *C.R. Acad. Sci. Paris*, **254** (1962), 2910–2912.
12. Algebraic and abstract simple groups, *Ann. of Math.*, **80** (1964), 313–329.
13. Géometries polyédriques finis, *Rend. Mat. e Appl.*, **23** (1964), 156–165.
14. Sur les systemes de Steiner associées aux trois 'grands' groupes de Mathieu, *Rend. Mat. e Appl.*, **23** (1964), 166–184.
15. Sur les constantes de structure et le théorème d'existence des algèbres de Lie semi-simples, *I.H.E.S. Publ. Math.*, **31** (1966).
16. Normalisateurs de tores I. Groupes de Coxeter étendus, *J. Algebra*, **4** (1966), 96–116.
17. Algèbres alternatives, algèbres de Jordan et algèbres de Lie exceptionnelles, *Indag. Math.*, **28** (1966), 223–237.
18. Classification of algebraic semisimple groups, *Proceedings of Symposia in Pure Mathematics, Vol. 9 Algebraic groups and discontinuous subgroups*, A.M.S. (1966), 33–62.
19. Tabellen zu den einfachen Lie Gruppen und ihren Darstellungen, *Lecture Notes in Mathematics*, **40** (1967), Springer.
20. Le groupe de Janko d'ordre 604,800, *Theory of finite groups (Harvard Symposium)*, 91–95, Benjamin (1969).
21. Groupes finis simples sporadiques, *Seminaire Bourbaki*, (1969–70), Exposé 375.

22. Notes on finite *BN*-pairs, to appear.

J. A. Todd and G. C. Shephard
1. Finite unitary reflection groups, *Canad. J. Math.*, **6** (1954), 274–304.

T. Tsuzuku
1. A characterisation of finite projective linear groups, *Proc. Japan Acad.*, **40** (1964), 155–156.

B. J. Veisfeiler
1. A class of unipotent subgroups of semisimple algebraic groups, *Uspehi Mat. Nauk.*, **21** (1966), 222–223.

B. L. van der Waerden
1. *Gruppen von linearen Transformationen*, Chelsea Publishing Co., New York (1948).

G. E. Wall
1. On the conjugacy classes in the unitary, symplectic and orthogonal groups, *J. Austral. Math. Soc.*, **3** (1963), 1–62.

H. N. Ward
1. On Ree's series of simple groups, *Trans. Amer. Math. Soc.*, **121** (1966), 62–89.

H. Weyl
1. Über die Darstellungen halbeinfacher Gruppen durch lineare Transformationen, I, II, *Math. Z.*, **23** (1925), 271–309, **24** (1926), 328–395.
2. The structure and representations of continuous groups, *I.A.S. notes*, Princeton (1934–5).
3. *The Classical Groups*, Princeton University Press (1946).

D. J. Winter
1. On groups of automorphisms of Lie algebras, *J. Algebra*, **8** (1968), 131–142.

E. Witt
1. Die 5-fach transitive Gruppe von Mathieu, *Abh. Math. Sem. Univ. Hamburg*, **12** (1938), 256–264.
2. Über Steinersche Systeme, *Abh. Math. Sem. Univ. Hamburg*, **12** (1938), 265–274.
3. Spiegelungsgruppen und Aufzählung halbeinfacher Liescher Ringe, *Abh. Math. Sem. Univ. Hamburg*, **14** (1941), 289–337.

T. Yokonuma
1. Sur la structure des anneaux de Hecke d'un groupe de Chevalley fini, *C. R. Acad. Sci. Paris*, **264** (1967), 344–347.
2. Sur la commutant d'une représentation d'un groupe de Chevalley fini, *C. R. Acad. Sci. Paris*, **264** (1967), 433–436; *J. Fac. Sci. Univ. Tokyo*, **15** (1968), 115–129.

Index of Notation

Symbol	Meaning	Page of definition
$^2D_l(q^2)$	The twisted orthogonal group of type D_l over $GF(q^2)$	251
$^3D_4(K)$	The triality twisted group of type D_4 over K	251
$^3D_4(q^3)$	The triality twisted group of type D_4 over $GF(q^3)$	251
e_r	A root vector in a simple Lie algebra	51
$e(Q)$	A multiplicative group isomorphic to the additive group Q	148
E_6	The type of a simple Lie algebra over $\mathbb{C}$	43
$E_6(K)$	The Chevalley group of type E_6 over K	64
$E_6(q)$	The Chevalley group of type E_6 over $GF(q)$	121
$^2E_6(K)$	The twisted group of type E_6 over K	251
$^2E_6(q^2)$	The twisted group of type E_6 over $GF(q^2)$	251
E_7	The type of a simple Lie algebra over $\mathbb{C}$	43
$E_7(K)$	The Chevalley group of type E_7 over K	64
$E_7(q)$	The Chevalley group of type E_7 over $GF(q)$	121
E_8	The type of a simple Lie algebra over $\mathbb{C}$	43
$E_8(K)$	The Chevalley group of type E_8 over K	64
$E_8(q)$	The Chevalley group of type E_8 over $GF(q)$	121
F_4	The type of a simple Lie algebra over $\mathbb{C}$	43
$F_4(K)$	The Chevalley group of type F_4 over K	64
$F_4(q)$	The Chevalley group of type F_4 over $GF(q)$	121
$^2F_4(K)$	The twisted Ree group of type F_4 over K	251
$^2F_4(2^{2m+1})$	The twisted Ree group of type F_4 over $GF(2^{2m+1})$	251
G	A group, usually a Chevalley group	68
$\bar{G}$	The universal Chevalley group associated with G	190
$\hat{G}$	The extension of G by its group of diagonal automorphisms	118
G'	The commutator subgroup (derived group) of G	170
G^1	A twisted group	226
G_2	The type of a simple Lie algebra over $\mathbb{C}$	43
$G_2(K)$	The Chevalley group of type G_2 over K	64
$G_2(q)$	The Chevalley group of type G_2 over $GF(q)$	121
$^2G_2(K)$	The twisted Ree group of type G_2 over K	251
$^2G_2(3^{2m+1})$	The twisted Ree group of type G_2 over $GF(3^{2m+1})$	251
$GF(q)$	The Galois field with q elements	2
$GL_n(K)$	The general linear group of degree n over K	2
H	The diagonal subgroup of a Chevalley group G	97
$\mathfrak{H}$	A Cartan subalgebra of a simple Lie algebra	35
h_r	The co-root associated to a root r	42
$\mathfrak{H}_r$	The 1-dimensional subspace of $\mathfrak{H}$ containing h_r	83
H_r	The hyperplane orthogonal to a root r	20
$h(\chi)$	The automorphism of the Lie algebra $\mathfrak{L}_K$ determined by a K-character χ of P	98
$\hat{H}$	The group of automorphisms $h(\chi)$ of $\mathfrak{L}_K$	98
$h_r(\lambda)$	The element of H associated with a root r and an element λ of K	92

Symbol	Meaning	Page of definition
$O_{2l}^+(q)$	The orthogonal group of degree $2l$ over $GF(q)$ leaving invariant a quadratic form of index l	6
$O_{2l}^-(q)$	The orthogonal group of degree $2l$ over $GF(q)$ leaving invariant a quadratic form of index $l-1$	6
$p_1, \ldots, p_l$	A system of fundamental roots	38
P	The additive group generated by $p_1, \ldots, p_l$	97
P_J	The parabolic subgroup BN_JB associated with a set J of fundamental roots	108
$P_W(t)$	The polynomial $\sum\limits_{w \in W} t^{l(w)}$	135
$P_{W^1}(t)$	The polynomial $\sum\limits_{w \in W^1} t^{l(w)}$	254
$PGL_n(K)$	The projective general linear group of degree n over K	2
$PSp_n(K)$	The projective symplectic group of degree n over K	4
$PSO_n(K, f)$	The projective special orthogonal group of degree n over K leaving invariant the quadratic form f	5
$P\Omega_n(K, f)$	The projective group of the commutator subgroup $\Omega_n(K, f)$ of $O_n(K, f)$	5
$PSU_n(K, f)$	The projective special unitary group of degree n over K leaving invariant the Hermitian form f	7
$\mathbb{Q}$	The field of rationals	153
$q_1, \ldots, q_l$	The fundamental weights of a simple Lie algebra	98
Q	The additive group generated by $q_1, \ldots, q_l$	99
$\mathbb{R}$	The real number field	12
R	The root of maximum height	156
S	The sum of the fundamental weights	145
S	An equivalence class in Φ	227
$\mathfrak{S}$	The algebra of polynomial functions on $\mathfrak{P}$	123
$\mathfrak{S}_n$	The set of homogeneous polynomials in $\mathfrak{S}$ of degree n	132
$SL_n(K)$	The special linear group of degree n over K	2
$Sp_n(K)$	The symplectic group of degree n over K	2
$SO_n(K, f)$	The special orthogonal group of degree n over K leaving invariant the quadratic form f	5
$SU_n(K, f)$	The special unitary group of degree n over K leaving invariant the Hermitian form f	7
St A	The star of an element A in a chamber complex	274
U	The unipotent subgroup of a Chevalley group G generated by the positive root subgroups	68
U_m	The subgroup of U generated by the root subgroups corresponding to roots of height at least m	78
U_r	The product of the root subgroups corresponding to positive roots other than r	104
U_w^+	The product of the root subgroups corresponding to positive roots transformed by w into positive roots	115
U_w^-	The product of the root subgroups corresponding to positive roots transformed by w into negative roots	115

Index

333